G. O. Mott
2183 McCarty

PATTERN ANALYSIS IN AGRICULTURAL SCIENCE

Pattern Analysis
in Agricultural Science

Edited by W. T. Williams

CSIRO

Melbourne

Elsevier Scientific Publishing Company

Amsterdam • Oxford • New York

Published jointly by

CSIRO
372 Albert Street, East Melbourne, Australia 3002

ISBN 0 643 00169 7

and

Elsevier Scientific Publishing Company
335 Jan van Galenstraat, P.O. Box 211, Amsterdam, The Netherlands
American Elsevier Publishing Company, Inc.
52 Vanderbilt Avenue, New York, N.Y., 10 017, U.S.A.

ISBN 0 444 99844 6

Library of Congress Catalog Card Number: 76–4711

Printed in Australia at The Dominion Press, North Blackburn, Vic. 3130

Contents

Foreword

When the then Officer-in-Charge of the CSIRO Davies Laboratory, Mr. Les Edye, first told me of his plans to organize a course on Pattern Analysis, I found the idea exciting and imaginative and I readily offered any assistance which the University could give. It is, of course, the policy of universities to make their facilities available for such activities. In this case, however, there were special reasons for my eagerness to see collaboration between James Cook University and the Davies Laboratory.

CSIRO scientists and University academics in Townsville share the problem of working in institutions situated 1500 kilometres from the nearest metropolitan city. The isolation is overcome to some extent by attending conferences at other universities and by inviting distinguished scientists and academicians to visit us. The most effective way, however, is to mount seminars and courses in Townsville which will, by virtue of the topic and the quality of the course leaders, attract participants from all parts of Australia. Not only do we benefit by intercourse with them, but they gain an insight into the special activities and problems of research and teaching in a remote tropical environment.

It was our very good fortune to have had in Townsville an outstanding scientist, Dr. Bill Williams, to spearhead the course. His unique contribution in Pattern Analysis techniques coupled with his gifted style of presentation— clear, concise and always entertaining—ensured the success of the course; a rewarding experience indeed for all who attended.

This book, then, is the product of the course on Pattern Analysis in Agricultural Research. It is, however, far more than a record of the course; it is an important contribution to scientific literature, filling as it does a major gap in the texts now available. I have no doubt that it will become a basic reference not only for agricultural scientists but also for biological scientists generally.

K. J. C. BACK

Vice-Chancellor, James Cook University of North Queensland

Preface

This book has been written for, and almost entirely by, agricultural research workers. Agricultural research, especially in the tropics, can be a very lonely profession, and it is not unusual for the nearest biometric help to be a thousand miles away; so the worker is often forced to be his own biometrician. He will have received some elementary statistical training in his agricultural course, which will see him through an orthogonal analysis of variance and a simple regression, but seldom much further. Yet it is increasingly common to record, not a single measurement, but a whole set of correlated measurements; for these it is often better to use multivariate statistical methods, in which almost no agriculturists—and by no means all statisticians—have received training. Moreover, his problem may not be truly statistical at all; there are occasions, notably in survey problems, when what is needed is an efficient summary of an extremely complex situation, and for such problems the 'pattern analysis' techniques of ordination and classification may be more appropriate. It is not necessary, or even desirable, that the agriculturist should attempt to master such methods himself. All powerful tools can be dangerous in inexperienced hands, and multivariate and pattern-seeking methods are no exception. It is, however, essential that he should know what *can* be done, and how to set about getting it done. This is particularly difficult in the pattern-seeking field, in which few statisticians have experience, the literature is widely scattered in unfamiliar journals, and new methods are appearing at an alarming rate. Elementary books are now beginning to appear, but even these may be inaccessible because of their use of matrix algebra or information statistics. As a result, many vital data are being wasted. It would be easy to cite examples where agricultural workers, often in physically extreme conditions, have devotedly collected data over a lengthy period, sometimes over years, only to find at the end that the results were not amenable to any form of analysis with which they were familiar. Only too often, the results have then been set aside and the information they contain lost.

This, then, is the purpose of this book. It is not intended to make every man into his own pattern analyst; it aims only to explain roughly how these methods work, when they are used, and what the ensuing results look like. The book falls into three sections. The first is a brief account of matrix algebra, using only agricultural examples; it is our hope that the agriculturist would discover that this subject—at least as far as he needs to take it—is surprisingly easy. The second section outlines the theory and present stage of development of pattern-seeking methods, and explores their relationship to the methods of conventional statistics. The last section is a set of twenty-one case histories. Agricultural research workers tend to be very practical people, who find a demonstration more convincing than elegant theory; in this section, therefore, a number of workers briefly describe problems they have elucidated with the aid of pattern-analysis techniques.

The book arose out of a course organized by members of the Davies Laboratory of the CSIRO Division of Tropical Agronomy, which is situated in Townsville, Queensland, in the dry tropics. The course was physically held in the James Cook University of North Queensland, to whose Administration we are much indebted for the use of their excellent facilities. We are indebted, too, to those members of the Davies Laboratory (R. L. Burt, L. A. Edye, J. B. F. Field, R. F. Isbell, R. L. McCown and J. G. McIvor) and the University (T. J. Rogers) who acted as an organizing committee; and particularly to John McIvor, the Secretary of the Committee, whose untiring and unobtrusive efficiency ensured that lecturers could concentrate on the task in hand, secure in the knowledge that nothing would go wrong. A few diagrams and segments of text have been previously published elsewhere; and we are grateful to Annual Reviews Inc. (*Annual Review of Ecology and Systematics*), The British Society of Soil Science (*Journal of Soil Science*) and The Elsevier Scientific Publishing Company (*Geoderma*) for permission to republish material which has already appeared in their Journals.

Finally, the Editor wishes to express his personal indebtedness to Les Edye, the then Officer-in-Charge of the Davies Laboratory, for his enthusiastic sponsorship of the course; to Nola Stokes, who typed the entire manuscript in its many recensions; to Dianne Field, who prepared the index; and to Margaret Walkom, of the CSIRO Editorial and Publications Service, for the skill and experience she has brought to seeing this volume into print.

W. T. WILLIAMS

Contributors

R. L. Burt, Ph.D. CSIRO Division of Tropical Agronomy, Davies Laboratory, Townsville, Queensland 4810.

H. T. Clifford, Ph.D., D.Sc., F.L.S. Botany Department, University of Queensland, St. Lucia, Queensland 4067.

M. B. Dale, Ph.D. CSIRO Division of Tropical Agronomy, Cunningham Laboratory, Mill Road, St. Lucia, Queensland 4067.

L. A. Edye, M.Agr.Sc. CSIRO Division of Tropical Agronomy, Davies Laboratory, Townsville, Queensland 4810.

P. Gillard, Ph.D. CSIRO Division of Tropical Agronomy, Davies Laboratory, Townsville, Queensland 4810.

D. A. Hedges, B.Sc.Agr. CSIRO Division of Animal Physiology, Pastoral Research Laboratory, Armidale, New South Wales 2350.

R. K. Jones, Ph.D. CSIRO Division of Tropical Agronomy, Cunningham Laboratory, Mill Road, St. Lucia, Queensland 4067.

G. N. Lance, M.Sc., Ph.D. CSIRO Division of Computing Research, P.O. Box 1800, Canberra City, A.C.T. 2601.

D. L. Lloyd, B.Agr.Sc. Department of Primary Industries, Queensland Wheat Research Institute, Toowoomba, Queensland 4350.

P. W. Milne, Dip.El.Eng. CSIRO Division of Computing Research, P.O. Box 1800, Canberra City, A.C.T. 2601.

A. W. Moore, M.Sc., Ph.D. CSIRO Division of Soils, Cunningham Laboratory, Mill Road, St. Lucia, Queensland 4067.

R. Reid, B.A. CSIRO Division of Tropical Agronomy, Davies Laboratory, Townsville, Queensland 4810.

P. J. Robinson, M.Sc., Ph.D. CSIRO Division of Tropical Agronomy, Davies Laboratory, Townsville, Queensland 4810.

J. S. Russell, M.Sc., Ph.D.   CSIRO Division of Tropical Agronomy, Cunningham Laboratory, Mill Road, St. Lucia, Queensland 4067.

J. P. Thompson, B.Agr.Sc.   Department of Primary Industries, Queensland Wheat Research Institute, Toowoomba, Queensland 4350.

J. C. Tothill, Ph.D.   CSIRO Division of Tropical Agronomy, Cunningham Laboratory, Mill Road, St. Lucia, Queensland 4067.

W. T. Williams, Ph.D., D.Sc., Hon. D.Sc., D.I.C., F.L.S., F.I.Biol.   CSIRO Division of Tropical Agronomy, Davies Laboratory, Townsville, Queensland 4810.

W. H. Winter, Ph.D.   CSIRO Division of Tropical Agronomy, Davies Laboratory, Townsville, Queensland 4810.

PART 1

Elements of Matrix Algebra

1

Matrix Terminology

W. T. Williams

Agronomists seem to be afraid of matrices, or at least of the *idea* of matrices. They tend to have a suspicion that matrix algebra is an esoteric and difficult branch of mathematics which no agronomist can hope to master; and it must be admitted that existing texts, even Searle's (1966) admirable 'Matrix Algebra for the Biological Sciences', do nothing to allay such suspicions. It is true that any branch of mathematics can be difficult; it is only necessary to browse through the opening pages of Frege's (1950) 'Foundations of Arithmetic' to discover that the statement $1 + 1 = 2$ conceals some formidable philosophical difficulties. Yet it is also true that, provided one does not wish to know too much, many branches of mathematics can be surprisingly easy. Matrix algebra is like this. If an agronomist wishes to indulge in original research into the methodology of pattern analysis, much work lies ahead; but if he only wishes to *use* the methods, he need know very little. His main difficulty, in fact, will be linguistic. Like every subject in the world, matrix algebra has developed its own technical language; once the agronomist has become familiar with this language and has learnt to carry out one or two simple arithmetic operations, he has all he needs to know to enable him to read and understand those agronomic papers which have used matrix methods. The purpose of these opening chapters is to provide just this basic knowledge.

Definitions: Matrix, Dimensions, Vector, Scalar

A matrix is simply a rectangular block of numbers; and such blocks arise in agriculture whenever more than one measurement is made on more than one animal or plant. The following data have been extracted from those used by Dr Winter in Part 3, Case 3.19; they represent the weight in kilograms of protein and fat in the carcasses of three sheep.

	Protein (kg)	Fat (kg)
Sheep No. 1	2·80	6·44
Sheep No. 2	3·26	9·64
Sheep No. 3	3·91	18·69

We call such a block of numbers a *data matrix*.

If we wish to make *general* statements about matrices we are not immediately concerned with the nature of the measurements or of the animal or plant on which they were made; we are interested only in the numbers themselves. We need, however, to show that they belong together in the same matrix, so we enclose them in square brackets:

$$\begin{bmatrix} 2\cdot80 & 6\cdot44 \\ 3\cdot26 & 9\cdot64 \\ 3\cdot91 & 18\cdot69 \end{bmatrix}. \tag{1.1}$$

In some earlier works ordinary brackets ( ) or curly brackets { } were used, but square brackets are now usual. It is, however, forbidden to use simple vertical lines on either side of the matrix, since, as we shall see later, this has a special meaning in matrix algebra. The individual entries—2·80, 6·44 and so on—are called the *elements* of the matrix.

This matrix has three rows and two columns; we say that it has *dimensions* (3 × 2), since in matrix work the rows are always specified first, the columns second.

If we had recorded only protein, the matrix would reduce to

$$\begin{bmatrix} 2\cdot80 \\ 3\cdot26 \\ 3\cdot91 \end{bmatrix}.$$

It would have only a single column, and dimensions (3 × 1). If we had carried out both measurements on a single sheep, the matrix would reduce to

$$[2\cdot80 \quad 6\cdot44].$$

It would now have only a single row, and dimensions (1 × 2). A matrix with only a single column or single row is called a *vector;* the protein values for the three sheep would constitute a *column vector*, the two measurements for the single sheep would constitute a *row vector*.

We sometimes need to carry out calculations using a number which is not itself part of a matrix; for example, if we wished to convert our measurements from kilograms to grams, we should need to multiply every element of the matrix by 1000. A number such as 1000, which is just a number and not part of a matrix, we call a *scalar*.

MATRIX TERMINOLOGY

Symbols

If we wish to be still more general, we must abandon our specific set of numbers and replace them by algebraic symbols, i.e. letters. We might choose to use the letter a. We then need to be able to show, for any given element, in which row and in which column it is to be found. For this we need to provide two subscripts, one for the row number and one for the column number; as always, the row subscript comes first. The general form of our (3×2) matrix will then be:

$$\begin{bmatrix} a_{11} & a_{12} \\ a_{21} & a_{22} \\ a_{31} & a_{32} \end{bmatrix}.$$

If we wish to be still more general, and identify the element in the ith row and jth column, we should write it as a_{ij}.

If we are referring to the matrix whose general term is a_{ij} we sometimes write it as (a_{ij}). More usually we require a single symbol, and we then use a single capital letter, e.g. **A**. 'Bold face' (black type) is traditionally used, but if this is not available an italic capital A is used instead. For a vector we use a lower-case letter, e.g. **a**; the context will decide whether this is a row or column vector.

Transpose of a Matrix

We might as easily have written our original matrix example (1.1) with the measurements (protein and fat) as rows and the three sheep as columns; the matrix would then appear as

$$\begin{bmatrix} 2 \cdot 80 & 3 \cdot 26 & 3 \cdot 91 \\ 6 \cdot 44 & 9 \cdot 64 & 18 \cdot 69 \end{bmatrix}.$$

The rows have become columns and the columns have become rows; the dimensions are now (2×3). We call this the *transpose* of the original matrix. If the latter had been symbolized by **A**, we symbolize the transpose by **A'**. If the operation were to be carried out twice we should, of course, be back where we started, i.e.

$$(\mathbf{A'})' = \mathbf{A}.$$

The Square Matrix

We shall now add a further measurement to our original two. The weight of water in kilograms for the three sheep carcasses was $10 \cdot 05$, $11 \cdot 96$ and $13 \cdot 16$ respectively. The new matrix will be:

$$\begin{bmatrix} 2 \cdot 80 & 6 \cdot 44 & 10 \cdot 05 \\ 3 \cdot 26 & 9 \cdot 64 & 11 \cdot 96 \\ 3 \cdot 91 & 18 \cdot 69 & 13 \cdot 16 \end{bmatrix}. \tag{1.2}$$

This matrix has as many columns as rows, and we describe it as a *square matrix*. Its dimensions are (3 × 3); but we more usually say that it is of *order* 3. An important part of this matrix is the diagonal string of numbers starting at the top left-hand corner and ending at the bottom right-hand corner, i.e. the numbers $2 \cdot 80$, $9 \cdot 64$, $13 \cdot 16$; we call this the *principal diagonal* of the matrix. The sum of the elements in the principal diagonal is called the *trace* of the matrix; in this case this is $2 \cdot 80 + 9 \cdot 64 + 13 \cdot 16$, i.e. $25 \cdot 60$.

We shall now suppose that Dr Winter had been interested in the inter-relationships between his three measurements, and that over the complete experimental set of 57 sheep he had calculated correlation coefficients between protein and fat ($+0 \cdot 642$), protein and water ($+0 \cdot 936$), and fat and water ($+0 \cdot 605$). The correlation between any measurement and itself is, of course, always $+1$; and we could write the complete set of correlation coefficients in the form of a *correlation matrix* as follows:

$$
\begin{array}{c}
\text{Protein} \\[4pt]
\text{Fat} \\[4pt]
\text{Water}
\end{array}
\begin{array}{ccc}
\text{Protein} & \text{Fat} & \text{Water} \\
\left[\begin{array}{ccc}
1 \cdot 000 & 0 \cdot 642 & 0 \cdot 936 \\
0 \cdot 642 & 1 \cdot 000 & 0 \cdot 605 \\
0 \cdot 936 & 0 \cdot 605 & 1 \cdot 000
\end{array}\right].
\end{array}
\qquad (1.3)
$$

This matrix has an important property. The correlation between protein and fat is the same as the correlation between fat and protein, and this is true for all pairs. It follows that the element in the ith row and jth column is the same as the element in the jth row and ith column, and we can write this as $a_{ij} = a_{ji}$. Such a matrix, which is symmetrical around the principal diagonal, is said to be a *symmetric* matrix. It follows that if such a matrix is transposed, so that the rows become columns and the columns become rows, the matrix is unaltered; in other words, a symmetric matrix equals its own transpose, or $\mathbf{A}' = \mathbf{A}$. In matrix (1.2) on the other hand, although it might *happen* that the fat content of sheep 1 was the same as the protein content of sheep 2, so that $a_{12} = a_{21}$, this would not necessarily—or even usually—be true; so that for a matrix such as (1.2) we have in general $a_{ij} \neq a_{ji}$ and $\mathbf{A}' \neq \mathbf{A}$. Such a matrix is said to be *asymmetric*.

The symmetric matrix we have examined here has all the elements in the principal diagonal identical, but this is not necessarily the case. Instead of calculating the correlation coefficients, we might have calculated the variance of each of the measurements and the covariance between all pairs of measurements. The resulting *variance–covariance matrix* is then

$$
\left[\begin{array}{ccc}
0 \cdot 196 & 1 \cdot 126 & 0 \cdot 638 \\
1 \cdot 126 & 15 \cdot 651 & 3 \cdot 679 \\
0 \cdot 638 & 3 \cdot 679 & 2 \cdot 362
\end{array}\right].
\qquad (1.4)
$$

This matrix is still a symmetric matrix, since everywhere $a_{ij} = a_{ji}$ (and, of course, of necessity $a_{ii} = a_{ii}$).

Finally, it should be noted that if a matrix is symmetric, the whole of its information is contained in the principal diagonal and the elements to its right (the *upper triangle*) or its left (the *lower triangle*). Use is made of this fact in preparing symmetric matrices for computing purposes, since it reduces the number of elements that have to be punched. In the case of a correlation matrix, there is no need to punch the principal diagonal, since it is known in advance that every element in it is 1, but this is not true of a variance–covariance matrix such as matrix (1.4) above. It is wise, if using a term such as 'upper triangle', to specify whether or not the principal diagonal is to be included.

Two special types of square matrix will occasionally be encountered. Suppose that in matrix (1.4) the three measurements had been completely uncorrelated. The matrix would then have appeared as

$$\begin{bmatrix} 0 \cdot 196 & 0 & 0 \\ 0 & 15 \cdot 651 & 0 \\ 0 & 0 & 2 \cdot 362 \end{bmatrix}.$$

Such a matrix, with non-zero elements in the principal diagonal but all off-diagonal elements zero, is called a *diagonal matrix*.

The last example is artificial. Consider the following matrix

$$\begin{bmatrix} 0 & 2 & 4 \\ -2 & 0 & 5 \\ -4 & -5 & 0 \end{bmatrix}.$$

Examination of this matrix will show that the element in the ith row and the jth column has the same absolute *value* as the element in the jth row and the ith column, but the opposite sign; in the algebraic symbolism, $a_{ij} = -a_{ji}$. But let us consider an element of the principal diagonal, in the ith row and ith column, that is to say a_{ii}. To be consistent over the matrix we shall want to say that $a_{ii} = -a_{ii}$; but the only number which remains unchanged if its sign is changed is zero. In algebraic symbolism, we may say that if $a_{ii} = -a_{ii}$ then $a_{ii} = 0$. The matrix must therefore have a string of zeros down the principal diagonal. Such a matrix is said to be *anti-symmetric* or *skew-symmetric*; the use of such matrices has been suggested in plant ecology, so they may yet appear in agronomic literature.

References

Frege, G. (1950). 'The Foundations of Arithmetic'. (Trans. J. L. Austin.) (Blackwell: Oxford.)

Searle, S. R. (1966). 'Matrix Algebra for the Biological Sciences'. (Academic Press: London.)

2

Manipulation of Matrices

W. T. Williams

The basic idea of using matrices is to enable us to deal with whole blocks of numbers at once instead of having to process them one at a time. It follows that we want to be able to carry out with matrices the same sort of arithmetic operations that we do with single numbers, i.e. addition, subtraction, multiplication and division. It is important to realize that matrices have no mysterious properties which determine such arithmetic rules for us; we are at liberty to devise any rules that we find convenient. However, there would be little point in each of us devising special rules for his own special purposes, because we all hope to be able to use standard computer programs for our matrix operations; in practice, therefore, we accept the rules that have been found convenient by other workers during the period since matrices were first defined well over a century ago. In this chapter we deal with the rules for addition, subtraction and multiplication; division poses some special problems, and we shall defer it till the next chapter.

Addition and Subtraction of Matrices

Of the three matrices in (2.1) below, consider matrices (i) and (ii). These represent a small part of the data from an experiment reported by Edye *et al.* (1975), in which a number of accessions of the legume *Stylosanthes* were grown at a number of climatically different sites over several years. The individual entries represent kilograms of dry matter produced in a year; the three rows in each case represent three species of *Stylosanthes;* and the two columns represent two different sites. The two matrices represent the values for (i) 1971 and (ii) 1972.

$$\begin{bmatrix} 139 & 213 \\ 38 & 220 \\ 557 & 1021 \end{bmatrix} + \begin{bmatrix} 956 & 913 \\ 212 & 3843 \\ 4046 & 2321 \end{bmatrix} = \begin{bmatrix} 1095 & 1126 \\ 250 & 4063 \\ 4603 & 3342 \end{bmatrix}. \qquad (2.1)$$

$$\text{(i)} \qquad\qquad\qquad \text{(ii)} \qquad\qquad\qquad \text{(iii)}$$

Suppose we now ask, what was the total dry-matter production of each species at each site over the two years? We shall obviously add the values for species 1 and site 1 in the two years; in other words, we add 139 to 956 and obtain the answer 1095. We do the same for the remaining five possibilities, and obtain the matrix (iii), where every element is the sum of the corresponding elements in the other two matrices. This is how we define matrix addition; we simply add corresponding elements in the two matrices.

There is, however, a possible difficulty. Suppose a fourth species had been grown in 1972 but not in 1971, because insufficient seed had been available in the first year. We should now have an extra row in matrix (ii); in matrix terms, the dimensions of matrix (i) would still be 3×2, but those of the new matrix (ii) would be 4×2. You cannot add two numbers together if one of them is missing, and it would be grossly misleading to pretend that the missing number was zero; if the species *had* been grown in 1971 it might have produced very well. It follows that two matrices cannot be added unless both have the same number of rows and the same number of columns, as in (2.1) above both matrices have three rows and two columns, so that both are of dimensions (3×2). Matrices can thus be added only if they have identical dimensions; two matrices which fulfil this condition are said to be *conformable for addition.*

Suppose we were given matrix (iii) of (2.1) above, and also matrix (ii)— the values for 1972—but we had contrived to mislay the values for 1971 and wanted them. We should obviously carry out the same process backwards: we should subtract every element in matrix (ii) from the corresponding element in matrix (iii), and so obtain matrix (i). Again, therefore, we simply subtract corresponding elements; and it follows that if two matrices are conformable for addition, they are also *conformable for subtraction.*

Multiplication of Matrices

Multiplication provides a trap for the unwary beginner: after so much talk of corresponding elements, he can be forgiven if he jumps to the conclusion that one simply multiplies corresponding elements. This is in fact occasionally done for very special purposes; but it has not been found generally useful, so this is not the way in which we define matrix multiplication. It will be convenient to consider three possibilities in order: multiplication of a matrix by a scalar, by a vector and by another matrix.

Multiplication by a scalar

Suppose, in (2.1) above, we wished to change the values from kilograms to grams. We should have to multiply every element in each matrix by 1000, which, because it is a number and not a block of numbers, is a scalar. It is obviously sensible to say that we have multiplied each *matrix* by the scalar 1000. This is therefore how we define multiplication of a matrix by a scalar: we simply multiply every element of the matrix by the scalar in question. In symbolic language, consider a matrix $\mathbf{A}$ with general term a_{ij}, and multiply it by the scalar n; the result, $n\mathbf{A}$, will itself be a matrix of the same dimensions as $\mathbf{A}$, but with general term na_{ij}.

Multiplication by a vector

The example we shall quote represents a small part of the data from an experiment reported by Williams *et al.* (1971), designed to investigate whether application of superphosphate to a paddock increased the conception rate of Droughtmaster cows. There were 3 phosphate treatments, P_0, P_1 and P_3; in each treatment there were 8 paddocks; and in each paddock there were 4 cows. In any 1 year, therefore, there could be 0, 1, 2, 3 or 4 calves born in each paddock (in the year of the data we shall use there was in fact no 0 record). In (2.2) the individual entries represent the number of paddocks (out of 8) in the treatment defined by the row which produced the number of calves defined by the columns:

$$
\begin{array}{lcccc}
\text{No. of calves:} & 1 & 2 & 3 & 4 \\
P_0 & 1 & 2 & 4 & 1 \\
P_1 & 1 & 3 & 1 & 3. \\
P_3 & 0 & 2 & 2 & 4
\end{array}
\tag{2.2}
$$

Suppose we now wish to find out how many calves were produced in each treatment in the year in question. We start with P_0. One paddock produced 1 calf, a total of (1×1) or 1 calf; 2 paddocks produced 2 calves each, a total of (2×2) or 4 calves; 4 paddocks produced 3 calves each, a total of (4×3) or 12 calves; and 1 paddock produced 4 calves, a total of (1×4) or 4 calves; grand total $1 + 4 + 12 + 4$, or 21 calves. Carrying out the same calculations for the other rows, we find that P_1 produced 22 calves and P_3 produced 26. Now, this type of calculation is something we often wish to do; so let us *define* matrix multiplication in such a way that it will do it for us. The entries of the above table can obviously be written as a matrix; and we shall write the calf numbers (1, 2, 3, 4), which were used as multipliers, in the form of a vector. We can then write the calculation as follows:

$$
\begin{bmatrix} 1 & 2 & 4 & 1 \\ 1 & 3 & 1 & 3 \\ 0 & 2 & 2 & 4 \end{bmatrix} \cdot \begin{bmatrix} 1 \\ 2 \\ 3 \\ 4 \end{bmatrix} = \begin{bmatrix} 21 \\ 22 \\ 26 \end{bmatrix}.
\tag{2.3}
$$

What we have done is to take the first element in the first *row* of the matrix and multiply it by the first element in the *column* which is the vector; take the second element in the first row of the matrix and multiply it by the second element in the column of the vector; and so on. We have in fact taken a *row* of the matrix and multiplied it by a *column* which is the vector; the elements which have to correspond are those in the row of the matrix and the column of the vector. Now, there are as many elements in a row of the matrix as there are columns; since the matrix has dimensions (3×4), each row contains four elements. There are as many elements in the column vector as it has rows; since the vector has dimensions (4×1), its column has four elements. We shall be unable to complete the multiplication unless there are as many columns in the matrix as there are rows in the vector; in this case the dimensions are respectively (3×4) and (4×1), so the multiplication is possible; we say that the matrix and the vector are *conformable for multiplication*. The result is a column vector of three elements, so we could summarize the change in dimensions by the sequence

$$(3 \times 4) \cdot (4 \times 1) \rightarrow (3 \times 1).$$

Note that the 4s 'cancel out', as it were; this often provides a useful check as to whether the multiplication has been carried out correctly.

Multiplication by another matrix

Suppose that, in the previous example, we were primarily interested not in the number of calves born but in the number of *failures* to calve (out of 4 in each paddock); this was in fact the case in the experiment in question. In (2.2) above, all that would change would be the heading and the numbers at the head of the columns; instead of 'number of calves: 1, 2, 3, 4' we should have 'failures to calve: 3, 2, 1, 0'. We should carry out the matrix multiplication (2.3) in the same way, but the multiplying vector would be

$$\begin{bmatrix} 3 \\ 2 \\ 1 \\ 0 \end{bmatrix}$$

and the result would be

$$\begin{bmatrix} 11 \\ 10 \\ 6 \end{bmatrix}.$$

Now suppose that when we undertook our calculations, we were not sure which way round was going to be more useful, so we decided to carry out both sets of calculations. We could obviously set out the result as two matrix $\times$ vector multiplications, both looking like equation (2.3), but with different multiplying vectors and different column vectors as answers. Why not write them out as a single equation?

$$\begin{bmatrix} 1 & 2 & 4 & 1 \\ 1 & 3 & 1 & 3 \\ 0 & 2 & 2 & 4 \end{bmatrix} \cdot \begin{bmatrix} 1 & 3 \\ 2 & 2 \\ 3 & 1 \\ 4 & 0 \end{bmatrix} = \begin{bmatrix} 21 & 11 \\ 22 & 10 \\ 26 & 6 \end{bmatrix}. \qquad (2.4)$$

In effect we have multiplied a matrix of dimensions (3×4) by a matrix of dimensions (4×2), and we again see that

$$(3 \times 4) \cdot (4 \times 2) \longrightarrow (3 \times 2).$$

The columns of the second matrix are simply regarded as separate vectors, and we have multiplied one matrix by another.

Although anybody who handles matrices frequently soon finds that this 'row by column' principle of multiplication becomes automatic, it must be admitted that workers new to the field sometimes encounter difficulties; at the critical moment they cannot remember which element is to be multiplied by which. The beginner is strongly advised to make up pairs of little matrices for himself, check that they are conformable for multiplication, multiply them, check that the dimensions of the resulting matrix are what they should be, and perhaps persuade a friendly biometrician to check his actual answers.

Pre- and Post-multiplication

If we are adding ordinary numbers (scalars) it is of no importance what order we add them in: $(a + b)$ is always the same as $(b + a)$. It will be clear from the rules for addition given above that the same is true of matrices: $(\mathbf{A} + \mathbf{B})$ is the same as $(\mathbf{B} + \mathbf{A})$. Again, with ordinary numbers the same is true of multiplication: ab is always the same as ba. When we are multiplying matrices, this happy situation breaks down. Consider equation (2.4) above, and suppose we tried to carry out the multiplication the other way round. The right-hand matrix has dimensions (4×2) and the left-hand matrix has dimensions (3×4). If we reverse these and try to multiply a matrix of dimensions (4×2) by a matrix of dimensions (3×4) we cannot do it: the number of columns of the first matrix is not equal to the number of rows of the second matrix, and the two matrices are not conformable for multiplication. The product, this way round, simply cannot exist. In other words, given two matrices $\mathbf{A}$ and $\mathbf{B}$, the product $\mathbf{AB}$ may exist, but the product $\mathbf{BA}$

may not. But let us consider a case in which *both* can exist. We postulate two matrices, **A** with dimensions (3×2) and **B** with dimensions (2×3). If we multiply **A** by **B**, the dimension sequence will be

$$(3 \times 2) . (2 \times 3) \longrightarrow (3 \times 3).$$

But if we multiply **B** by **A** the sequence will be

$$(2 \times 3) . (3 \times 2) \longrightarrow (2 \times 2).$$

It is obvious that the answers cannot be the same, since the dimensions of the two answers are different (the reader might invent a simple case and try it for himself). In other words even if **AB** and **BA** both exist they are in general unequal; or, in symbolic language, **AB** $\neq$ **BA**. In technical language, matrix multiplication is *non-commutative*. This means that we must always specify the *order* in which matrix multiplication is to be taken; if the product is **AB**, we say that **A** is *post-multiplied* by **B** or that **B** is *pre-multiplied* by **A**.

We shall now consider a simple statistical example, which in this case is fictitious. In (2.5) the values of x and y represent lengths and breadths of five leaves; and we wish to know whether length and breadth are correlated (nobody is likely to pay much attention to a correlation coefficient based on five observations, but we shall let that pass).

$$
\begin{array}{cc}
x & y \\
5 & 2 \\
3 & 1 \\
2 & 1 \\
6 & 4 \\
4 & 2
\end{array}
\qquad (2.5)
$$

In order to calculate a correlation coefficient we first need the sum of the squares of all the x values $(\sum x^2)$, that is $5^2 + 3^2 + 2^2 + 6^2 + 4^2$, which is 90. Next we need $\sum y^2 : 2^2 + 1^2 + 1^2 + 4^2 + 2^2$, or 26; and then the sum of the cross-products $(\sum xy)$, or $(5 \times 2) + (3 \times 1) + (2 \times 1) + (6 \times 4) + (4 \times 2)$, or 47. Can we do this calculation by matrix manipulation?

In fact we can. Take the matrix of observations in (2.5) and *pre-multiply it by its own transpose*. The multiplication is

$$
\begin{bmatrix}
5 & 3 & 2 & 6 & 4 \\
2 & 1 & 1 & 4 & 2
\end{bmatrix}
\begin{bmatrix}
5 & 2 \\
3 & 1 \\
2 & 1 \\
6 & 4 \\
4 & 2
\end{bmatrix}.
$$

We are to multiply a matrix of dimensions (2×5) by one of dimensions (5×2), so the answer must have dimensions (2×2). If we call the observations in (2.5) the matrix $\mathbf{A}$, then its transpose—the left-hand matrix in our multiplication—will be $\mathbf{A}'$. We begin by multiplying the first row of $\mathbf{A}'$ by the first column of $\mathbf{A}$, to give us the first element in the first column of the answer; we have $(5 \times 5) + (3 \times 3) + (2 \times 2) + (6 \times 6) + (4 \times 4)$, which is 90, or $\sum x^2$. To obtain the second element in the first column of our answer, we multiply the second row of $\mathbf{A}'$ by the first column of $\mathbf{A}$; this gives us $(5 \times 2) + (3 \times 1) + (2 \times 1) + (6 \times 4) + (4 \times 2)$, which is 47, or $\sum xy$. To get the second column of our answer we repeat the process using the second column of $\mathbf{A}$; the complete answer is

$$\begin{bmatrix} 90 & 47 \\ 47 & 26 \end{bmatrix} \text{ or } \begin{bmatrix} \sum x^2 & \sum xy \\ \sum xy & \sum y^2 \end{bmatrix}.$$

This trick, of pre-multiplying a matrix of observations by its own transpose, to give a matrix of squares and cross-products—what is called in the trade a *dispersion matrix*—is a basic operation in multivariate statistics and in most modern ordination procedures.

Finally, as an exercise, the reader might care to verify that, if $\mathbf{A}$ and $\mathbf{B}$ are any two matrices conformable for multiplication, the following is always true:

$$(\mathbf{AB})' = \mathbf{B}'\mathbf{A}'.$$

In other words, the transpose of the product of two matrices is the product of the two transposes, taken in the reverse order.

The Identity Matrix

Consider the diagonal matrix, of any order, which has unities (the number 1) all down its principal diagonal and zero everywhere else. We shall find out what happens if we pre- or post-multiply other matrices by a matrix of this nature. The *Stylosanthes* matrix (2.1 (i)) will serve as an example. The matrix is

$$\begin{bmatrix} 139 & 213 \\ 38 & 220 \\ 557 & 1021 \end{bmatrix}. \tag{2.6}$$

Since this has 2 columns, we can only post-multiply it by a matrix which has 2 rows; we are going to multiply it by a square matrix (2×2), so that our dimension sequence will be

$$(3 \times 2) \cdot (2 \times 2) \longrightarrow (3 \times 2).$$

The reader should now have no difficulty in verifying the following equation:

$$\begin{bmatrix} 139 & 213 \\ 38 & 220 \\ 557 & 1021 \end{bmatrix} \cdot \begin{bmatrix} 1 & 0 \\ 0 & 1 \end{bmatrix} = \begin{bmatrix} 139 & 213 \\ 38 & 220 \\ 557 & 1021 \end{bmatrix}.$$

Now, if we wish to pre-multiply matrix (2.6) by such a matrix, since (2.6) has 3 rows, the matrix we use must have 3 columns, so that the dimension sequence will be

$$(3 \times 3) \cdot (3 \times 2) \longrightarrow (3 \times 2).$$

Again, the reader will find that

$$\begin{bmatrix} 1 & 0 & 0 \\ 0 & 1 & 0 \\ 0 & 0 & 1 \end{bmatrix} \cdot \begin{bmatrix} 139 & 213 \\ 38 & 220 \\ 557 & 1021 \end{bmatrix} = \begin{bmatrix} 139 & 213 \\ 38 & 220 \\ 557 & 1021 \end{bmatrix}.$$

In other words, this remarkable matrix is such that we can make it of whatever order we choose; if a matrix is either post-multiplied or pre-multiplied by such a matrix, it is unchanged. This matrix occupies exactly the same position in matrix arithmetic as the number 1 does in ordinary arithmetic—multiplying by it leaves things exactly as they were before. For this reason it is called the *identity matrix*, and is always represented by $\mathbf{I}$; in some earlier books it is called the 'unit matrix', but this term is now not usual.

References

Edye, L. A., Williams, W. T., Anning, P., Holm, A. McR., Miller, C. P., Page, M., and Winter, W. H. (1975). Sward tests of some morphological–agronomic groups of *Stylosanthes* accessions in dry-tropical environments. *Aust. J. Agric. Res.* **26**, 481–96.

Williams, W. T., Haydock, K. P., Edye, L. A., and Ritson, J. B. (1971). Analysis of a fertility trial with Droughtmaster cows. *Aust. J. Agric. Res.* **22**, 979–91.

3

Determinants: The Inverse of a Matrix

W. T. Williams

The Origin of Determinants

At some stage in his early youth, every agronomist will have been caused to learn how to solve simultaneous equations in two unknowns, x and y. He will be presented with equations such as the following:

$$\left.\begin{aligned} ax + by &= p, \\ cx + dy &= q. \end{aligned}\right\} \qquad (3.1)$$

We begin by finding x. To do this we must *eliminate y;* we multiply the first equation by d and the second by b, obtaining:

$$adx + bdy = dp,$$
$$bcx + bdy = bq.$$

Subtracting we obtain

$$x(ad - bc) = dp - bq,$$

whence

$$x = (dp - bq)/(ad - bc).$$

Similarly, we find that

$$y = (aq - cp)/(ad - bc).$$

It is the denominator of these two fractions that interests us. If we examine the coefficients of x and y in the original pair of equations (3.1), we see that we should obtain the value of the denominator by a process of cross-multiplication:

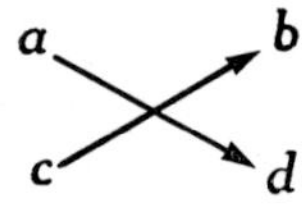

16

To avoid the inconvenience of having to print diagonal arrows, we symbolize the process as follows

$$\begin{vmatrix} a & b \\ c & d \end{vmatrix}. \tag{3.2}$$

Since this arrangement arises from eliminating either x or y from the original equations, (3.2) was originally called an *eliminant*; but the term has dropped out of use, and (3.2) is now always known as a *determinant*. The reader must be careful to distinguish between (i) and (ii) below:

$$\begin{bmatrix} a & b \\ c & d \end{bmatrix} \qquad \begin{vmatrix} a & b \\ c & d \end{vmatrix}$$
$$\text{(i)} \qquad\qquad \text{(ii)}$$

(i) is a *matrix*: a 2×2 block of numbers, with no necessary connexion with each other apart from the experimental conditions from which they have arisen. (ii) is a *determinant*; it is a single number, a scalar, the value of which can be found by *evaluating* the determinant, i.e. by carrying out the calculation symbolized by $(ad - bc)$. Nevertheless, we are at liberty to say that (ii) is the determinant *associated with* the matrix (i); and if we write the matrix (i) as $\mathbf{A}$, we could write the associated determinant (ii) as $|\mathbf{A}|$. We shall see shortly that the determinant associated with a matrix can tell us something important about the matrix that we may need to know.

Consider the two (2×2) matrices below:

$$\begin{bmatrix} 6 & 3 \\ 8 & 4 \end{bmatrix} \qquad \begin{bmatrix} 7 & 3 \\ 1 & 4 \end{bmatrix}$$
$$\text{(i)} \qquad\qquad \text{(ii)}$$

Let us evaluate their associated determinants. That of (i) is $(6 \times 4 - 8 \times 3)$, which $= 0$; that of (ii) is $(7 \times 4 - 3 \times 1)$, which $= 25$. A matrix like (i), of which the associated determinant is 0, is said to be *singular*; and singular matrices, which are very common in ecology and not uncommon in agronomy, can cause difficulties. In particular, should the reader ever find himself involved in multivariate statistics (such as the multivariate analysis of variance), he will find that the significance tests depend on the evaluation of determinants; and that, if the matrices concerned happen to be singular, the tests will automatically fail. A similar difficulty arises in certain ordination procedures. In contrast, matrix (ii) above is said to be *non-singular*.

Higher-order Determinants

Suppose we had, instead of the pair of equations (3.1) in x and y, a set of three equations, in x, y and z:

$$ax + by + cz = p,$$
$$dx + ey + fz = q,$$
$$gx + hy + kz = r.$$

By analogy from the pair of equations (3.1) we should expect that when we have found the fractions for x, y and z, they will have a common denominator; and we should obviously like to generalize the system we have already set up, and represent this denominator by the three-rowed determinant:

$$\begin{vmatrix} a & b & c \\ d & e & f \\ g & h & k \end{vmatrix}. \tag{3.3}$$

The problem is, how are we to evaluate this? The only way for us is to do it the hard way—to work laboriously through the algebra, find out what the answer ought to be, and then see how we can get it from (3.3). When we have slogged through the algebra, we find that the answer is an expression of six terms:

$$aek - afh - bdk + bfg + cdh - ceg.$$

We could rewrite this as follows:

$$a(ek - fh) - b(dk - fg) + c(dh - eg).$$

But the expressions in the brackets could obviously be rewritten as two-rowed determinants:

$$a\begin{vmatrix} e & f \\ h & k \end{vmatrix} - b\begin{vmatrix} d & f \\ g & k \end{vmatrix} + c\begin{vmatrix} d & e \\ g & h \end{vmatrix}. \tag{3.4}$$

If we now carefully compare this expression with (3.3), we can see how to turn one into the other. Consider the first row of (3.3) and take the first element, a. Strike out the rest of the row and column in which it occurs, and express what is left as a two-rowed determinant; this is the *two-rowed minor* corresponding to a. Now move along to the second element, b, change the sign, and do the same thing. Finally, move along to the third element, c, change the sign again, and repeat the process; we finish with expression (3.4) above, and from this we can evaluate the determinant. The process is very general; we have used the first row; but (providing we are careful about sign changes) *any* row or column we care to choose will give the same result. In practice, there are ways of simplifying the computation; but since no agronomist is likely to have to evaluate determinants by hand, we shall not consider them here.

Finally, it will be noted that we have tacitly assumed that all determinants are square. In fact they are. It would no doubt be possible to assign a meaning to the concept of a non-square determinant if anybody had found the idea useful; but nobody ever has, so determinants remain irrevocably square. It follows that only a square matrix has an associated determinant.

Arithmetic Manipulation of Determinants

It will be remembered that when dealing with matrices, we were free to define any rules of arithmetic that we found useful; but this is not true of determinants. For each determinant ultimately represents a single number, and any operation we undertake on two determinants must produce, when the result has been evaluated, the same answer as would have been produced by operating on the single numbers that the original determinants represented. As a result, the rules for the arithmetic manipulation of determinants are quite different from the corresponding rules for matrices. We shall not discuss the rules here, because agronomists are unlikely ever to need them. It is not necessary for an agronomist to be able to manipulate determinants; it is only necessary that he should know what they are.

The Problem of Division

Suppose we wished to divide an ordinary number—*any* number—by, say, 7, and we had no rules for doing it. The first thing we should want to do would be to *generalize* the process; we should not want the rules for dividing 33 by 7 to be quite different from the rules for dividing 192 by 7. We can get over this—we frequently do it on desk-calculators—by *multiplying* the number in question by 1/7. Whatever the number we wish to divide, we have now reduced the problem to devising rules for dividing 1 by 7. We have already assumed that we know how to multiply, though not to divide; so we have to recast the problem of dividing 1 by 7 into a form of multiplication. What we say is this: let there be a quantity x which satisfies the equation $7x = 1$; i.e. we are to find a number x such that, when it is multiplied by 7, the answer will be 1. If we can find it we shall call it the *reciprocal* of 7, and write it as 7^{-1}. In general, to divide anything by a number a, we need to find a quantity x such that $ax = 1$; and we shall call x the reciprocal of a, and write it as a^{-1}.

Let us consider two special cases. Suppose $a = 0$. There is then *no* finite value of x which will satisfy the equation $ax = 1$; in other words, the reciprocal of a does not exist. But now let us suppose that a, though non-zero, is very small; perhaps it represents the parts per million of a trace element in some plant dry matter. The answer might be 0·001 or

$0 \cdot 002$, but the distinction between these might be within the known errors of our analytical determination. Nevertheless, though the difference between the two actual values is negligible, the discrepancy between their reciprocals is enormous—1000 and 500. In a case like this, when a tiny change in the input data—likely to be due to experimental error—produces a very large change in the answer, we say that the calculation (in this case the calculation of the reciprocal) is *ill conditioned*. Note that it is the *calculation* which is ill conditioned, not the number or its reciprocal.

We have so far been considering division by a scalar; but now suppose we wish to divide one matrix by another square matrix $\mathbf{A}$. $\mathbf{A}$ now corresponds exactly to our scalar number a in the previous paragraph, and the matrix corresponding to the number 1 is the identity matrix $\mathbf{I}$. It follows that to be able to divide by a matrix $\mathbf{A}$ we need to find a matrix $\mathbf{X}$ which satisfies the matrix equation $\mathbf{AX} = \mathbf{I}$. If we can find such a matrix we usually call it not the reciprocal, but the *inverse* of $\mathbf{A}$; and again we write it as $\mathbf{A}^{-1}$. We note in passing (but shall not attempt to prove) that if the matrix can be found, it will satisfy not only the equation $\mathbf{AX} = \mathbf{I}$, but also the equation $\mathbf{XA} = \mathbf{I}$. In other words, it does not matter whether a square matrix $\mathbf{A}$ is pre- or post-multiplied by its inverse $\mathbf{A}^{-1}$; the result is still $\mathbf{I}$. We shall not here be concerned with the computation of such a matrix; nowadays there are always computer programs to carry out the task (which by hand is excessively laborious). We are, however, concerned to ascertain whether the matrix $\mathbf{X}$ always exists.

In fact, sometimes it does not, and for a reason very similar to that of the previous scalar example. If $|\mathbf{A}| = 0$, i.e. if $\mathbf{A}$ is singular, there is *no* matrix $\mathbf{X}$ that will satisfy the equation $\mathbf{AX} = \mathbf{I}$. Furthermore, suppose that $|\mathbf{A}|$, the determinant associated with $\mathbf{A}$, is very small; we shall again find that very small changes in the elements of $\mathbf{A}$—probably due to experimental error—produce very large changes in the elements of $\mathbf{X}$. Again, we say that the calculation of the inverse $\mathbf{A}^{-1}$ is *ill conditioned*. You may occasionally encounter the statement that 'the matrix $\mathbf{A}$ is ill conditioned'. This statement is erroneous, since only a calculation can be ill conditioned, and not all calculations concerned with $\mathbf{A}$ fall into this category; for example, the calculation of the determinant $|\mathbf{A}|$ is never ill conditioned, since small changes in the elements of $\mathbf{A}$ produce only small changes in $|\mathbf{A}|$. What people mean when they say that a matrix is ill conditioned is that the calculation of its *inverse* is ill conditioned.

Extensions of the 'Inverse' Concept

In the foregoing section we have tacitly assumed that $\mathbf{A}$ is always square. In fact, it need not be; even though $\mathbf{A}$ is not square, there still may be a matrix $\mathbf{X}$ which satisfies the equation $\mathbf{AX} = \mathbf{I}$. However, the solution may not now be unique; there may be a whole family of matrices $\mathbf{X}$ all of which satisfy the equation. Moreover, the matrix $\mathbf{X}$ which satisfies the equation $\mathbf{AX} = \mathbf{I}$ may not be the same matrix $\mathbf{X}$ as that which satisfies the equation $\mathbf{XA} = \mathbf{I}$. The rules governing the inverses of rectangular (non-square) matrices are in fact more complex and recondite than those we have outlined here; but they have not yet appeared in any agronomic work known to me, so for the moment they can safely be neglected.

The concept of the *generalized inverse* is different; it has already begun to appear in ecology, and so may yet rear its head in agricultural work. It is therefore advisable that agronomists should at least know what it is about. Let us consider the basic 'inverse' equation:

$$\mathbf{AX} = \mathbf{I}. \tag{3.5}$$

We already know that if $|\mathbf{A}| = 0$, there is no matrix $\mathbf{X}$ that will satisfy the equation. Now let us post-multiply both sides of equation (3.5) by $\mathbf{A}$, so that we obtain

$$\mathbf{AXA} = \mathbf{IA} = \mathbf{A}. \tag{3.6}$$

A matrix $\mathbf{X}$ can always be found to satisfy this equation; but we encounter a new difficulty. If $|\mathbf{A}| \neq 0$, i.e. if $\mathbf{A}$ is non-singular, $\mathbf{X}$ is unique and is the ordinary inverse $\mathbf{A}^{-1}$; but if $|\mathbf{A}| = 0$, the only case in which we are really interested, the solution is not unique; a whole family of matrices $\mathbf{X}$ will satisfy the equation. However, it has now been shown that it is possible to add further constraints to equation (3.6), such that there is now only one matrix $\mathbf{X}$ that satisfies all the conditions, and this matrix exists whether $\mathbf{A}$ is singular or not. This matrix, sometimes called the 'unique generalized inverse', is not easy to compute, but appropriate computer programs are beginning to appear.

4

Latent Roots and Vectors

W. T. Williams

Origin of the Concept

We have been unable to find a simple agronomic example for our next matrix concept, so we shall be forced to invent one. We postulate an animal which invariably produces two offspring in its first year and three in its second, after which it is sent to market. Suppose we decide to enter this field by buying x first-year stock and y second-year stock. How many of each age shall we have the following year?

The x first-year animals will each produce two offspring, so from these there will be $2x$ first-year animals next year; similarly, the y second-year animals will produce an additional $3y$ offspring, so we shall have a total of $(2x + 3y)$ first-year offspring. The original x first-year animals will all become second-year animals; but the y second-year animals will have been sent to market, so next year there will be a total of only x second-year animals. We could write this as a matrix multiplication as follows:

$$\begin{bmatrix} 2 & 3 \\ 1 & 0 \end{bmatrix} \cdot \begin{bmatrix} x \\ y \end{bmatrix} = \begin{bmatrix} 2x + 3y \\ x \end{bmatrix}.$$

Suppose we now say that we want the proportions of first- and second-year animals to remain the same, though we expect the total numbers to increase. We could write the condition for this situation as follows:

$$\begin{bmatrix} 2 & 3 \\ 1 & 0 \end{bmatrix} \cdot \begin{bmatrix} x \\ y \end{bmatrix} = \lambda \begin{bmatrix} x \\ y \end{bmatrix}. \tag{4.1}$$

In this case λ is a scalar multiplier; and the problem we have to solve is, can values of λ, x and y be found that will satisfy the equation? Our example is, of course, trivial; but the problem is an important one in the study of natural populations where, knowing something of the fertility and probabilities of survival of different age groups, we want to know whether a stable age structure can be maintained.

Let us put the problem in more general terms. For the matrix $\begin{bmatrix} 2 & 3 \\ 1 & 0 \end{bmatrix}$ we shall write $\mathbf{A}$, and we can now let $\mathbf{A}$ be *any* square matrix. For the vector $\begin{bmatrix} x \\ y \end{bmatrix}$ we shall write $\mathbf{x}$, where $\mathbf{x}$ must be a vector with as many rows as $\mathbf{A}$ has columns. The equation now reads

$$\mathbf{Ax} = \lambda\mathbf{x}, \tag{4.2}$$

and we have to solve this equation for the scalar λ and the vector $\mathbf{x}$. Any value of λ which satisfies this equation we shall call a *latent root* or *eigenvalue* (the terms are synonymous) of the matrix $\mathbf{A}$; the vector $\mathbf{x}$ which corresponds with a given value of λ we shall call the corresponding *latent vector* or *eigenvector* of the matrix $\mathbf{A}$. We now need to be able to evaluate these quantities.

Calculation of Roots and Vectors

We can obviously rewrite equation (4.2) in the form

$$\mathbf{Ax} - \lambda\mathbf{x} = 0$$

and at first sight it looks as though we might be able to take the $\mathbf{x}$ outside a bracket. However, we cannot; because inside the bracket we should be left with $(\mathbf{A} - \lambda)$, which is nonsense, since there is no way of subtracting a single number from a block of numbers. However, we already know that a matrix or vector is unchanged if it is pre-multiplied or post-multiplied by the identity matrix $\mathbf{I}$, which can be of any order we please. So we can rewrite our equation as

$$\mathbf{Ax} - \lambda\mathbf{Ix} = 0$$

and we can now write

$$(\mathbf{A} - \lambda\mathbf{I})\,\mathbf{x} = 0. \tag{4.3}$$

We cannot solve this as it stands; but $(\mathbf{A} - \lambda\mathbf{I})$ is itself a square matrix, and therefore has a determinant; and a proof exists that any value of λ which satisfies

$$|\mathbf{A} - \lambda\mathbf{I}| = 0$$

will also satisfy equation (4.3).

Let us now solve our artificial example (4.1) above. Here the matrix $\mathbf{A}$ is

$$\begin{vmatrix} 2 & 3 \\ 1 & 0 \end{vmatrix}$$

and $\lambda\mathbf{I}$ is

$$\lambda\begin{vmatrix} 1 & 0 \\ 0 & 1 \end{vmatrix}.$$

So $(\mathbf{A} - \lambda\mathbf{I})$ in this case is

$$\begin{bmatrix} 2 & 3 \\ 1 & 0 \end{bmatrix} - \lambda \begin{bmatrix} 1 & 0 \\ 0 & 1 \end{bmatrix} = \begin{bmatrix} 2 & 3 \\ 1 & 0 \end{bmatrix} - \begin{bmatrix} \lambda & 0 \\ 0 & \lambda \end{bmatrix} = \begin{bmatrix} 2 - \lambda & 3 \\ 1 & -\lambda \end{bmatrix}.$$

So we have to solve the *determinantal* equation

$$\begin{vmatrix} (2 - \lambda) & 3 \\ 1 & -\lambda \end{vmatrix} = 0.$$

But we know how to multiply out this simple determinant; it is

$$- \lambda(2 - \lambda) - 3,$$
$$\text{i.e. } \lambda^2 - 2\lambda - 3.$$

So we merely have to solve the quadratic equation $\lambda^2 - 2\lambda - 3 = 0$ whence, virtually by inspection, $\lambda = 3$ or -1. Observe that there are *two* values of λ which will satisfy the equation; in general, if the order of the matrix $\mathbf{A}$ is n, there will be n latent roots, though some of these may be zero. In our case we are not interested in the value $\lambda = -1$, since we cannot give a useful meaning to the concept of minus animals; our problem has therefore only a single agronomically meaningful solution.

We now have to find the latent vector which corresponds to the solution $\lambda = 3$. Equation (4.1) has now become

$$\begin{bmatrix} 2 & 3 \\ 1 & 0 \end{bmatrix} \begin{bmatrix} x \\ y \end{bmatrix} = 3 \begin{bmatrix} x \\ y \end{bmatrix}.$$

The left-hand multiplies out, as we know, to

$$\begin{bmatrix} 2x + 3y \\ x \end{bmatrix}$$

and the right-hand to

$$\begin{bmatrix} 3x \\ 3y \end{bmatrix}$$

and since these are equal, corresponding elements must be equal. Take the element in the first row; we have $2x + 3y = 3x$, or $x = 3y$. But the second element gives us exactly the same answer: $x = 3y$. In other words, the latent vector does not tell you the *absolute* number of animals to buy, only the *proportions*. Providing you always, at the start of your experiment, buy three times as many first-year stock as second-year stock, these proportions will remain unchanged indefinitely (verify this for a few values of x and y).

This 'proportional' property poses a computational problem. If $\mathbf{A}$ is of order 2, the resulting quadratic equation in λ is easy to solve by hand by algebraic means. If the order is 3, we are confronted with an equation in λ^3, which is much more difficult to solve; an equation in λ^4 is more difficult still, and equations in λ^5 or of higher order cannot be solved at all by algebraic means. More recondite computer methods have to be used; and in practice, nowadays, latent roots are always extracted by computer programs. Every scientific computer possesses at least one subroutine for the extraction of latent roots and vectors. But since a latent vector possesses no absolute values, what is the computer to print out?

Evidently we need to make the elements of the latent vector **x** fulfil some additional condition; we must, in fact, impose some additional *constraint* on its elements. The roots-and-vectors concept lays down no laws as to what this constraint shall be, and we are at liberty to impose any constraint we find convenient. In practice, this means that the constraint is decided by whoever writes the computer program we are going to use. In some older programs the element with the largest absolute value of the proportion (whether this was positive or negative) was set equal to $+1$, and everything else scaled and signed accordingly. This has not proved widely useful, and it is now usual to standardize the elements so that the sum of their squares is 1; in our artificial case this would be equivalent to requiring that not only must $x = 3y$, but also $x^2 + y^2 = 1$. The vector printed out would then be, to three decimal places,

$$\begin{bmatrix} 0 \cdot 949 \\ 0 \cdot 316 \end{bmatrix}.$$

For certain purposes, it is necessary to make the sum of squares equal to the corresponding latent root; in our case this would impose the constraint $x^2 + y^2 = 3$, and the vector would be printed out as

$$\begin{bmatrix} 1 \cdot 643 \\ 0 \cdot 548 \end{bmatrix}.$$

Signs sometimes puzzle the beginner. In our artificial example, both elements of the vector in which we are interested are positive; but they need not have been. If we had pursued our second latent root ($\lambda = -1$) we should immediately find that the requirement for the two elements of the latent vector was simply $x = -y$. Let us take a slightly more complicated case; suppose our matrix had been of order 3, and that one of the latent vectors had elements x_1, x_2 and x_3 such that the proportions were

$$x_1 : x_2 : x_3 = +1 : -2 : +3$$

and suppose we impose the usual constraint that $x_1^2 + x_2^2 + x_3^2 = 1$, we shall find that *either* of the two following vectors will fulfil both conditions:

$$\begin{bmatrix} +0 \cdot 267 \\ -0 \cdot 535 \\ +0 \cdot 802 \end{bmatrix} \qquad \begin{bmatrix} -0 \cdot 267 \\ +0 \cdot 535 \\ -0 \cdot 802 \end{bmatrix}.$$

Which version you get depends on the idiosyncrasies of your own particular computer program. In other words, the overall sign of a latent vector is arbitrary; if the set of signs you get is inconvenient, you are at liberty to change them—provided you are consistent and change them *all*.

Properties of Latent Roots

There are two simple properties of latent roots you should know; they concern the sum of the roots (which we might write as $\sum_i \lambda_i$) and the product of the roots (which we might write as $\prod_i \lambda_i$). *The sum of the roots is always equal to the trace of the matrix.* For example, in our simple matrix $\left[\begin{smallmatrix} 2 & 3 \\ 1 & 0 \end{smallmatrix}\right]$ the trace (the sum of the elements in the principal diagonal) is $2 + 0$, or 2. We have already found that its latent roots are 3 and -1, whose sum is also 2. The second property is that *the product of the roots is always equal to the determinant of the matrix.* In our case the determinant is $(2 \times 0) - (3 \times 1)$, or -3; the product of the roots is 3×-1, or again -3. It follows that if a matrix has one or more zero latent roots, it must be singular; and conversely, if a matrix is known to be singular it must have at least one latent root of zero.

A Fundamental Relationship

Suppose, for a given matrix, we had extracted all its latent roots and calculated all the corresponding latent vectors; it might be useful, and it would certainly be aesthetically satisfying, if we could find some relationship that joined all these elements together. Such a relationship does in fact exist; we shall not attempt to prove it, but shall demonstrate it by means of a simple example.

We again begin with our simple matrix $\mathbf{A}$, $\left[\begin{smallmatrix} 2 & 3 \\ 1 & 0 \end{smallmatrix}\right]$. We already know that the elements of the first vector are in the proportion of $3 : 1$ and those of the second vector $1 : -1$. The relationship we seek must be unaffected by vagaries of standardization, and we shall use these proportions as actual values. We can then define a matrix of vectors, $\mathbf{V}$, such that $\mathbf{V} = \left[\begin{smallmatrix} 3 & 1 \\ 1 & -1 \end{smallmatrix}\right]$. We now need the inverse of this matrix, $\mathbf{V}^{-1}$. It is in fact

$$\begin{bmatrix} \frac{1}{4} & \frac{1}{4} \\ \frac{1}{4} & -\frac{3}{4} \end{bmatrix}$$

(the reader might care to confirm this by both pre- and post-multiplying $\mathbf{V}$ by this matrix). Now carry out the triple multiplication $\mathbf{V}^{-1}\mathbf{A}\mathbf{V}$:

$$\begin{bmatrix} \frac{1}{4} & \frac{1}{4} \\ \frac{1}{4} & -\frac{3}{4} \end{bmatrix} \begin{bmatrix} 2 & 3 \\ 1 & 0 \end{bmatrix} \begin{bmatrix} 3 & 1 \\ 1 & -1 \end{bmatrix} = \begin{bmatrix} 3 & 0 \\ 0 & -1 \end{bmatrix}.$$

But 3 and -1 are the latent roots of $\mathbf{A}$; and it can be shown that this will always be the case. The result is a diagonal matrix with the latent roots of $\mathbf{A}$ down the principal diagonal and zero everywhere else. We often symbolize such a matrix by Λ, so that our relationship can be expressed as $\mathbf{V^{-1}AV} = \Lambda$. An agronomist is most unlikely ever to need to use this relationship; but if he reads modern review papers on pattern analysis he is likely to find it quoted, and he might as well know what it means.

Rank and Order

Let us again suppose that we have made two measurements on three leaves, and obtained the following little matrix, where the leaves are rows and the measurements are columns:

$$\begin{bmatrix} 2 & 3 \\ 0 & 1 \\ 1 & 0 \end{bmatrix}.$$

We shall, as in Chapter 2, obtain the uncentred dispersion matrix by pre-multiplying this matrix by its own transpose, obtaining the matrix

$$\begin{bmatrix} 5 & 6 \\ 6 & 10 \end{bmatrix}.$$

The reader might care to set up and multiply out the determinantal equation which will give us the latent roots; he will find it is

$$\lambda^2 - 15\lambda + 14 = 0,$$

whence the roots are 14 and 1.

But now suppose we had made *three* measurements on *two* leaves, so that the matrix now appeared as

$$\begin{bmatrix} 2 & 0 & 1 \\ 3 & 1 & 0 \end{bmatrix}.$$

If we pre-multiply this matrix by its transpose, we obtain

$$\begin{bmatrix} 13 & 3 & 2 \\ 3 & 1 & 0 \\ 2 & 0 & 1 \end{bmatrix}. \tag{4.4}$$

But this is a matrix of order 3, so it must have three latent roots; it looks as though we have created an extra bit of information simply by turning the data matrix on its side.

However, the reader might again care to work out the determinantal equation for matrix (4.4); he will find that it is

$$\lambda(\lambda^2 - 15\lambda + 14) = 0.$$

In other words, we have the same two roots as before, plus an additional zero root. In effect, the calculation is informing us that we really have only two completely independent sets of information. You can check that this is true by examining matrix (4.4) more carefully: you will find that the first column is equal to three times the second column plus twice the third column. There is therefore one redundant column, and consequently one zero root. We say that the matrix (4.4) is of *order* 3 but of *rank* 2. In general, if we measure m quantities on n things, and m is greater than n, the $m \times m$ dispersion matrix cannot have rank greater than n (it can be less). We say that the n things have been *over-defined;* the $m \times m$ dispersion matrix has a rank lower than its order, and is of necessity singular. Since modern devices such as the amino-acid analyser and the quantometer enable the experimenter to measure a number of properties simultaneously, over-defined matrices are increasingly common in agricultural experimentation and may invalidate subsequent multivariate statistical procedures.

Postscript

We believe that the agronomist who has mastered these four chapters will know enough matrix algebra to enable him to read with understanding papers that use matrix terminology. There is much more that could be learnt; but where a subject is ancillary to one's own discipline, the important thing is not how much, but how little, one needs to learn.

PART 2

Techniques and Principles of Pattern Analysis

5

Attributes

W. T. Williams

The General Concept

Pattern analysis almost always begins with a set of things, which may be paddocks, people, cows, soils, anything that can conceivably be ordinated or classified. The things, the elements to be ordinated or classified, are now usually simply called *individuals* or *entities*. Sneath introduced the rather more cumbersome term *operational taxonomic units* (usually abbreviated as O.T.U.s), though this term is now less used than formerly. About each of these individuals we have a number of items of information, which we call *attributes*. In pattern analysis, the term 'attribute' is used in a wider sense than in classical statistics. In the latter discipline, distinction is made between *variables* (which are continuously varying quantities), *variates* (which are variables that can be associated with a statistical distribution) and *attributes* (*sensu stricto*), which are qualities possessed or not possessed by an individual. Variates have no place in pattern analysis; but the term 'attributes' is used to cover variables, attributes *sensu stricto* and indeed any other forms of information. Attributes can take many forms and the classification of types of attributes is the subject of this chapter.

The Three Basic Attribute Types

Nominal attributes

A nominal attribute is a description that falls into discrete states, such that a given individual must be in one, and only one, state. It might be colour (red, white or blue); rock type (shale, sandstone, granite, basalt); leaf shape (simple, pinnate, palmate); and so on. The essential feature of a nominal attribute is that although the states may be numbered for computational convenience, no meaning is to be assigned to the order in which the states are to be taken. In other words, even though the states might be numbered 1, 2, 3 and 4, it is assumed that the difference between 1 and 4 need be no greater than the difference between 1 and 2 or between 2 and 4. Nominal attributes are sometimes called *disordered multistate attributes* (for 'non-exclusive' nominal attributes, see the section on 'linked attributes' below).

A special case of the nominal attribute is that with only two states (present or absent, yes or no). These are often called *binary* or *qualitative* attributes; it is convenient to handle them separately, since they can be 'packed' into the essentially binary-oriented word of a computer, when they take less space and can be handled quickly and efficiently. In some computer programs (none, so far as I know, in Australia) a further distinction is made within the general class of binary attributes, which are divided into *symmetrical* and *asymmetrical*. The distinction arises as follows. Suppose we are carrying out a plant ecological survey in an area where, in all, some 50 species are represented. Further, suppose that two of our sample sites are floristically very poor, so that each has only a single species and this is different in the two cases. Do we wish these two sites to look very alike because they are both floristically so impoverished, or very different because they do not possess a single species in common? This is a question for the ecologist to decide, not the pattern analyst. However, if the presence of a species is coded as 1 and its absence as 0, then if they are to look alike two 0s (a 'double-zero match') will count as strongly as two 1s; but if they are to look unlike, two 1s will count towards similarity but two 0s will not. The former is the symmetrical case, the latter the asymmetrical. It is more usual in Australian programs to deal with this problem outside the main classificatory program, by defining symmetric or asymmetric measures.

Numeric attributes

These are quantities which can in principle vary continuously, such as weight in grams or length in centimetres. They were formerly called *metric* attributes; but since the term 'metric' has a restricted meaning in topology, some more neutral term is usually preferred. They are also sometimes called *quantitative* attributes.

Ordinal attributes

An ordinal attribute, like a nominal attribute, is a description which falls into discrete states, such that a given individual can be in one and one only. In this case, however, the *order* of the states is meaningful, though the distances between states are undefined. Examples might be: seeds small, intermediate or large; stem glabrous, downy, softly hairy or densely hairy; and so on. Here, if the states are numbered, there *is* more difference between states 1 and 4 than between 1 and 2. Ordinal attributes, which are in effect ranks or scores, are very common; they are also an unmitigated nuisance. This is because no really satisfactory way of handling them numerically is known. In some programs they have to be treated as if they were real numbers; this is equivalent to assuming that the states are equidistant, which they usually are not. In other, information-statistic, programs a specific algorithm for handling ordinals exists, but is suspected to be suboptimal. Statistical transformations exist which are aimed at making the scores approximate to a normal distribution; but the populations with which pattern analysis deals are commonly too small for these to have much validity or to be worth the trouble. They are therefore best avoided if possible; but it is not always possible.

Meristic Attributes

These are essentially counts, and can only take integral values; examples are the number of petals in a flower or the number of specimens of a given crustacean in a marine grab sample. No special algorithm exists for these, but care must be taken in the decision as to how they are to be coded. If a group of plants contains some with 2 and others with 4 petals, it is not valid to calculate a mean number of petals and regard the whole group as having, on average, 3 petals; this would imply a transfer from dicotyledons to monocotyledons. In such a case the attribute should be coded as nominal, so that the states remain separate. However, if the range of values is large (the number of crustacea might range from zero to several hundred) a mean value is interpretable, and the attribute can safely be coded as numeric, being regarded as continuously varying. The criteria are, in fact, (i) the range of values that the meristic attribute can take and (ii) whether the concept of a mean value is interpretable, or at least not misleading.

Linked Attributes

The concept of linked attributes is relatively recent and has arisen in two distinct contexts. The first type for consideration is the *linked binary attribute*, which particularly arises in higher-plant taxonomy. In an actual case, there were 13 different types of hair on record in the group of species under examination, and any one species could bear almost any subset out of the whole 13. This can obviously be handled as a set of 13 binary attributes; but the users considered that this would unduly weight the hair character, and wished that the complete set should only contribute the equivalent of a single attribute. In effect, the user wished to average the contributions from all 13 types. A linked set such as this could also be regarded as a nominal attribute, in this case of 13 states, but such that any individual could appear in more than one state at a time. This in fact is how they are dealt with in modern programs; in most program specifications they are called *non-exclusive disordered multistates*.

The second type is the *linked numeric attribute*, which originally arose in the classification of soils. A soil sample may be the subject of a number of, for example, chemical determinations; and these determinations may be repeated at successive depths down the soil profile. Suppose there were 10 different determinations made at 6 successive levels. There are then two things we could do. First, we could regard this as a set of 10×6, i.e. 600 distinct attributes. Alternatively, we could for each determination calculate similarities at each of the levels and average these, leaving us with only 10 attributes; this is the 'linked' solution. With most measures, if there are no missing values, these produce identical results; but if an appreciable proportion of values is missing (because some determinations could not be made at some levels) the two methods produce different answers. There is in the literature some limited and purely heuristic evidence that in these circumstances the linked alternative is better; the algebraic reason for this—and even whether it is generally true—is unknown. The method has also been extensively used in recent years in the analysis of field trials of plant introduction material at a range of sites over several years. In such experiments climatic or administrative difficulties, or shortage of seed, may result in some introductions not being planted at some sites in some years, so that an appreciable proportion of data is missing. In these cases the 'linked' algorithm—the other attributes being regarded as linked over years—appears to be effective.

Serially Dependent Attributes

Suppose we were classifying species of insects, and an important binary character was whether the species possessed or did not possess wings. We shall assume three insect species, A, B and C, such that A and B both have wings and C has not. Now, if an insect has wings, there are potentially a considerable number of attributes concerning wings that can be scored for that insect; but if it has no wings, these additional questions cannot be answered and the additional attributes cannot be scored. Such attributes are known as *serially dependent;* because only if the answer to a particular question (does it have wings?) is 'yes', can these attributes be scored; if it is 'no' they cannot, and the further questions are *inapplicable*. All existing programs score such attributes as 'missing'.

This produces a paradox, first pointed out by Kendrick (1965) in connexion with classificatory work on the Fungi Imperfecti. In our insect case, we shall begin by calculating some measure of similarity between A and B, A and C, and B and C. It is obvious that the fact that A and B differ in a number of wing characters that cannot be scored for C does not make either of them any more like C. Kendrick showed, however, that this is precisely what happens. If A and B differ in enough dependent characters, they will both appear more similar to C than they do to each other; the dependent characters will override the primary characters on which they are dependent.

Williams (1969) later showed that a single dependent character cannot override, but that any larger number can and usually will. Numerical methods of overcoming the problem are known in theory, but they are not implemented by any production program known to us; and even these will fail if there is higher-order dependence, i.e. if already dependent attributes have further sets dependent on them. Our advice at present is to avoid serially dependent attributes as far as possible. If it is not possible, the strings of such attributes should be kept short.

Intractable Attributes

There are a few attributes for which no satisfactory algorithm is known. Perhaps the best-known example is that of chromosome number, because of the possibility of ploidy. If we have three plants with chromosome numbers 14, 27 and 28, existing programs will regard those with 27 and 28 as most alike; but it is not improbable that the 14 and 28 are diploid and tetraploid versions of the same species, whereas the 27 may be an altogether unrelated species. Worse, of two plants, both with chromosome numbers of 24, one might be a diploid with $n = 12$, the other a triploid with $n = 8$. As a result, chromosome numbers are never included in numerical classifications; they are used only to assist in the interpretation of final results (for an example, see Edye *et al.* 1970).

False Attributes: Judgments

There are certain fields of activity in which pattern analysis frequently encounters difficulty, notably studies in criminology (e.g. delinquents) or social medicine (e.g. neurotics); in these cases the attributes recorded may show little or no pattern. The reason is that to state that somebody is 'delinquent' or 'neurotic' is not to record an attribute which could be confirmed by an independent observer weighing or measuring him, or closely examining him with a hand-lens; it is to record a *judgment* on him made by somebody else. There is now ample evidence that the pattern of judgments lies in the people who make them, not in the people upon whom they are made. To take a trivial example, if the minimum age for legal drinking varies between areas, the pattern of convictions for under-age drinking reflects the area difference, and genuine attributes recorded for the individuals are largely irrelevant.

Value judgments for performance are not unknown in agronomy; analysts must be alert to the possibility that such attributes may reflect not the characteristics of the plant material evaluated, but the predilections of the agronomist who made the evaluation.

Mixtures of Attributes

Most modern classificatory programs will accept mixtures of attributes, i.e. an individual may be defined by any or all of the types of attribute here described. However, it is implicitly assumed that *all* attributes are concerned with a single underlying pattern. For example, it is common in terrestrial ecological survey to collect both floristic and environmental information. If the floristic information takes the form of the presence or absence of all vascular plants recorded for the area, the floristic data set can be exhausted completely; there is no more information to collect. However, the environmental data can only be a small part of the enormous mass of environmental information that could in theory be collected. Now, it is usually not known in advance what particular type of environmental factors are controlling the floristic pattern; if it *were* known, there would be no point in the survey. But it would be quite possible to collect the wrong environmental information; such information might well be patterned, but reflecting a pattern that had no relation to the floristic pattern. In such a case the classification will be vainly attempting to extract two conflicting patterns simultaneously, and the results would probably be uninterpretable. Analysts are advised never to mix data types in this manner unless they have very good and explicit reasons for so doing.

References

Edye, L. A., Williams, W. T., and Pritchard, A. J. (1970). A numerical analysis of variation patterns in Australian introductions of *Glycine wightii* (*G. javanica*). *Aust. J. Agric. Res.* **21**, 57–70.

Kendrick, W. B. (1965). Complexity and dependence in computer taxonomy. *Taxon* **14**, 141–54.

Williams, W. T. (1969). The problem of attribute-weighting in numerical classifications. *Taxon* **18**, 369–74.

6

Similarity Measures

H. T. Clifford and W. T. Williams

Careful writers make a distinction between *similarity* measures (which increase with increasing likeness) and *dissimilarity* measures (which decrease with increasing likeness); however, the term 'similarity measures', or sometimes 'similarity coefficients', is often used indiscriminately for both. In practice, most modern computer programs operate on dissimilarity measures. The literature contains a bewildering array of such measures, but only relatively few of these are in regular use. Those that have found general acceptance have done so because they possess properties that are desirable from the point of view of ease of calculation, and are readily interpretable, at least in the sense that they are suitable for generating hypotheses concerning the structure of the data. Some may be applied only to particular types of data, others only to particular situations. The correct choice of measure is extremely important, and careful consideration should be given to the problem of defining similarity before an analysis is begun. It will be convenient, before considering such general principles as can be enunciated in this respect, first to describe the more important measures in common use.

Matching Coefficients

These are confined to binary attributes; as originally defined they are true similarity measures, and they are the oldest in the literature. Their use immediately raises the question of symmetry which has been briefly mentioned in the previous chapter. In ecology, the two states of such an attribute usually represent the presence or absence of a species in a site; but in taxonomy the two states may be two distinct properties of an organ, such as 'fruits fleshy' or 'fruits dry'. In the latter case the two states are, as it were, interchangeable, and the problem of 'double-zero matches' does not apply.

We shall illustrate three simple matching coefficients by means of an artificial example. In the following table, A to J represent 10 plant species, which may be present (1) or absent (0) in either of 2 sites:

	Species									
	A	B	C	D	E	F	G	H	I	J
Site 1	1	0	1	0	1	0	1	0	0	1
Site 2	1	1	0	1	1	0	0	1	1	1

We could summarize this situation in a 2 × 2 contingency table as follows:

	No. of species present in site 1	No. of species absent from site 1
No. of species present in site 2	3	4
No. of species absent from site 2	2	1

In other words, there are 3 species in both sites (1, 1 match), 1 species in neither (0, 0 match), 2 in site 1 which are absent from site 2, and 4 in site 2 which are absent from site 1. We could set this out in general form as follows:

a	b	$(a + b)$
c	d	$(c + d)$
$(a + c)$	$(b + d)$	$(a + b + c + d)$

The *simple matching coefficient* is defined as the total number of matches, both (1, 1) and (0, 0), divided by the total number of comparisons, in this case 4/10, or 0·40. In algebraic symbols we can write this as $(a + d)/(a + b + c + d)$. The *Jaccard coefficient* is defined as the number of (1, 1) matches divided by the total number of comparisons, excluding (0, 0) matches. There are 3 (1, 1) matches, and when the single (0, 0) match has been excluded, 9 comparisons remain; so the coefficient is 3/9, or 0·33. We can write this in the general form $a/(a + b + c)$. An alternative form of this measure is one usually attributed to Czechanowski. It is defined as the number of (1, 1) matches divided by the arithmetic mean of the number of presences in the two sites. In algebraic symbols, there are $(a + c)$ species in site 1 and $(a + b)$ in site 2; the arithmetic mean of these two quantities is $\frac{1}{2}(2a + b + c)$. The number of (1, 1) matches is a, so that the measure becomes $2a/(2a + b + c)$, or 6/12, or 0·50. Each of these measures possesses the property that it is constrained between 0 and 1; and each can consequently be turned into a dissimilarity measure by subtracting it from 1. The simple matching coefficient then becomes $(b + c)/(a + b + c + d)$, the Jaccard becomes $(b + c)/(a + b + c)$ and the Czechanowski becomes $(b + c)/(2a + b + c)$.

How are we to choose between these? There is virtually no practical difference between the Jaccard and Czechanowski measures; they become 0 together, they become 1 together, and they are *jointly monotonic*, in that they go up and down together. The important difference is that between the simple matching coefficient (S.M.C.) and the other two. The S.M.C. counts double-zero matches towards similarity; it is a *symmetric* measure, and so is commonly used in taxonomy. The Jaccard and Czechanowski measures are used in ecology when it is not desired that double-zero matches should count towards similarity.

Euclidean Distance

When the attributes measured on each entity are continuous variables, such as the *amount* of each species in each site, matching coefficients are of limited value. Each variable could, of course, be converted into a binary attribute by the simple device of regarding all values above a given value as 1 and all those equal or below as 0; but this discards a great deal of information about the entities being classified. Measures are therefore needed which can exploit the whole of the available information. The simplest, and oldest, of these is Euclidean distance. To calculate this measure we have first to assume that the measurements we have made can be regarded as coordinates on a set of axes in Euclidean space. Suppose we had measured the lengths and breadths of three leaves, A, B and C; we shall call their lengths x_1, x_2 and x_3, and their breadths y_1, y_2 and y_3. We could plot these on a graph as in Fig. 6.1. We can calculate the distance between, say, A and C by means of the theorem of Pythagoras; if we call this distance $d\,(A, C)$, we have

$$d\,(A, C) = \{(x_3 - x_1)^2 + (y_3 - y_1)^2\}^{\frac{1}{2}}. \tag{6.1}$$

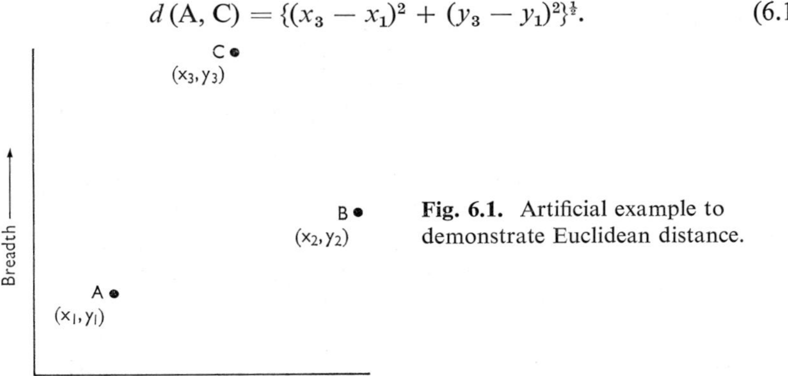

Fig. 6.1. Artificial example to demonstrate Euclidean distance.

In practice we should normally use the square of this distance rather than the distance itself; this is because it is regarded as a virtue in similarity measures if they are additive over attributes, which d^2 is, but d is not. We can generalize this into the case for many attributes and therefore for many dimensions. If we had a total of s measured attributes, the squared distance between the ith and jth entity would be given by

$$d_{ij}^2 = \sum_{k=1}^{s} (x_{ik} - x_{jk})^2. \tag{6.2}$$

To save this quantity becoming uncomfortably large, it is common practice to divide it by s, the number of attributes. This is always necessary if (as is common in taxonomy) there are missing data, for in such a case s, the number of possible attribute comparisons, will differ between different pairs of individuals.

We now encounter a new difficulty. Suppose one of our attributes was weight in grams and another was height in centimetres. We cannot add grams to centimetres in any meaningful way, so that as things stand we cannot calculate Euclidean distance unless all our measurements are in the same units. It will not suffice for them to have the same physical dimensions; kilograms and milligrams have the same physical dimensions, but we should not wish to add them together. We must divide the value for each attribute by some quantity which shall be measured in the same units as is the attribute, and it will usually be appropriate to select for this quantity some property of the whole population under study. The values of our attributes after such a transformation will then represent pure numbers; they will be *dimensionless* and will have no units of measurement. What quantity are we to use for this process of *standardization*? One that is often used is the *range*, the difference between the largest and smallest values of the given attribute in the whole population. This ensures that the contribution of any single attribute to the overall distance must lie between 0 and 1. More commonly we use the *standard deviation*, in the form

$$\left(\frac{\sum (x - \bar{x})^2}{n} \right)^{\frac{1}{2}}.$$

This ensures that each attribute is reduced to *unit variance*. If the term 'standardized Euclidean distance' is used without further explanation, it is this particular form of standardization that is implied.

Finally, it should be noted that the use of Euclidean distance is not confined to continuous variables. Burr (1968) has shown that comparable transformations can be defined for nominal attributes, so that Euclidean distance can still be calculated even if the attribute set is a mixture of numeric and nominal attributes. No comparable solution is known for ordinal attributes; all that can be done in this case is to regard the state labels (1, 2, 3 etc.) as real numbers, i.e. to assume that the states are *equidistant*.

Manhattan Metric Measures

Let us reconsider Fig. 6.1. Suppose that the two points A and C are objects situated in a town at opposite corners of an island block of buildings, as in Fig. 6.2.   The Euclidean distance between A and C implies that we can

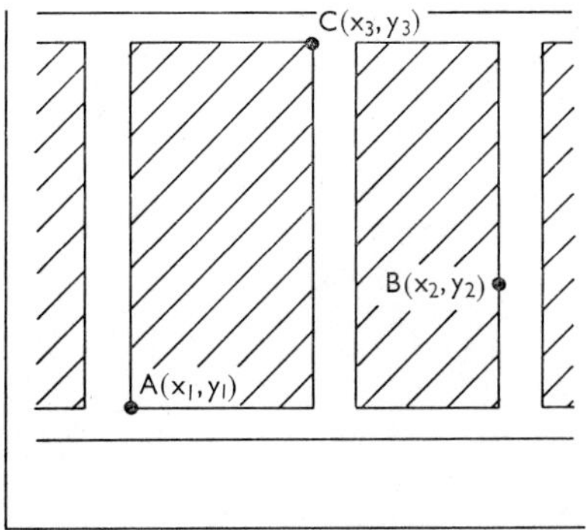

Fig. 6.2. Artificial example to demonstrate the Manhattan metric.

plunge straight through the middle of the block. In practice we normally cannot; we must walk along one street for a distance of $(x_3 - x_1)$ units, then along another street at right angles for a distance $(y_3 - y_1)$ units. However, we cannot represent the distance along the first street by the algebraic symbol $(x_3 - x_1)$, since this would become negative if we walked in the opposite direction; we must use the *absolute* value or *modulus* of this quantity, which we write as $\left| x_3 - x_1 \right|$. Expression (6.1) then becomes

$$d\,(\mathrm{A}, \mathrm{C}) = \left| x_3 - x_1 \right| + \left| y_3 - y_1 \right|$$

and the more general expression (6.2) becomes

$$d_{ij} = \sum_{k=1}^{s} \left| x_{ik} - x_{jk} \right|. \tag{6.3}$$

It is because of this city-block analogy that the quantity (6.3) is usually known as the Manhattan, city-block or lattice metric; it has a considerable history of use in pattern analysis. As with Euclidean distance, it needs to be standardized by attributes; but in this case standardization by the standard deviation has always been considered inappropriate. Three forms of standardization are in use, and the three standardized forms have been given different names. In the *Gower metric*, whose properties were first investigated by J. C. Gower, it is standardized by range. The *Bray–Curtis measure*, so called because a version of it was used in a classical ecological study by J. R. Bray and J. T. Curtis, is standardized as follows:

$$\frac{\sum_k \left| x_{ik} - x_{jk} \right|}{\sum_k \left(x_{ik} + x_{jk} \right)}.$$

Finally, the *Canberra metric*, so called because it was first used in a Canberra classificatory program, is

$$\sum_k \frac{|x_{ik} - x_{jk}|}{x_{ik} + x_{jk}} . \qquad (6.4)$$

The Bray–Curtis and Canberra metrics should only be used where the data are everywhere non-negative; if the Gower and Canberra metrics are divided by s, all three measures are then constrained between 0 and 1. Relationships exist between these measures and the matching coefficients. The range-standardized Euclidean distance and the Gower metric collapse in the binary case into the simple matching coefficient; the Bray–Curtis measure becomes the Czechanowski measure; and the Canberra metric becomes the Jaccard measure.

Choice of Metric Measure

These measures are not equally suitable for all types of data. In particular, they differ markedly in their response to outlying values and to zero values. 'Outlying values' are isolated large values, out of scale with the bulk of the data matrix; they are common, for example, in rainfall data and in faunistic data from marine ecological surveys. Consider the following artificial example:

	Species			
	1	2	3	4
Site A	2	3	4	5
B	5	4	6	12
C	5	40	6	12

It is obvious that B and C are identical except for the rather large value of species 2 for site C. If we were to use unstandardized Euclidean distance, this large value would completely dominate the situation; but the main interest lies in the differences between variance-standardized Euclidean distance and the three Manhattan metric measures. The values for the measures for (A, B), (A, C) and (B, C) are given in the following table:

	Standardized Euclidean	Gower	Bray–Curtis	Canberra
A, B	3·375	0·751	0·317	0·296
A, C	4·530	1·000	0·636	0·475
B, C	1·094	0·243	0·400	0·205

All except the Bray–Curtis measure imply that B and C are the most alike sites, despite the relatively large value for C(2); this value has, however, dominated the Bray–Curtis measure to the extent that it nominates the A, B pair as most alike. If we now increase the value of C(2) from 40 to 400, the values become:

A, B	3·375	0·751	0·317	0·296
A, C	4·503	1·000	0·936	0·506
B, C	1·122	0·249	0·880	0·245

The tenfold increase in C(2) has made a very large change in the Bray–Curtis measure, C now appearing still more remote from A and B; but it has had little effect on the remaining measures. The Bray–Curtis measure is evidently extremely sensitive to outlying values, the remainder are not.

The main zero-value problem arises with the Canberra metric. It will be seen from expression (6.4) that if x_{ik} is 0, the metric takes its maximum value of 1 irrespective of the value of x_{jk}. This is occasionally countered in practice by trapping any zero/non-zero comparison and replacing the zero, for that comparison only, by a small positive value. Finally, it should be noted that double-zero matches make no contribution to the Bray–Curtis measure; with the other three, specific steps must be taken to suppress these if they are unwanted.

Information-statistic Measures

These are now widely used in pattern analysis. They are particularly applicable to nominal (especially binary) attributes; and although variants capable of handling ordinal or numeric attributes exist, they are relatively weak. Their existence makes it possible to use these measures with data of mixed attribute types; but they should not be used unless the data are predominantly nominal. We first need to distinguish between *information content* and *information change*. The information content of a population is a measure of the heterogeneity, or disorder, of that population; the more heterogeneous, the higher the information content. It follows that a population of identical individuals has zero information content. However, if two populations are combined, there will normally be an increase in disorder over and above that present in both the original populations. If we had a population A with information content I_A, and another population B with information content I_B, we might fuse these into a new combined population C, with information content I_C. I_C cannot be less than the sum of I_A and I_B, and will usually be greater; so we define an *information gain*, ΔI, as

$$\Delta I = I_C - (I_A + I_B) .$$

ΔI can now be used as a measure of dissimilarity between the populations A and B. It is worth noting that 2ΔI is distributed substantially as a χ^2; this tempted a number of earlier workers (including the editor of this book) into using this property to define significance tests that were later shown to be invalid.

A detailed consideration of information theory is beyond the scope of this book; we note only that an information content normally takes the form of minus the logarithm of a probability. We shall now use this relationship to show that both symmetric and asymmetric information measures can be devised.

The Symmetric Measure

Suppose we have a group of n quadrats defined by the presence or absence of s plant species, and suppose that a_j of the quadrats contain the jth species. For the moment we consider only this single species. If we pick a quadrat at random, the best estimate we can get (actually, the maximum-likelihood estimate) of the probability that it will contain the species is a_j/n. Similarly, the best estimate that it does *not* contain the species is $(n - a_j)/n$. There are a_j quadrats containing the species; and the probability that they all do contain it must be the product of their individual probabilities, i.e. $(a_j/n)^{a_j}$. Similarly, the probability that all those without it do not contain it must be $[(n - a_j)/n]^{n-a_j}$. Therefore the probability of the whole population (strictly, the likelihood of the population) must be the product of these quantities, i.e. $(a_j/n)^{a_j} . [(n - a_j)/n]^{n-a_j}$. But this is for only one of the species; and we shall have to multiply together a whole set of such expressions, one for each species. So we can symbolize the probability that the entire population is as it is in the form

$$L = \prod_{j=1}^{s} (a_j/n)^{a_j} \left[(n - a_j/n)\right]^{n-a_j}.$$

The information content, I, that we are seeking will be minus the logarithm of this quantity; for most purposes the base of the logarithm is immaterial, and it is usual to use 'natural' logarithms to base e, since this is the standard computer form. We therefore have

$$I = -\ln L = -\sum_{j=1}^{s} \left\{ a_j \ln (a_j/n) + (n - a_j) \ln \left[(n - a_j)/n\right] \right\} .$$

Rearranging and simplifying, this reduces to

$$I = sn \ln n - \sum_{j=1}^{s} \left\{ a_j \ln a_j + (n - a_j) \ln (n - a_j) \right\} .$$

Quantities of the form $x \ln x$ are ubiquitous in information measures; tables of them exist (it should be noted that $0 \ln 0 = 0$) though it is usually more convenient to generate the tables within the computer as required. Here are the first 10 values:

x	$x \ln x$
1	0·0000
2	1·3863
3	3·2958
4	5·5452
5	8·0472
6	10·7506
7	13·6214
8	16·6355
9	19·7750
10	23·0259

Consider a set of 10 quadrats defined by the presence or absence of 4 species. Suppose that 2 contain the first species, 5 the second, 4 the third, and all contain the fourth. The reader might care to use the table above to show that the information content of the population is 18·6657. The unit of measurement, if it is needed, is the 'natural bel'.

The Asymmetric Measure

We are not here concerned with quadrats, but with a single sample; it might be a marine grab sample, or perhaps a collection of insects caught in a light-trap. We will suppose that T individual animals have been caught in all, and we want a measure of the heterogeneity of the collection. If all were of the same species, the heterogeneity would clearly be zero; if every individual belonged to a different species, the heterogeneity would be as great as possible. We call the measure of the actual heterogeneity the *diversity* of the population, and proceed as follows.

Suppose that among our T individuals there are representatives of s different species, and that there are a_j individuals of the jth species. As before, if we picked an individual out at random, the probability that it is of this species is a_j/T. We know that there are a_j of this species, and the probability of having the whole group is $(a_j/T)^{a_j}$. Arguing as we did in the previous case, the likelihood of the entire population is given by

$$L = \prod_{j=1}^{s} (a_j/T)^{a_j}.$$

The corresponding information content, I, will equal $-\ln L$, so we have

$$I = \sum_{j=1}^{s} (a_j \ln T - a_j \ln a_j) .$$

But since $\sum_{j=1}^{s} a_j$ represents the whole population, and therefore $= T$, this reduces to

$$I = T \ln T - \sum_{j=1}^{s} a_j \ln a_j .$$

Apart from its uses in animal ecology, this form of measure is used in the processing of multistate nominal attributes. As a final example, the reader might consider a catch of 10 animals, 1 of which is of species A, 1 of species B, 3 of species C and 5 of species D: using the above table, he could show that the diversity of this population is $11 \cdot 6829$.

Reference

Burr, E. J. (1968). Cluster sorting with mixed character types. I. Standardisation of character values. *Aust. Comput. J.* **1**, 97–9.

7

Ordination: Principal Component Analysis

W. T. Williams

The Basic Problem of Ordination

Ordination is essentially a geometrical problem. The problem arises because most biologists are visually oriented; they find a pattern which can be visually displayed easier to grasp than one defined by a set of numbers or symbols. Consider the following artificial example, purporting to be 2 sets of measurements made on 10 objects:

Measure 1	2·0	3·5	5·0	6·5	8·0	9·0	8·0	15·0	14·0	16·0
Measure 2	6·4	6·0	7·1	7·6	9·8	9·1	10·0	10·5	10·6	12·4

We can display this visually by pretending—it is really no more—that measurement 1 defines an *X*-axis and measurement 2 a *Y*-axis at right angles to it. Each pair of measurements now becomes a pair of (x,y) coordinates in two-dimensional space, and we obtain a figure such as Fig. 7.1.

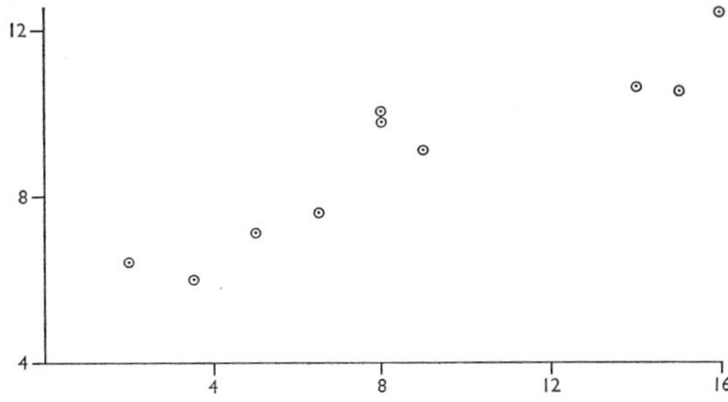

Fig. **7.1.** Artificial example of a two-dimensional ordination.

We immediately see that all the points lie roughly on a straight line, so that we could say that the two measurements are substantially 'linearly related'; and we also note that three of the points are separated from the remaining seven, suggesting that these three objects are in some sense

47

different from the rest (larger? older?). The important point is that we have displayed the *whole* of the data, and can quickly grasp any pattern it may contain.

But now suppose we had taken three measurements. To construct a similar model we should now need three axes, X, Y and Z. We now have a choice of three methods of displaying these. We could plot them as three pairs—X against Y, X against Z, and Y against Z—and try to imagine them as put together.

Alternatively, by a cunning use of perspective, we could display the whole three-dimensional system on a two-dimensional sheet of paper; but this is not easy to do, and the results are not always satisfactory. Lastly, we could build a three-dimensional model, using small balls suspended in space by wires to represent the objects. This is a time-consuming exercise, but at least we could in this way once more examine the whole of the data simultaneously. But now consider Table 7.1, which is an artificial data matrix, representing *four* measurements on eight objects.

Table 7.1

7·0	3·6	1·0	2·2
2·7	4·0	2·1	3·9
4·7	2·9	1·5	3·5
2·5	4·5	0·9	2·8
8·8	3·0	0·8	1·9
8·0	3·8	1·2	2·0
1·5	3·5	1·1	3·7
4·5	5·2	0·7	1·8

We now need four axes, X, Y, Z and W. If we decide to examine these pairwise, we shall finish up with six diagrams. If we decide to take them in threes and build three-dimensional models, we shall finish up with four such models. The one thing we cannot do in this universe is to build a four-dimensional model using all the data at once. However, if we *were* able to do so we should find that the points representing the objects in Table 7.1 all fell precisely on a straight line, plunging as it were through the middle of the four-dimensional space. In other words, *if we could find this line* we should be able to display the whole of our data in a single dimension; we should, in fact, have 'reduced the dimensionality' of the system without loss of information; though we should, of course, need to know in what way the measurements along our single dimension were related to the measurements along the original four. If the measurements had been subject to small errors, so that the points deviated slightly from a single straight line, we might still hope to find a 'best' straight line which accounted for the greater part of the original information; the simplicity of presentation would more than repay us for any slight loss of information we had incurred.

This is what ordination is about. If we have a set of individuals defined by a set of attributes, we cannot display the whole of the data if there are more than three attributes—which there usually are. We need to find some means of reducing the dimensionality of the system with as little loss of information as possible, so that we can examine its configuration.

Principal Component Analysis

The analytical method which underlies almost all modern methods of ordination is now universally known as 'principal component analysis'. It is sometimes shortened to 'principal components' or 'component analysis' or simply 'P.C.A.' In earlier works it was often called 'principal axes', and up to about 20 years ago it was commonly called 'factor analysis'. For example, Goodall's (1954) paper, 'An essay in the use of factor analysis', is in fact about component analysis; but this was not an error on Goodall's part—the distinction now invariably made between 'component' and 'factor' analysis (see Chapter 8) had not been rigidly established when Goodall was writing.

There is more than one way of approaching P.C.A.; it can be thought of as a problem in statistics, in formal algebra, or in coordinate geometry. Since ordination is a geometrical problem, we shall approach P.C.A. geometrically; and instead of setting out formal solutions, we shall illustrate the process by an example so simple that it can be carried completely through by hand, so that the reader can see what is happening at every stage. We shall frequently have recourse to the time-honoured formula 'It can be shown that . . .'; formal proofs naturally exist for every step, but the agronomist does not need them. The agronomist's basic question is, simply, 'What do we have to *do*?'

The problem and the 'best straight line'

Table 7.2 purports to represent a set of two measurements taken on six

Table 7.2

Measurement 1	Measurement 2
−3	11
−2	8
−4	7
−2	6
−8	2
−5	2

objects; when these objects are plotted as before in a two-dimensional space, we obtain the configuration shown in Fig. 7.2.

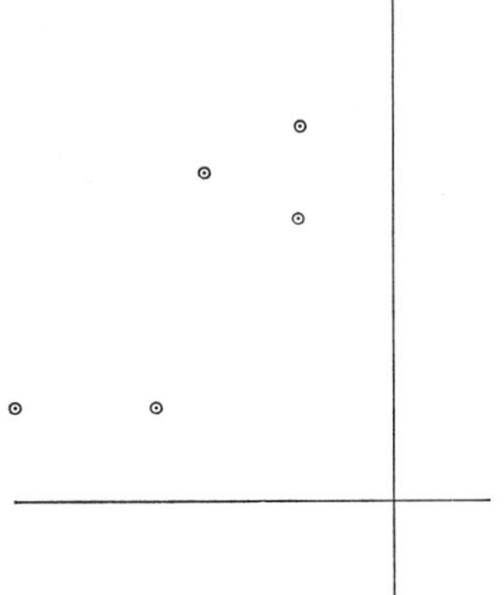

Fig. 7.2. Initial configuration of data set in two dimensions.

We shall now suppose that we are one-dimensional creatures, space-worms capable of seeing only forwards and back, but not sideways; so that two dimensions are too many for us, just as four are in the real world. We shall further suppose that we can only travel in a straight line; so we need to find the single best straight line through these points. But the word 'best' represents a value judgment; it has no absolute meaning until we give it one. The meaning we choose to give it is that, if we drop perpendiculars from the points onto the line, these shall be collectively as short as possible. Since we shall be calculating these, not measuring them with a ruler, we might run into difficulties with signs (since the points will not be all on the same side of the line); so the requirement we actually make is that the sum of the squares of the perpendiculars shall be as small as possible. Note that this is a quite different situation from regression. In regression we assume that one of the measurements is exact and that all the error lies in the other. Since in the two-dimensional case there are two ways in which we can do this, there are in general two regression lines. In the present case we are looking for the *single* best straight line, and there can be only one.

In the two-dimensional case it is not particularly difficult to find such a line; but as the number of dimensions increases beyond two, the necessary algebra becomes increasingly cumbersome and intractable, aud quite out of the question as a routine method. About a century ago, however, it was found that if we are prepared to accept one further constraint there is a relatively simple solution. The constraint is that the line must not only minimize the sums of squares of the perpendiculars; it must also pass through the origin. But if we look at Fig. 7.2 we see that a line through the origin and our six points is hardly the line we are looking for; it is going in quite the wrong direction. We want a line that goes somewhere along the long axis of our cloud, not one that cuts it across the middle. We can do nothing about the constraint itself, but we *can* do something about the origin: we can move it, for moving the origin will not affect our overall pattern.

But P.C.A. gives us no instructions as to where to move it to, that is a subjective decision we must take ourselves. We obviously want it to be somewhere in the middle of our cloud of points; and in practice we normally move it to the common mean, the point $(\bar{x},\bar{y})$, because this is computationally simple and usually effective. Every x value has now been replaced by $(x - \bar{x})$ and every y by $(y - \bar{y})$. The table is now as follows and Fig. 7.3 shows the corresponding diagram.

Measurement 1	Measurement 2
1	5
2	2
0	1
2	0
−4	−4
−1	−4

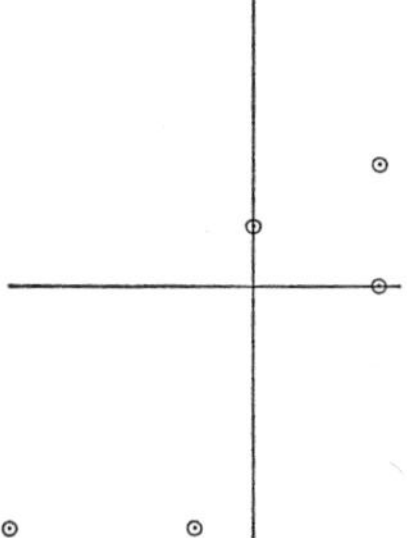

Fig. 7.3. The origin shifted to the common mean.

Now, our original axes naturally pass through the origin, because this is what we mean by the origin; and we know that the line we are seeking must also pass through the same origin. It follows that all we need to find is the *angle* that the new line makes with both the old. In our two-dimensional case if we know the angle it makes with one of the original axes, we also know the angle it makes with the other, since the two must add up to 90°; but if there were more dimensions than two, we should need to know all the separate angles concerned.

The method of computation

The reader must accept the fact that what follows is only an illustration, a recipe if you like, and not an explanation; anybody who wishes to know *why* we proceed in this manner must seek the knowledge in more formal texts. We begin by writing Table 7.2 as a matrix:

$$\begin{bmatrix} 1 & 5 \\ 2 & 2 \\ 0 & 1 \\ 2 & 0 \\ -4 & -4 \\ -1 & -4 \end{bmatrix}. \tag{7.1}$$

We now need the sums of squares and cross products of our two measurements, $\sum x^2$, $\sum y^2$ and $\sum xy$, together constituting the *dispersion matrix*. In Chapter 2 it was shown that we obtain this by pre-multiplying our matrix by its own tranpose; when we do this we obtain the matrix

$$\begin{bmatrix} 26 & 29 \\ 29 & 62 \end{bmatrix}.$$

We now require the latent roots of this matrix. As usual, we set up the determinantal equation, and we find it to be

$$\lambda^2 - 88\lambda + 771 = 0$$

and the roots of this equation are $78 \cdot 13$ and $9 \cdot 87$. Remember that the sum of these must equal the trace of the matrix, i.e. $26 + 62$, or 88; so that the sum of the two roots is also equal to the total sum of squares, of both measurements, that we have to account for.

We now want the latent vectors corresponding to these two roots standardized so that the sum of the squares of the elements of each equals 1. The two vectors are in fact

$$\begin{bmatrix} 0 \cdot 486 \\ 0 \cdot 875 \end{bmatrix} \text{ and } \begin{bmatrix} -0 \cdot 875 \\ 0 \cdot 486 \end{bmatrix}.$$

We shall begin by considering the first of these, which corresponds to the larger latent root. It can be shown that the two values, $0 \cdot 486$ and $0 \cdot 875$, are the cosines of the angles that the new line makes with the old X- and Y-axes respectively. $0 \cdot 486$ is the cosine of $61°$ and $0 \cdot 875$ is the cosine of $29°$; the new line must therefore make an angle of $61°$ with the old X-axis and $29°$ with the old Y-axis. This line (labelled I) has been drawn in Fig. 7.4.

There is a second root and a corresponding latent vector, which must define a second line. The value $-0 \cdot 875$ is the cosine of $(180° -29°)$, and again $0 \cdot 486$ is the cosine of $61°$. The line so defined has been drawn in the diagram and is labelled II; we have now exhausted our information.

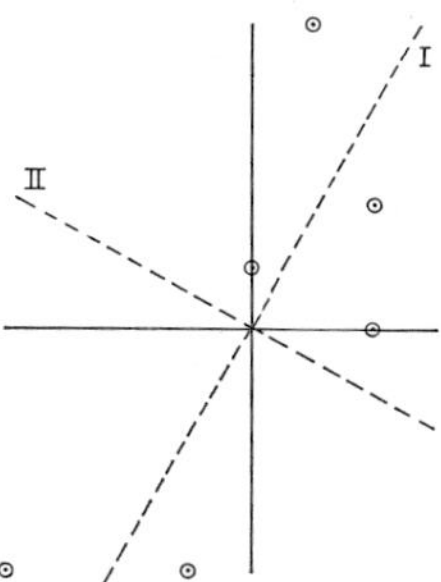

Fig. 7.4. The first (I) and second (II) principal axes shown as dotted lines.

However, we now need to know the coordinates of our six points on the new axes; we obtain these by post-multiplying the data matrix by the matrix of latent vectors. We therefore carry out the multiplication:

$$
\begin{bmatrix} 1 & 5 \\ 2 & 2 \\ 0 & 1 \\ 2 & 0 \\ -4 & -4 \\ -1 & -4 \end{bmatrix} \cdot \begin{bmatrix} 0\cdot486 & -0\cdot875 \\ 0\cdot875 & 0\cdot486 \end{bmatrix} = \begin{bmatrix} 4\cdot86 & 1\cdot55 \\ 2\cdot72 & -0\cdot78 \\ 0\cdot87 & 0\cdot49 \\ 0\cdot97 & -1\cdot75 \\ -5\cdot44 & 1\cdot56 \\ -3\cdot98 & -1\cdot07 \end{bmatrix}
$$

In Fig. 7.5 the points are plotted with respect to the new axes. Now, we know that axis I is our 'best' straight line; it minimizes the sum of squares of the perpendiculars and takes out the maximum amount of information that can be removed by a single line. But how much information has it taken out? We can estimate this by calculating the sum of squares of the coordinates on the axis (this is also the sum of squares of the deviations from the mean; for the mean is zero, having been made so). The result is $78\cdot15$, which is equal to the latent root within the accuracy of rounding-off errors. Similarly, the sum of squares of the coordinates on axis II is $9\cdot89$, the second latent root. It should now be clear that the successive latent vectors define the position in space of the successive 'best' axes; the successive latent roots define the amount of information associated with each line.

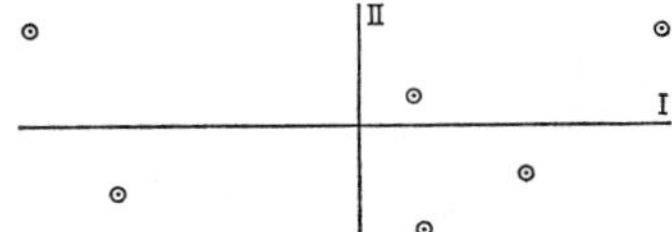

Fig. 7.5. The points plotted with respect to the principal axes.

Before leaving the calculation, we must admit that we have so far evaded the problem of the signs of the latent vectors. We know from Chapter 4 that the overall sign of a latent vector is arbitrary; so what happens if we change the sign of one or more of our vectors? All that can happen is that we may produce a mirror image of our original pattern instead of the pattern itself. In real life this is of no importance; it is the interrelationships of the various axes that most interest us. It is only in a case such as the present, where we are concerned to illustrate the extraction of principal components without changing the original space in any way, that we need to be careful about signs.

Properties and Problems

How many components?

In our little example, the first latent root accounted for 78·13 units of the sums of squares out of a total of 88 units, i.e. the first component accounted for 89% of the information, leaving only 11% for the second component. This is the sort of thing one hopes for: our original dispersion matrix might have been based on 10 or even 20 measurements, but we should hope that the first few components would take out so much that we could safely neglect the remainder. The biologist with a little statistical knowledge naturally asks at this point whether there is a significance test that will enable him to make an objective judgment as to how much can safely be discarded. This raises a new problem. The reader may have noticed that in the whole of the foregoing we have not mentioned the word 'distribution', nor was there any need to do so; so far as the efficient reduction of the dimensionality of the space is concerned, the process is completely distribution-free and is applicable to *any* set of numbers that can be thought of as defined in a Euclidean space. But if we are to ask for a significance test, the data must be substantially multivariate normal; that is to say, the individual variates must be approximately normally distributed and the correlations between them must be substantially linear. Given such a case, one would like a test that assigned a probability to the hypothesis that all subsequent roots were zero. No such test is known; but there *is* a test—due to Bartlett—that all subsequent roots are equal. Since in practice this would usually mean that they were all too small to bother about, the test more or less fulfils the same purpose. However, multivariate normality is a condition few agricultural data matrices achieve. Worse, if the dispersion matrix is singular, the test fails completely, since the statistic to be calculated has the determinant of the dispersion matrix in its denominator. In practice, therefore, the test is of little use, and the investigator retains only as many components as he can interpret, or at least handle.

Occasions arise when the roots do not fall sharply, but drift downwards with no obvious cutting-point; the first few roots may then take out only a disappointingly small proportion of the information. There are two possible causes of this. The first, which is not uncommon, arises when the variances of the original measurements are approximately equal (or, for reasons we discuss below, have been made so), and the measurements are only very weakly correlated. In such a case no really efficient reduction of the space is possible. The second cause is that the correlations between the original measurements may be strongly curvilinear. Principal component analysis takes out only the best *straight* lines and is relatively helpless when confronted with non-linear configurations. This situation is common in plant ecology, since (except over a very narrow range) the relationship between the amount of a species and almost any environmental parameter is bell-shaped, not linear.

Orthogonality

Cross-multiply the final coordinates in our example (i.e. obtain $(4 \cdot 86 \times 1 \cdot 55) + (2 \cdot 72 \times -0 \cdot 78) + $ etc.); the answer, within the limits of our rounding-off errors, will be zero. The components are, in fact, *orthogonal* to one another; that is, they are completely uncorrelated. This is an extremely useful property of principal components. A biometrician not infrequently wishes to undertake a multiple regression; but stepwise multiple regression of a highly internally correlated system is a treacherous business, and there is no certainty that the result will be optimal. However, if the system is reduced to its principal components these, since they are independent of each other, can be separately processed by univariate regressions. An exactly similar situation exists with a multivariate analysis of variance; it is often more satisfactory to extract the principal components and then carry out univariate analyses on these. However, if statistical operations of this sort are intended, it must always be remembered that all the normal distributional limitations will now apply.

Non-invariance

Consider our original data matrix (7.1), and suppose that by an error in recording one of the two sets of values had been divided by 2, so that the matrix should really have been

$$\begin{bmatrix} 2 & 5 \\ 4 & 2 \\ 0 & 1 \\ 4 & 0 \\ -8 & -4 \\ -2 & -4 \end{bmatrix} \qquad (7.2)$$

In a regression situation this would not matter. The regression equation we obtained would *look* different, but we could always transform it back to the same regression line. This is not the case in P.C.A. If we carried out a P.C.A. on matrix (7.2) the result would not be the same as the one we have carried out on matrix (7.1); and the axes from matrix (7.2) *cannot be transformed back into those of matrix* (7.1). We say that P.C.A. is 'not invariant under change of scale'. The units of measurement are now an integral part of the system and must be chosen with great care.

Standardization

In our artificial example, both our variables had much the same range. But suppose we had taken a number of measurements, all in the same units, such that one of them had a range of values greatly exceeding that of any of the others. It will be obvious from the geometry of such a situation that most of the information must lie along this particular axis, and the first principal component must run very close to it. In such a case there is little point in carrying out a P.C.A., since we know before we start approximately where the first axis must lie. An obvious way out of this difficulty is to *standardize* the variables before we start, so as to bring them to equal range. We could, for example, divide all the values for a given variable by the range of that variable; or we could bring all the variables to equal (unit) variance by dividing throughout by the standard deviation of each variable. Both methods are in use, but it is most usual to use the equal-variance standardization.

We shall then find that our dispersion matrix is closely related to a correlation matrix. In fact, if there were n individuals and the correlation coefficient between the ith and jth variable was r_{ij}, the entry in the dispersion matrix would be nr_{ij}. This suggests that if we wanted to standardize, we might as well start with the correlation matrix. This is in fact often done, but there is one point to be watched. If we tried to do what we have done on our artificial example, but used the correlation matrix, we should find that the coordinates on the principal axes no longer fitted the original diagram; the two diagrams would differ in scale by a factor of $\sqrt{n}$. Some 'canned' programs can be a nuisance because of this; they may assume that you have used the correlation matrix but want to re-scale the answer to what it would have been if you had used the dispersion matrix. If you *had* used the dispersion matrix in such a case, you would again find the scale wrong.

Inconsistent dimensions

We have so far tacitly assumed that all our measurements have the same dimensions—they are all in centimetres, or grams, or what you will. But suppose some were in grams, others in centimetres and still others in degrees

Celsius? There is obviously no way in which these can be added as they stand. However, if you have standardized to equal range or variance, the variables will also have been made *dimensionless*, and a P.C.A. is possible. Some writers take a 'purist' approach and maintain that this should never be done, but this is a counsel of despair. You have admittedly distorted your data by rendering them dimensionless, but they may still provide you with profitable information as to the interrelationships of your standardized attributes.

Interpretation

The 'meaning' of a component

Undertaking a P.C.A. is tantamount to saying something like, 'I postulate that these data have been generated by a small number of independent linear factors. *This postulate being accepted*, find me the most efficient set of factors which will suffice.' The answer does *not* tell you that the system is of this nature; it tells you that *if* it happened to be of this nature, this is what the structure would look like. This is why we say that P.C.A. is a system for *generating* hypotheses, not for testing them. We shall suppose that you are prepared to take the answer at its face value and want to ascribe a meaning, or an interpretation, to your hypothetical factors. This can be done by examining the latent vectors, since the elements on these vectors are the cosines of the angles between the components and the original attribute axes. If a component lies almost perfectly along one of the original axes, so that it reflects virtually nothing else, the cosine will be close to 1; if there is an original axis to which it is completely unrelated, the cosine will be close to 0. You can regard the vector elements as 'weighting' coefficients, telling you to what extent a component reflects each original attribute, and whether the relationship is positive or negative. You will be able to say that a particular component is 'made up of' so much of this original attribute, so much of that attribute, so much of the other attribute, and so on. This combination of characters may suggest to you an agricultural meaning, which may form the basis of a useful hypothesis. But it must never be forgotten that this *is* only a hypothesis; avoid the trap of saying that you have *identified* the factors.

Rotation

Users sometimes complain that the pattern of the elements on the latent vectors defies any simple biological interpretation. On inquiring further, the analyst may find that the user hoped that each component would be associated with only a small number of the original attributes, which would greatly simplify interpretation. Unfortunately, there is no reason why the P.C.A. solution should satisfy this additional requirement. Methods have therefore been devised whereby, after the components have been extracted, the component axes can again be swung round in space in such a way that each has, as far as possible, high associations with some original axes and low associations with others—a situation sometimes described as 'simple structure'. There is a subjectivity about this approach, since there is no longer a unique answer; the answer will depend on the particular form of simplicity that has been specified. Moreover, if the axes are to be rotated at all, it does not follow that they need remain orthogonal; a still simpler structure might be found if the axes were swung quite independently, so that they finished up at oblique angles with each other. Rotation programs exist, particularly for orthogonal rotation; but the interpretation of the results they produce needs great care, and it is wise for the beginner to take advice from somebody experienced in their use.

Reference

Goodall, D. W. (1954). Objective methods for the classification of vegetation. III. An essay in the use of factor analysis. *Aust. J. Bot.* **2**, 304–24.

Suggestions for Further Reading

Blackith, R. E., and Reyment, R. A. (1971). 'Multivariate Morphometrics'. (Academic Press: London.)
Kendall, M. G. (1957). 'A Course in Multivariate Analysis'. (Griffin: London.)
Seal, H. (1964). 'Multivariate Statistical Analysis for Biologists'. (Wiley: New York.)

8

Other Ordination Procedures

W. T. Williams

Factor Analysis

Factor analysis is among the oldest of all ordination procedures; like P.C.A., it was developed at the beginning of the century. Originally developed by Spearman for use in psychology, its use is still very largely confined within that subject. Although there have been sporadic attempts to apply it to, for example, ecological or horticultural data, it has not become widely popular. There are probably three reasons for this. First, it is computationally more demanding than component analysis, so it is not to be undertaken lightly. Secondly, there is a greater element of subjectivity in factor analysis, in that certain important decisions have to be taken by the user; as a result, its successful application calls for considerable experience, which few workers outside psychology ever obtain. Lastly, in Australia at any rate, efficient and easy-to-use computer programs are hard to come by. This may be because the centre of factor-analysis work has long been the U.S.A., so that most of the basic texts and programs are American.

The basic problem

Suppose we proposed to undertake a component analysis of five attributes (A, B, C, D and E) and had elected to begin with a correlation matrix; and suppose we obtained a matrix like the following artificial example:

	A	B	C	D	E
A	1·00	0·83	−0·62	0·02	0·21
B	0·83	1·00	−0·51	−0·13	0·29
C	−0·62	−0·51	1·00	0·25	0·01
D	0·02	−0·13	0·25	1·00	−0·11
E	0·21	0·29	0·01	−0·11	1·00

It is obvious that A, B and C are quite highly correlated; but D and E are not, either with the ABC group or with each other. It is as though A, B and C are reflecting some basic pattern in the data as a whole; but D and E are behaving quite independently, each going its own separate way unconnected with any pattern elsewhere in the system. If we are seeking a general pattern, we shall obviously not find it in D and E, which might as well be discarded from the analysis. Now, the correlation matrix is a variance–covariance matrix, and all the variance has to be accounted for. But we have made all the variances equal, so that all our original axes are equally important. The result of this will simply be that D and E cannot take part in any appreciable reduction of dimensions; each will need a separate latent root and vector to accommodate it. It is this situation that factor analysis attempts to prevent.

We postulate that each of our original measurements is really made up of two parts, a *common-factor* part and a *unique* part, and that all the unique parts are independent of everything else, including each other. What we have really done (if we can do this) is to *increase* the dimensionality of our space; because where we originally had 5 axes, we now have 10. However, this does not matter, since we are not interested in the unique axes. As a result, each of our variances will also be divisible into a common-factor variance and a unique variance; we call the common-factor variance the *communality*. In our artificial example, the communalities for A, B and C would be high, those for D and E would be low. Now if instead of the values of $1 \cdot 00$ in our principal diagonal we inserted the communalities, we should have reduced the rank of the matrix, so that some of the latent roots (the ones we are not interested in, because they are associated primarily with the unique variances) would become zero or nearly so. In the matrix above, the latent roots are, in order, $2 \cdot 41$, $1 \cdot 04$, $1 \cdot 01$, $0 \cdot 40$ and $0 \cdot 13$. If instead of filling the principal diagonal with the value $1 \cdot 00$, we replace it by the string of values $0 \cdot 93$, $0 \cdot 82$, $0 \cdot 87$, $0 \cdot 38$ and $0 \cdot 26$, we obtain for our latent roots the values $2 \cdot 26$, $0 \cdot 58$, $0 \cdot 42$, $0 \cdot 03$ and $-0 \cdot 02$. The last two roots have all but vanished, and we are left with only the common-factor pattern we are seeking.

The search for communalities

If the order of our original matrix was n, it can be shown that there is an exact solution for communalities only if $(8n + 1)$ is a perfect square; but this is quite useless in practice, and we proceed as follows. We guess an initial set of communalities, and then *decide in advance* how many roots and vectors we are interested in; we extract the roots and vectors of the matrix, terminating the extraction of roots at the predetermined number and standardizing the elements of the vectors so that the sum of their squares equals the corresponding latent root. There is then a simple calculation which provides an improved estimate of the communalities. This revised estimate is entered in the original matrix, and we go round again; and we continue going round until our communalities are substantially constant.

But if we do not know in advance the pattern we are seeking, how can we know how many axes will be needed to accommodate it? This is a very real problem; the standard texts provide a number of conflicting 'rules of thumb', and this remains a personal decision and somewhat a matter of luck.

Once the number of axes has been decided, and the communalities calculated, we have tacitly assumed that the roots and vectors will be extracted by the methods described in the earlier chapters. This would usually be the case nowadays; but a newcomer to the field who ventures into the literature will find frequent references to a 'centroid solution'. This is based on a numerical model quite different from that of the standard P.C.A. model; but it is far more amenable to hand calculation, and provides surprisingly good approximations to the roots and vectors.

There are further difficulties ahead. If we require the coordinates of the individuals on the new set of factor axes, we can no longer obtain these by simple matrix multiplication; more complex computation is necessary. Moreover, since the whole purpose of factor analysis is to obtain the simplest possible underlying pattern, it is common practice to carry out a subsequent rotation to obtain new, simpler axes, which may be orthogonal or oblique.

It will be clear, even from this sketchy account, that factor analysis does not provide the simple, unique and unequivocal answer produced by component analysis. Attempts have been made to remove the subjectivity inherent in factor analysis by defining a more rigorous mathematical model. The most successful of these is undoubtedly the maximum-likelihood solution of Lawley and Maxwell (1963); but despite its mathematical elegance, this has not proved widely useful. The computational procedure requires that the initial dispersion matrix shall not be singular, and the associated significance tests require that the data shall be multivariate normal; moreover, even if these conditions can be met, convergence to the optimum solution tends to be slow. We may suspect that the older, simpler, centroid solution will still be the most suitable for everyday use. Factor analysis is not yet part of the necessary armoury for a pattern analyst working with agricultural data; but our advice to anybody confronted with a factor-analysis problem would be to try the centroid solution first.

Principal Coordinate Analysis

This is the single most important ordination procedure for the pattern analyst. It is simple, powerful and unequivocal, easy to program and widely applicable. Curiously enough, it arose historically out of a factor-analysis problem. In ordinary factor analysis, starting with a correlation matrix, the correlations are between all possible pairs of attributes; such an analysis was usually known as an R-analysis. Occasionally, however, factor analysts would begin with a matrix of correlations between all possible pairs of *individuals* over all attributes: this was known as a Q-analysis. The procedure worried mathematicians, since it appeared to contravene every reasonable mathematical principle underlying component analysis. The trouble was that it worked; it gave analysts meaningful and profitable solutions. Now, it is a sound assumption that if a method consistently works, it is almost certainly based on a sound mathematical principle, even if the principle is unknown or mistaken for something else. John Gower, at Rothamsted, elected to re-examine the problem, and solved it in his classic (1966) paper. (Biologists will probably find his 1967 simplified exposition more accessible.) In this paper he showed conclusively that factor analysts carrying out Q-analyses were producing the right result for the wrong reason; he severed the connexion with factor analysis and developed the elegant and very general procedure he called *principal coordinate* analysis.

Q-analysis as a mapping procedure

We suppose that *any* measure has been calculated between all pairs of individuals; we shall symbolize the measure between the ith and jth individuals as a_{ij}, and we shall assume (as will normally be the case) that the measure is symmetrical, so that $a_{ij} = a_{ji}$. We shall denote the matrix of such measures by (a_{ij}). Now suppose we treat this matrix as for a Q-factor analysis; that is, we extract the roots and vectors, and standardize the vectors so that the sum of the squares of the elements equals the corresponding latent root. Gower showed that the vectors now define a Euclidean space (which is naturally tending to become known as a 'Gower space'); the vectors define successive orthogonal axes, and the elements of the vectors are the coordinates of the individuals in the new space. Subject to conditions we shall consider below, the new system is Euclidean, even if the original measures were not. How do the individuals—the points in the new space—stand in relation to one another? Let us symbolize the distance between the ith and jth individuals in the new space as d_{ij}; then Gower showed that this distance is related to the original values in the matrix (a_{ij}) in the following manner:

$$d_{ij}^2 = a_{ii} + a_{jj} - 2a_{ij}. \tag{8.1}$$

Now suppose that before we extracted the roots and vectors, we had made certain changes to the matrix; we shall put unities all down the principal diagonal, so that $a_{ii} = a_{jj} = 1$, and we shall replace all the individual a_{ij} values by their 'one-complements', i.e. by $(1 - a_{ij})$.

Expression (8.1) now becomes

$$d_{ij}^2 = 1 + 1 - 2(1 - a_{ij}),$$
$$= 2a_{ij},$$

so we have preserved the square roots of our original values. It should be easy to show that had we again entered the principal diagonal with unities, but replaced every a_{ij} value with $(1 - \frac{1}{2}a_{ij}^2)$, we should preserve the original values themselves.

For all this to be strictly true, the matrix (a_{ij}) must fulfil two conditions. First, it must be symmetric, since no relationship corresponding to expression (8.1) is known for an asymmetric matrix. Fortunately, almost all real-life similarity or dissimilarity matrices are symmetrical. The asymmetric case has, however, turned up in studies of forest regeneration. If a_{ij} represents the number of seedlings of tree-species A occurring under mature individuals of tree-species B, then a_{ji} will represent the number of seedlings of tree-species B which occur under mature individuals of tree-species A; and in such a case we cannot expect a_{ij} and a_{ji} to be equal. Williams *et al.* (1971) have suggested an extension of the Gower procedure to handle this case, but it is too early to say whether it will prove generally applicable.

The second requirement is that the matrix (a_{ij}) must have no negative latent roots; some of its roots may be zero, but none must be negative. Such a matrix is said to be *positive semi-definite*, often abbreviated to 'p.s.d.' Even if (a_{ij}) is not p.s.d., the algebraic relation (8.1) will still hold; but part of the space will now be imaginary, and therefore no longer Euclidean. Very small negative roots do not matter; but a sizable negative root can cause the information in the real part of the space to be seriously distorted. Fortunately, all similarity and dissimilarity measures in common use appear to define p.s.d. matrices, though proofs that this must be so are known in only very few cases. However, missing values in the original data matrix can destroy the p.s.d. property of the dissimilarity matrix; if there is a high proportion of missing values, it is a wise precaution to examine *all* the roots of the matrix before proceeding.

The Gower adjustment

We have now mapped our original dissimilarity measures into a Euclidean space; but this new space will have as many axes as there are individuals, which will normally be too many to handle. However, there is no reason why we should not now reduce the dimensionality of the space by carrying out an ordinary principal component analysis. In his early work (before his paper was published) Gower actually did this, so the total computation involved *two* root-and-vector extractions. However, Gower then discovered a simpler solution. We take the original (a_{ij}) matrix, and calculate the means of all the rows (we shall denote a row mean by r_i), the means of all the columns (c_j), and the grand mean of the whole matrix (g). From every element of (a_{ij}) we subtract the appropriate row mean and the appropriate column mean; then we add the grand mean. As a result any single value a_{ij} is replaced by

$$a_{ij} - r_i - c_j + g.$$

Gower showed that if this modified matrix is submitted to a principal coordinate analysis (i.e. roots and vectors extracted, vectors standardized so that the sum of squares equals the root), the Euclidean space which results has already been reduced as if by principal component analysis; we have obtained two root-and-vector extractions for the price of one. This is the procedure that is always followed.

Back-correlation of Gower vectors

When we have obtained our Gower vectors, we can take as many as we wish, plot them in pairs, and so examine the configuration of our individuals in the Gower space. However, in order to explain this configuration we shall want to 'interpret' the Gower axes in terms of the original attributes. Every vector has an element for each individual, and each individual has a number of attributes. We can therefore calculate, for each vector in turn, the correlation coefficient between that vector and each of the original attributes; we sort these into descending order of their absolute values (since high negative coefficients are as informative as high positive ones), and thus discover, for every vector, which of the original attributes it most strongly reflects.

Applications of principal coordinate analysis

We can usefully distinguish two main uses of principal coordinate analysis. In the first it is used solely as an ordination procedure, and such an application needs no further comment. In the second it is used primarily as a prior transformation for conventional statistical analysis. It must be admitted that its use for this purpose arouses misgivings among statisticians; the main difficulty is that although the procedures to be described appear effective and even powerful, their algebraic rationale is completely unknown. In this chapter we shall merely describe the procedures without considering their propriety; a fuller examination of the problems involved is given in Chapter 18. We begin by assuming we have laid out a factorial experiment, and on each element of the design we have recorded a number of measurements, which might be highly inter-correlated. We could analyse each of our measurements separately by univariate analyses of variance; but this tends to be uninformative, or at least difficult to interpret, if the measurements are indeed highly correlated. We might consider carrying out a multivariate analysis of variance; but the significance tests for such an analysis are less robust than those of its univariate counterpart, and the analysis cannot be carried out at all if the system is highly non-orthogonal or over-defined. In such a case we can proceed as follows: we take the original data matrix and carry out a principal coordinate analysis, then interpret the vectors by means of back-correlation. Now, Gower vectors are latent vectors, and therefore completely independent of each other; we can therefore analyse each of them—or as many as we choose to retain— separately by univariate analyses of variance. An interesting feature of this process is that if the original treatments are orthogonal, the main effects tend to dispose themselves on separate Gower vectors. A particularly valuable feature of this process arises if the system has a very large error contribution; for this may itself be patterned, and separate out as a distinct vector. As a result, the true main effects can be tested against an error term from which an unwanted pattern has been partitioned out.

Another valuable application of principal coordinate analysis is its use in the analysis of sequential data. For example, an agronomist might have set up a factorial experiment whose results issued in the form of long sequences of cumulative liveweight gains. In the past there has been no completely satisfactory way of carrying out an analysis of variance on a set of sequences; often all that has been done is to analyse a single terminal value, or perhaps a set of annual values, but this is to discard all or most of the sequential information. The earliest solution in the literature is that of Yates *et al.* (1964), who fitted quartic regressions to the sequences and analysed the regression coefficients. It is doubtful whether such a method would be of universal application; the interpretation of high-order regression coefficients may be difficult or impossible, and their calculation tends to be ill conditioned. Dale *et al.* (1970) have devised an alternative approach. Let us take our sequences and 'chop' each into a number of discrete states; anything between 3 and 8 states is commonly used, and there are simple computer programs which will do this automatically.

By chopping a continuous variable into a stepped one, we have admittedly discarded information, but not so much as if we had ignored the sequence altogether. Our new, 'chopped' sequences will now each define a *transition matrix*, which tells us how many times the state defined by the column follows the state defined by the row. Such a matrix can be associated with a numerical value, its *information content;* and every pair of sequences can jointly be associated with an *information gain*. If there are n units in the experimental design, and therefore n sequences, there will be $\frac{1}{2}n(n-1)$ information gains. These are in effect dissimilarity measures between whole sequences, and can be mapped into Euclidean space by principal coordinate analysis. The resulting Gower vectors can then be subjected to univariate analyses of variance. There is only one drawback; when we have discovered which of our treatments or interactions are significant, we should like to know how these have affected the sequences. We should therefore like in some way to back-correlate our Gower vectors with the original sequences; unfortunately, at present no way of doing so is known.

Canonical Analysis; Canonical Coordinate Analysis

Occasions not infrequently arise when we have two sets of data for the same set of individuals. In ecology, a set of sample sites may have records of both plant species and soil measurements; in animal nutrition experiments there may be input and output measurements. Our interest is then very often to ascertain the extent to which one set of measurements is correlated with the other, and the particular attributes which have been responsible for these correlations. This is the province of *canonical analysis*, which we can regard as an extension of component analysis. Geometrically, we again regard the individuals as points in a space defined by attribute axes; but some of these axes relate to one set of variables (say, the X variables), the remainder to the other (Y) set. We could take the system all together, and by component analysis find the single best line which summarizes all the axes, both X and Y; but this is not what we want to do, since we do not want to mix our two sets of variables. We therefore find the best line among the X axes and the best line among the Y axes, subject to the additional constraint that these two lines *must be as highly correlated as possible*. We finish up with a set of *pairs* of lines, each pair having an X-member and a Y-member; and we impose the constraint that each line must be independent of all other lines on its own side, and of all except one on the other side. The first pair possess the largest *canonical correlation coefficient*, the second pair the next largest, and so on. If R_{xx} is the correlation matrix between all the X variables, R_{yy} that for the Y variables, and R_{xy} is the matrix of correlations between all the X values and all the Y values, we have the uncomfortable task of solving the determinantal equation

$$\left| R_{xx}^{-1} \cdot R_{xy} \cdot R_{yy}^{-1} \cdot R_{yx} - \rho^2 I \right| = 0. \tag{8.2}$$

The canonical correlation coefficients we are seeking are the successive values of ρ, the roots of this equation.

It cannot be said that in this form the method has been useful in agriculture, or indeed in biology in general: the new axes almost always prove extremely difficult to interpret. This is believed to be because, in most biological data sets, there is a high proportion of unique variances which effectively confuses or even conceals the common-factor pattern we are seeking.

It is therefore common practice, at least in agricultural experimentation in Australia, to use a modified system which we have come to call *canonical coordinate analysis*. The two data sets are separately mapped in Euclidean space by means of principal coordinate analysis, so that we obtain two sets of vectors which we may call the Gower X-vectors and the Gower Y-vectors. Now, if as usual we retain only the first few of these we shall have discarded most of the unique 'rubbish', since it will be left behind in the vectors we have discarded. What will happen if we now carry out a canonical analysis on the Gower vectors? The first thing that happens is a remarkable simplification of equation (8.2). For all the Gower X-vectors are orthogonal to each other, i.e. they are uncorrelated; so that R_{xx} has unities down the principal diagonal and zeros elsewhere. In other words, it becomes the identity matrix I. But the identity matrix is its own inverse, just as 1 is its own reciprocal; so R_{xx}^{-1} is replaced by I, and so is R_{yy}^{-1}. Equation (8.2) now reduces to the much simpler equation

$$\left| R_{xy} . R_{yx} - \rho^2 I \right| = 0.$$

This version has invariably proved informative; the roots and vectors are easy to extract, and the coordinates of the individuals on the paired canonical axes can be obtained by simple matrix multiplication. The Canberra program CANONGO accepts two sets of Gower vectors, and carries through the complete canonical analysis automatically; examples of this procedure will be found in Part III.

Canonical Variate Analysis

This is not a pattern-analysis technique at all, but a true statistical technique. It is only mentioned here because it is sometimes, unhappily, referred to simply as 'canonical analysis', and the unwary reader may therefore confuse it with the methods of the previous section. It arises when a classification already exists, and when it is desired to know whether groups of this classification differ significantly from each other over all attributes. It requires that the within-group dispersion matrix shall be constant over all the groups and shall not be singular; and the data must be substantially multivariate normal. It has found little use in agricultural experimentation, where these conditions can rarely be fulfilled.

Computational Considerations

It will be clear that a pattern analyst needs a computer program which, given a matrix of measures between attributes or individuals, can carry out either a principal component or a principal coordinate analysis. Unlike the situation we shall be exploring in classificatory programs, such a program is easy to write, given only a minimum knowledge of FORTRAN programming. The basic requirement is a subroutine that will extract the roots and vectors of a symmetric matrix, with the roots in descending order of magnitude, and will leave the roots and vectors in arrays which can then be accessed by the programmer. All scientific computers of any size possess such a subroutine; many computers possess more than one, and the main problem here may be one of choice.

There are then four basic requirements:

(i) A *control-card* which defines the analysis to be undertaken and whose instructions can be appropriately stored. This must specify five things:

(*a*) The order of the matrix to be expected (either an upper or lower triangle can be used; most programs take the upper triangle).

(*b*) The nature of the principal diagonal. For a principal coordinate analysis or the component analysis of a correlation matrix, this will be everywhere 1; and since this is the commonest case, this can be used as a default option. However, if a dispersion matrix is to be used, a separate principal diagonal will have to be read in, and the program needs warning of this.

(*c*) Transformation of off-diagonal entries. For most principal coordinate analyses the value a_{ij} will need to be replaced by $(1 - a_{ij})$; for a principal component analysis the values will be left unchanged. Other transformations can be built in for special purposes, but will be seldom if ever used.

(*d*) Whether the Gower adjustment (subtraction of row and column means, addition of the grand mean) is to be carried out; it is needed for coordinate analyses but not for component analyses.

(*e*) Whether the vectors are to be standardized so that the sums of squares of elements equal 1 (most subroutines leave the vectors in this condition) for component analysis, or standardized so that the sums of squares equal the corresponding latent roots; this is required for coordinate analysis.

(ii) An *input routine* for the off-diagonal entries. The format can be fixed by the programmer; the Canberra program GOWER uses a free-format subroutine INCOEF, which uses a comma as separator between elements and an asterisk as terminator. This subroutine will also be used for reading in the principal diagonal if this is required.

(iii) A *setting-up routine* for the Gower input matrix. This must complete the matrix from the triangle read in, carry out any transformation specified, and carry out the Gower adjustment if required. At this point the roots-and-vectors subroutine is called in.

(iv) A *restandardization routine* to leave the latent vectors in whichever form of standardization has been specified. These, and their corresponding roots, are then printed out.

The program can, of course, be more complex than this; the Canberra program GOWER provides facilities for plotting and punching the vectors, but these are luxuries.

References

Dale, M. B., Macnaughton-Smith, P., Williams, W. T., and Lance, G. N. (1970). Numerical classification of sequences. *Aust. Comput. J.* **2**, 9–13.

Gower, J. C. (1966). Some distance properties of latent root and vector methods used in multivariate analysis. *Biometrika* **53**, 325–38.

Gower, J. C. (1967). Multivariate analysis and multi-dimensional geometry. *Statistician* **17**, 13–28.

Lawley, D. N., and Maxwell, A. E. (1963). 'Factor Analysis as a Statistical Method'. (Butterworth: London.)

Williams, W. T., Dale, M. B., and Lance, G. N. (1971). Two outstanding ordination problems. *Aust. J. Bot.* **19**, 251–8.

Yates, J. J., Edye, L. A., Davies, J. G., and Haydock, K. P. (1964). Animal production from a *Sorghum almum* pasture in south-east Queensland. *Aust. J. Exp. Agric. Anim. Husb.* **4**, 326–35.

Suggestions for Further Reading

Cattell, R. B. (1952). 'Factor Analysis'. (Harper: New York.)

Thomson, G. (1951). 'The Factorial Analysis of Human Ability'. (Univ. of London Press.)

9

Ordination: Recent Developments and Future Possibilities

M. B. Dale

The Basic Problem

There is currently available a viable collection of ordination methods which, judiciously applied, can provide a good deal of information. They are heuristic methods, that is they *serve to discover*, and it is rare for confirmation of the discovery to be possible without further data collection. One characteristic of heuristic methods is that they can normally be improved by the introduction of 'problem-specific' information. This constrains the area of interest and provides for a more detailed examination of a smaller problem area. Thus one possibility of development for ordination methods is simply the introduction of more problem knowledge. An analogy is possible with an architect who will suffer an *embarras de richesse* if simply asked to build a dwelling, and will ask for considerable detail concerning the purpose of the building and the habits of the dwellers. Such problem-specific information must be elicited from experts in the subject matter and incorporated as constraints in the techniques used for pattern analysis.

An alternative to this line of development is to examine the methods themselves more closely to determine their basic assumptions, and hence the constraints that they themselves impose on any analysis. It may then be possible to seek developments that relax these constraints. This is a search for increased generality and, to continue the architectural analogy, is rather like modular construction, with standard units which can be combined to provide a variety of structures, suitable for many different purposes.

Both these lines of development have their dangers. If too much problem specificity is demanded then each problem requires a new method with considerable waste of effort in redesigning for minor variations. On the other hand there does seem to be an inverse relationship between interpretability of results and generality of method, for in the limit a method capable of coping with everything will probably fail to discriminate between any

70

of the possibilities. A judicious mixture of the two components seems necessary, and this must be combined with an appreciation of the feasibility of actually computing the desired analysis.

The Underlying Assumptions

Most ordination methods are concerned primarily to simplify the relationships between items, or between the attributes which describe them, so that they can be presented as a simple visual representation of points which can be plotted. This is not the only function of ordination, but it is the function fulfilled by the most widely used methods, principal component analysis (Chapter 7) and principal coordinate analysis (Chapter 8). Both possess optimal properties which have been briefly described in the previous chapters; but it is important to realize that these properties themselves depend on certain assumptions. Both seek to define new variables or axes which are linear additive combinations of the old ones, in such a way that distances in the new representation are related to similarities in the old. Such a process makes three assumptions. The first is *metricity*, the assumption that the attributes can be treated as measured quantities like distances, which can be added or subtracted; the second is *additivity*, the assumption that the underlying units need only be added or subtracted to generate the data actually observed; and the last is *linearity*, the assumption that the raw data can be generated by linear (i.e. straight-line) functions of a set of underlying units. It is doubtful whether these assumptions are ever completely justified; indeed, Noy-Meir and Austin (1970) have demonstrated that the linearity assumption can never hold in ecology. The additivity assumption is possibly the least considered of these assumptions. Apart from some unpublished work by Gower showing the extreme computational intractability of complex non-additive models, the only non-additive model for ordination I know is that of Gilbert and Wells (1966). It is disturbing to find that with vegetation data this model seems to work much too well, with 90% or more of the variation on one axis. In most principal component analysis results 50% on one axis would be more usual. It is unfortunate that interpreting the results of such a non-additive analysis has so far always proved extremely difficult.

The fact that principal component analysis and principal coordinate analysis have been used with success in a very wide variety of applications suggests that they are numerically robust and may be expected to yield interpretable results providing the underlying assumptions are not grossly violated. However, interpretation itself is a largely subjective process. The axes produced by classical methods are uncorrelated and ordered in terms of an importance measure, so that some can be rejected as either being 'noise' or bearing so little signal that this can be regarded as submerged in noise and disregarded. With care, and some luck, it is often possible to interpret the remaining axes, but the dangers of too-easy reification must be remembered. Human observers are extremely adept at providing explanations of even the most strange results.

Scaling Methods and Unfolding

The assumptions of linearity and metricity are related, and methods are available which avoid both assumptions. Instead of using a least-squares function to relate distance in new and old representations, other functions can be used, notably monotonic regression functions. Instead of relying on the actual values of similarity measures these rely on their order with respect to each other.

The first practical method for accomplishing such an ordination was due to Kruskal (1964), and Anderson (1971) has shown that the results of applying such methods to vegetation data can be useful. However, you must pay for the generality with a more complex computational route and the possibility of convergence to local rather than global optima.

The usefulness of such multidimensional scaling methods can be extended in several ways. First consider the case where several sets of similarity measures are available for the same set of items but derived from several different sets of variables. Carroll and Chang (1970) have proposed an 'individual differences' scaling model where it is assumed that there is a common dimensionality for all the sets of similarities but that each set of attributes will differ in its weighting on these underlying dimensions, some of the weights possibly being zero-valued. A simple example is colour-blindness, where individuals differ in their ability to differentiate between either red and green or yellow and blue. The model seems appropriate to much work on human, and perhaps animal, perception (Bannister and Mair 1968). Recent work here includes the interesting studies of Srinivasan and Shocker (1973) where linear programming methods are used to relate preference data to measured variables.

The second extension is termed *unfolding* (Hays and Bennett 1961). Here it is argued that similarities measure only the magnitude of relationship between items and not its direction. Consider an environmental 'space' in which there are points indicating the positions of certain stands, and also points indicating the 'ideal' points for species (these ideal points represent hypothetical stands to which the species would be perfectly adapted). The model assumes that the abundance of a species in a stand is related to the distance between the stand point and the ideal species point. Thus for each species measurement is outwards from its 'ideal' but the direction is not given. Unfolding attempts to recover stand and species points directly, although in practice it is more likely that only a region containing the points will be identified. It is noteworthy that if all the 'ideal' points are peripheral then principal component (or coordinate) analysis will in fact perform a creditable approximation to the unfolding.

Scaling methods in effect recover a metric representation from non-metric data although some points are still left fixed in rank order only. There has been much confusion in the ecological literature because methods producing rank order solutions have been regarded as producing metric results (this is especially common with so-called gradient methods). However, scaling seems attractive because for some extra cost of computation and what seems a fairly small likelihood of obtaining a local optimum, the constraints of metricity and linearity can in part be avoided.

Seriation

The origins of ordination in ecology lie in the work of Curtis and McIntosh (1951). However, in later studies (Bray and Curtis 1957) the techniques adumbrated led directly towards the PCA methods of Goodall (1954). This is perhaps unfortunate, for until recently (Beals 1973; Hill 1973; Hatheway 1971) it concealed the fact that the original method was not a simplification method at all; instead it was a seriation method. Seriation developed in archaeology (from work by Flinders Petrie in the late nineteenth century) as a means of arranging items in a sequence such that similar items were adjacent or nearly so. The end members of such a sequence might have little or nothing in common provided there was continuity through the remainder of the items. This means that the emphasis is on local similarity, not on global minimization.

This is close to the concept of topological dimensionality introduced by Trunk (1968) who suggested an approximate maximum likelihood solution to the problem. Thus seriation starts with an advantage over PCA in that an acceptable dimensionality estimate can be obtained. Finding the axes themselves is, of course, quite complex computationally. Various sorting methods and similarity measures have been used to produce such series from raw data, but most of them are laborious to the extent of being infeasible for large amounts of data. Shepard and Carroll (1965) extended the scaling model by using a measure of 'continuity' in place of the monotonicity measure, while Lieth *et al.* (1973) resorted to using solutions to the travelling salesman problem for their ordering. However, Hill (1973) started by assuming that both stands and species need ordering and showed how an elegant use of eigenvalues could provide such a solution, with problems of the order of 1000×1000.

There are, of course, problems still remaining with these methods. The generality of the model often leads to problems of interpretation and the impact of error variance can be considerable. Hill's method seems entirely adequate when a single sequence is present but where branching and looping occur it is less helpful. Sneath (1966) presented one method for finding linear sequences in complex data which relies on local linearity and might with some development provide a useful method. Unpublished work by O'Callaghan indicates that automatic selection of neighbourhood size might be possible with such a method which would certainly improve on present line-finding methods.

Factor Analysis

The future of factor analysis in agricultural experimentation is far from clear. The concept of separating unique and common-factor variances is undoubtedly attractive; moreover, the use of oblique rotation implies a freedom from the otherwise almost universal constraint of orthogonality.

While non-orthogonality has always been a serious debating point in factor analysis, non-linearity and non-metricity have received less attention. Wood *et al.* (1966) proposed a quadratic component analysis method which could also be used for factor analysis, while much of the work in multi-dimensional scaling provides a basis for rank-order factor analysis. However, the major difficulty is simply the great variety of methods which have been proposed and which would require examination in an agricultural context. Any such examination would now represent a major research project.

Conclusion

The methods indicated here by no means exhaust all possible futures. To take only a single example, all the methods are based ultimately on a Euclidean-type space, in which the measures of distance are constant throughout the space. Rao (1948) has discussed the possibility of using a Riemannian space representation, in which the measure of distance itself changes with position in the space. However, it seems that little can be said about the curvature of a space when only a point set of samples from it is available. I can only hope that enough has been said here to show that developments are still possible in ordination, and that such developments may give rise to new methods which may prove useful in future pattern-analytic studies.

References

Anderson, A. J. B. (1971). Ordination methods in ecology. *J. Ecol.* **59**, 713–26.
Bannister, D., and Mair, J. M. M. (1968). 'The Evaluation of Personal Constructs'. (Academic Press: London.)
Beals, E. W. (1973). Ordination: mathematical elegance and ecological naiveté. *J. Ecol.* **61**, 23–6.

Bray, J. R., and Curtis, J. T. (1957). An ordination of the upland forest communities of southern Wisconsin. *Ecol. Monogr.* **27**, 325–49.

Carroll, J. D., and Chang, J. J. (1970). Analysis of individual differences in multidimensional scaling *via* an *n*-way generalization of the Eckart–Young decomposition. *Psychometrika* **35**, 283–319.

Curtis, J. T., and McIntosh, R. O. (1951). An upland forest continuum of the prairie-forest border region of Wisconsin. *Ecology* **32**, 476–96.

Gilbert, N., and Wells, T. C. E. (1966). The analysis of quadrat data. *J. Ecol.* **54**, 675–86.

Goodall, D. W. (1954). Objective methods for the classification of vegetation. III. An essay in the use of factor analysis. *Aust. J. Bot.* **2**, 304–24.

Hatheway, W. H. (1971). Contingency table analysis of rain forest vegetation. In 'Statistical Ecology'. (Eds G. P. Patil and E. C. Pielou.) Vol. 3, pp. 271–313. (Penn. State Univ. Press.)

Hays, W. L., and Bennett, J. F. (1961). Multidimensional unfolding: determining configuration from complete rank order preference data. *Psychometrika* **26**, 221–38.

Hill, M. O. (1973). Reciprocal averaging: an eigenvector method of ordination. *J. Ecol.* **61**, 237–50.

Kruskal, J. B. (1964). Multidimensional scaling by optimizing goodness of fit to a nonmetric hypothesis. *Psychometrika* **29**, 1–27, 115–29.

Lieth, H., Numata, M., and Suganuma, T. (1973). Studies in the grassland vegetation in the Kawatabi special research area of the Japanese I.B.P. *Vegetatio* **28**, 41–56.

Noy-Meir, I., and Austin, M. P. (1970). Principal components ordination and simulated vegetation data. *Ecology* **51**, 551–2.

Rao, C. R. (1948). On the distance between two populations. *Sankhya* **9**, 246.

Shepard, R. N., and Carroll, J. D. (1965). Parametric representation of nonlinear data structures. In 'Multivariate Analysis'. (Ed. P. R. Krishnaiah.) (Academic Press: New York.)

Sneath, P. H. A. (1966). A method for curve-seeking from scattered points. *Comput. J.* **8**, 383–91.

Srinivasan, V., and Shocker, A. D. (1973). Linear programming techniques for multidimensional scaling of preferences. *Psychometrika* **38**, 337–69.

Trunk, G. V. (1968). Statistical estimation of the intrinsic dimensionality of data collections. *Inf. Control* **12**, 508–25.

Wood, K. R., McCormack, R. L., and Villone, L. T. (1966). Non-linear factor analysis program A–78A. TM–1764 System Development Corporation. (Santa Monica, Calif.)

10

Types of Classification

W. T. Williams

The Concept of Purpose

The first question to be asked concerning any classification is simply, why is one undertaking it at all? Perhaps the commonest answer is that it eases communication between human beings. If you were thinking of taking a holiday in Italy, and asked me what the people were like, it would not help you to be told that they are warm-blooded, bilaterally symmetrical, with fur rather than feathers, and so on; this is information you already possess and is insufficiently detailed to fulfil the purpose behind your question. If, on the other hand, I supplied you with a complete list of the inhabitants of Italy, together with detailed descriptions and photographs, the information you wanted would be there, but buried in such a mass of unnecessary detail that you could never extract it. What you really wanted was to be told that the inhabitants usually had these physical characteristics, spoke this language, ate this type of food, and so on. In other words, you wanted to define an *intermediate* level of information, sufficiently economical to be quickly grasped by the human mind. This was exactly the problem that faced Mr Edye and Dr Burt with their *Stylosanthes* introductions: to know that all were species of *Stylosanthes* was too general; but a list of some 230 introductions, each specified by some 40 pieces of morphological and agronomic information, was too voluminous to be communicated meaningfully. This is the origin of the 'morphologic–agronomic groups' of Case 3.1.

Non-uniqueness

Unfortunately, different purposes usually imply different classifications. Suppose a mother had 10 children. Suppose, too, she was asked how many she had in the context of a discussion on the family finances. She might well reply that two were infants, four still at school, two boys were still living at home but were now working, and two girls had married and were off her hands. On the other hand, if she had been asked whether she had plans to enter any for the Conservatorium, she might equally well reply that three appeared to be musically talented, but seven did not. Each classification would be appropriate for its stated purpose. But if she were casually asked the question in no particular specified context, she would most likely say 'five boys and five girls'. The reason for this is that this particular classification implies a whole collection of attributes in addition to the one on which the classification has been overtly made; it is, in fact, a *predictive* classification, and so may prove useful in a wide variety of contexts. Such a classification is known as a *general-purpose* classification, and is best defined as a classification erected at a time when the uses to which it is to be put are not known. Most, though not quite all, numerical classifications are of this type; but it must always be remembered that classifications are not unique; and that, if a precise purpose *is* known in advance, a general-purpose classification might be completely unsuitable.

Typology of Classifications

Purpose is not the only consideration when choosing a classification; there are also computational and mathematical considerations. A user might know precisely the classificatory strategy he wished to use, but this might imply the use of a program that was computationally slow and consequently expensive. In such a case a user might decide that he could accept a slightly suboptimal solution in return for a much reduced computing time. In the early days of numerical classification, a common difficulty was that no algorithm was known which could process the type of data the user had collected; but this is seldom a problem today.

It will be convenient to describe the strategies as presenting a series of dichotomous choices; these have been set out in Fig. 10.1 in which the choices have been numbered for reference in the text.

Choice 1: exclusive or non-exclusive. An *exclusive* classification is one in which a given element occurs in one class and one class only; the population is divided into a set of mutually exclusive subclasses which nowhere overlap in their membership. This type of classification is normally seen in the taxonomy of living organisms, though even here there are exceptions: Bor (1960), unable to decide for a group of nine grass species whether they should be ascribed to the genus *Erianthus* or *Saccharum*, firmly put them in both. Exclusive classifications also arise in problems of land survey, in which the individuals are sample plots of land which it is desired to separate by meaningful boundaries.

In a *non-exclusive* classification any given element may appear simultaneously in more than one subclass; this situation usually occurs if a group is defined solely by its attributes; we may say of an individual, 'if it possesses these attributes it belongs to this group'. Such a group is open-ended; we

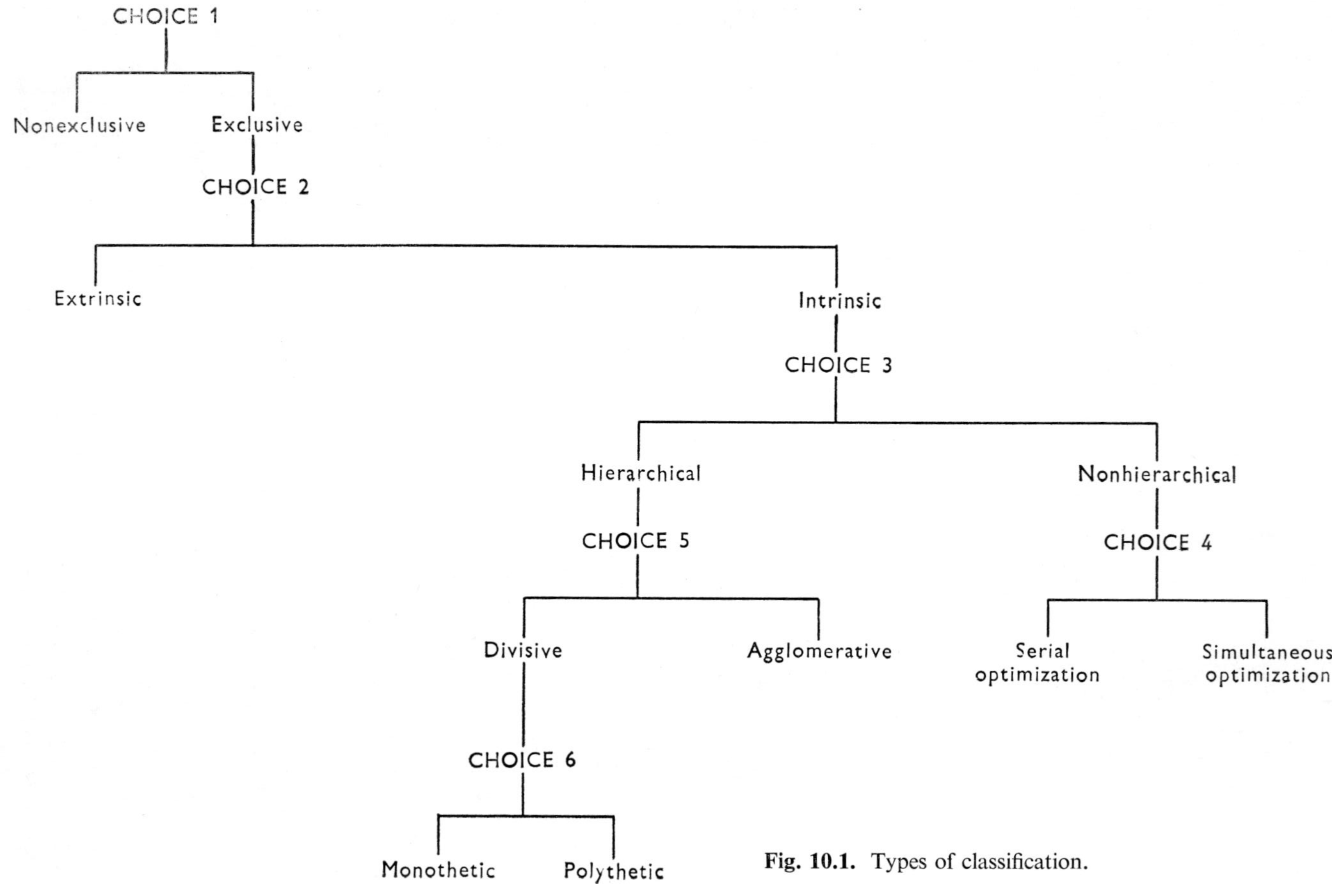

Fig. 10.1. Types of classification.

do not need to know how many elements it contains, nor anything about them other than that they possess the specified attributes. As an example we might classify diseases into three groups: A, those that produce fever; B, those that produce skin eruptions or a rash; and C, those that produce muscular pains. Tinea would be in group B; but measles would be in both groups A and B, and influenza would be in groups A and C. If we were to define sufficient groups of this type we might be able to identify a disease by enumerating the groups in which it occurs, and in fact classifications of this type have found their main use in medical diagnosis.

A simple example of the second type arises in forest survey, where for certain purposes it has been found expedient to regard each tree as the centre of a group consisting of itself and its m nearest neighbors. All, or almost all, trees will then be in several of these groups. The main use of such classifications, however, has been in the enumeration of groups of related words for purposes of library retrieval. Non-exclusive numerical classifications, in fact, have always been special-purpose systems erected with a particular end in view.

Choice 2: exclusive classifications—intrinsic or extrinsic. In an *intrinsic* classification, all attributes used are regarded as equivalent. The resulting groups may be of interest in their own right, as in pure taxonomy; or the boundaries between them may be examined to find out if they reflect discontinuities in some external attribute, as in land survey problems.

In the latter case an *intrinsic* classification is being used extrinsically, since the nature of the discontinuity sought is not known in advance. In an *extrinsic* classification, however, the external attribute is declared in advance, and the resulting groups, though based entirely on the internal attributes, are required to reflect discontinuities in the external attribute as closely as possible, the intrinsic pattern being overridden if need be. It is important to note the distinction between this approach and that of dividing the population on the external attribute in the first place and then examining the resulting pattern of the internal attributes. In such a situation the groups will, of course, reflect the external attribute with the greatest possible fidelity; but the resulting configuration of internal attributes may be such that it cannot be obtained by any operation on the internal attributes alone, so that the classification has no predictive value. True extrinsic classificatory methods are very rare; I know only those devised by Macnaughton-Smith (1965).

Choice 3: exclusive intrinsic classification — hierarchical or non-hierarchical. The previous choices have been concerned with the use to which the classification is to be put; once the user's requirements have been clearly defined, the strategy is determined. All exclusive intrinsic classifications, however, are used for much the same sort of purpose, so that the strategies within the general class are largely interchangeable and the choice will largely be determined by pragmatic considerations, mainly of a computational nature.

The most important of these choices is that between hierarchical and non-hierarchical strategies. A hierarchical strategy always optimizes a *route* between the entire population and the set of individuals of which it is composed. The route may be defined by progressive fusions, beginning with individuals and ending with the complete population; or by progressive divisions, beginning with the population and decomposing it into individuals or at least into subgroups.

Fusions and fissions are in practice always dichotomous; trichotomous fissions have been used but have not been found profitable. A hierarchical fusion strategy must necessarily optimize some function between the two groups to be fused next (termed i, j functions by Lance and Williams 1967) or between these two groups jointly and their fusion product (ij, k functions); similar considerations apply to the fission strategy. As a result the groups through which the process passes are not themselves necessarily optimal; the best route may be obtained at the expense of some slight reduction in homogeneity of individual groups. With a non-hierarchical strategy, on the other hand, it is the structure of the individual *groups* which is optimized, since these are made as homogeneous as possible. No route is defined between groups and their constituent individuals, so that the infrastructure of a group cannot be examined. Similarly, no route is defined between groups and the complete population. Despite these advantages, for those applications in which homogeneity of groups is of prime importance the non-hierarchical strategies are in principle very attractive; unhappily their current state of development lags far behind that of their hierarchical counterparts, which at their best are more flexible, provide a wider range of facilities, are numerically better understood and are computationally faster.

Choice 4: exclusive intrinsic non-hierarchical classifications—serial or simultaneous optimization. Non-hierarchical strategies are of two types. In the first, groups are obtained serially: that is to say, a group is defined and removed from the population; within the remainder a second group is defined and removed; and the process continues until all the population is accounted for. The process may end with a final reallocation, but the basic principle is that the groups are initially defined one at a time. Such general methodological principles as can be discerned are discussed by Lance and Williams (1967). Unfortunately, at present all strategies of this type appear open to criticism on numerical or computational grounds.

The second type of non-hierarchical strategy is that in which the groups are obtained simultaneously. All existing strategies of this type are in effect relatively minor variants of a single strategy: the population is partitioned in some way and the groups simultaneously optimized by an iterative process (for examples, see Ball and Hall 1965; Jancey 1966; MacQueen 1966). The model is always strictly Euclidean, the process is convergent (Needham 1966) and computer programs are in a satisfactory state of development; the system has not been extended to applications involving mixed data with missing values. Unfortunately, there is some reason to believe that the basic model is lacking in power, in that the type of classification produced is not that which a user commonly requires; the future of these strategies is consequently in some doubt.

Choices 5 and 6: exclusive intrinsic hierarchical classifications—agglomerative or divisive, polythetic or monothetic. It is convenient to take these two choices together, since they are not completely independent: all agglomerative classifications are polythetic and the most commonly used divisive classifications are monothetic. An *agglomerative* strategy is one that proceeds by progressive fusion, beginning with the individuals and ending with the complete population. A *divisive* strategy progressively splits the population into smaller and smaller groups.

All agglomerative strategies suffer from two disadvantages, the first of which is computational. The user's interest is normally concentrated in the higher levels of the hierarchy, so that it is almost invariably necessary to establish the complete hierarchy from individuals to population. For a population of n individuals, this may require the calculations appropriate to at least $(n - 1)^2$ fusions; and if n is large this may require an unacceptably long computation time. In contrast, if a user is interested only in the establishment of a small number of major subgroups, a divisive classification can be terminated at the appropriate level. Second, an agglomerative system is inherently prone to a small amount of misclassification, the ultimate cause of which is that the process begins at the inter-individual level, where the possibility of error is greatest. Divisive classifications are free of this disadvantage, but the reason they have not thereby completely displaced agglomerative systems lies in the monothetic–polythetic distinction. A *monothetic* division is one based on a single attribute; if the attribute is binary the population will be divided into those that do and those that do not possess the attribute, and comparable rules have been devised for other types of attribute. The division attribute cannot be chosen more or less indiscriminately, unless all that is required is an artificial key to individuals; it must be optimum in some sense, dividing the population into two groups as unlike as possible. Since its selection must therefore depend on the properties of the population under study, it follows that a monothetic agglomerative classification is impossible save in a trivial sense. In contrast, a *polythetic* system is one based on a measure of similarity or dissimilarity applied over all attributes, so that an individual is grouped with those individuals which, on the average, it most resembles. A polythetic division may *happen* to be monothetic as a result of the configuration of a particular set of data, but it is never so defined.

The outstanding advantage of monothetic classifications is the simplicity and clarity of the subgroup definitions; it enables new individuals to be allocated quickly and unambiguously to an existing classification. They are, however, liable in their turn to a troublesome form of misclassification. We suppose two groups, X and Y, to have been separated monothetically on a binary attribute, possessed by X and lacked by Y. Consider then an individual A, which on the whole most closely resembles the members of Y but happens to possess, or at least is recorded as possessing, the division attribute. A will appear in X and, since most users are concerned with overall (i.e. polythetic) resemblance, will appear to the user to have been misclassified. At a later stage in the analysis A will almost certainly be separated from X, and it is characteristic of divisive monothetic systems that they tend to produce by this means an unduly large number of fragmentary groups.

It would appear to follow that the ideal hierarchical system is divisive polythetic. Until very recently this was the great unsolved problem of numerical taxonomy in that, despite many attempts, no successful algorithm had been devised. A satisfactory solution now appears to have been found, and it may well be that this will displace many of the existing agglomerative programs for all except small data sets.

Inverse Classification; One- and Two-parameter Models

In all the foregoing we have tacitly assumed that we wish to classify a set of individuals by reference to a set of attributes, which is now known as a *normal* classification. But any data matrix can in principle be transposed, and we might then attempt to classify the attributes by reference to the individuals which they specify; such a classification is said to be *inverse*. No satisfactory classification of mixed attribute types is known, but no difficulty arises if the attributes are all binary (the commonest case) or all numerical. The classifications, one normal and the other inverse, of the same data set can then be synthesized into a *two-way table*, which purports to show which blocks of attributes define which groups of individuals.

This simple statement conceals a more profound problem, which is most easily grasped in the context of the scores obtained from intelligence tests in psychology. The score obtained by a given individual for a given test is a function both of the ability of the individual and of the difficulty of the test. However, if we carry out a numerical classification of the individuals over all tests, what we are doing—though we may not realize it—is to 'even out' the differences between the tests by reducing them all to the same variance (or some equivalent standardization). If the individuals are rows and the tests are columns, we are standardizing by columns, and thus assuming that all differences reside in the row totals. Similarly, in an inverse classification we bring all individuals to a common level of ability; we are standardizing the rows, and assuming that all the differences reside in the column totals. The two-way table thus represents an attempt to redress these two arbitrary procedures. Since all ordinary classifications are based either on the row or the column totals, these are said to be *one-parameter models*. Macnaughton-Smith (1965) suggested that it should be possible to devise a method of classification which did indeed express the score as some function of both the row and column totals; such a classification would use a *two-parameter model*. He suggested a method by which such a classification might in theory be approached. So far as I know, the suggestion remained purely theoretical until the recent work by Dr Dale (e.g. Dale and Anderson 1973) who is the only investigator to attempt to implement the suggestion. The methods in use are still relatively crude, but already show promise. The problem is of some agricultural interest, because it also arises in ecological survey; the probability of finding a given species in a given quadrat is a function both of the fertility (and therefore floristic richness) of the quadrat and of the abundance of the species in the area. If the two-parameter type of model is susceptible of development to a level comparable to that of the one-parameter models, it may assume great importance in the future.

References

Ball, G. H., and Hall, D. I. (1965). ISODATA, a novel method of data analysis and pattern classification. Tech. Commun. Stanford Res. Inst., Menlo Park, Calif.

Bor, N. (1960). 'Grasses of Burma, Ceylon, India and Pakistan'. (Pergamon: Oxford.)

Dale, M. B., and Anderson, D. J. (1973). Inosculate analysis of vegetation data. *Aust. J. Bot.* **21**, 253–86.

Jancey, R. C. (1966). Multidimensional group analysis. *Aust. J. Bot.* **14**, 127–30.

Lance, G. N., and Williams, W. T. (1967). A general theory of classificatory sorting strategies. I. Hierarchical systems. *Comput. J.* **9**, 373–80.

Macnaughton-Smith, P. (1965). Some statistical and other numerical techniques for classifying individuals. Home Office Res. Rep. No. 6. (H.M.S.O.: London.)

MacQueen, J. B. (1966). Some methods for classification and analysis of multivariate observations. West. Manage. Sci. Inst. Work Pap. No. 96. (Univ. of California.)

Needham, R. M. (1966). The termination of certain iterative processes. Rand Corp., Santa Monica, Calif., Rep. RM–5188–PR.

Suggestions for Further Reading

Clifford, H. T., and Stephenson, W. (1975). 'Introduction to Numerical Taxonomy'. (Academic Press.)

Sneath, P. H. A., and Sokal, R. R. (1973). 'Numerical Taxonomy'. (Freeman: San Francisco.)

11

Hierarchical Agglomerative Strategies

W. T. Williams

The General Problem

We shall defer the question of the choice between agglomerative and divisive strategies until the next chapter; for the moment we shall assume that it has been decided to use an agglomerative strategy. In the standard case we then begin with a data matrix of individuals described by a number of attributes; and our first task is to calculate some similarity or dissimilarity measure between all pairs of individuals. Again, for the moment we shall assume that the measure to be used has already been decided. We shall then have an inter-individual matrix of measures; if there are n individuals and s attributes, our matrix will have dimensions $(n \times n)$. Such a matrix is now commonly known as a *D-matrix* ('D' is here short for 'distance', since all existing programs operate on distance, or dissimilarity, measures). In a few mathematical texts (notably Jardine and Sibson 1971) the D-matrix is unfortunately referred to as the 'data matrix', but this use of the term is confusing and should be avoided.

We next have to identify that pair of individuals which are most like, i.e. that pair for which the dissimilarity measure is the smallest in the table (methods for resolving ambiguities exist in certain programs, but we shall not discuss them here). This pair of individuals is the first to be fused. Two of our original individuals have now disappeared from the system, and have been replaced by a new, composite individual—a *group* of two individuals. Our next task must be to calculate a measure of dissimilarity between this group and all other individuals still remaining in the system; and later, as other groups form, we shall need to know how dissimilar these are from each other. In other words, immediately after the first fusion we need to define a measure, compatible with our original inter-individual measure, between a group and an individual; and at some later stage we shall need to define a similar measure between two groups. It is the nature of the individual/group and group/group measures that determines the subsequent course of the analysis; as we say, it determines the *fusion strategy* which we have elected to adopt.

Except in the most trivial cases, different fusion strategies will produce quite different ultimate classifications; the choice of an appropriate fusion strategy is thus extremely important. Experience is a great help, but extensive experience can only be obtained by those few analysts who are continually processing data from a wide spectrum of users. The intention of this chapter is to provide the isolated user with at least some guide as to the decisions that need to be taken; it may assist him to prevent an analyst from providing a convincing and elegant answer to the wrong question.

Types of Agglomerative Strategy

Combinatorial and non-combinatorial strategies

It will be recalled that very early in the proceedings we need to fuse two (and later more than two) individuals together. We can obviously do this within the computer; for a numerical attribute we could store the total over all individuals in the group, and for an ordinal or nominal attribute we could store the number of individuals in each state of the attribute. We should, of course, also have to store the number of individuals in the group.

All early classificatory programs worked like that, by physically fusing individual records within the computer, and some still do. However, about 10 years ago John Gower invented a strategy of fusion that he called the *median* strategy. For reasons given below, this strategy is now virtually obsolete; but it is of great historical importance. Its novel feature was that it calculated individual/group and group/group measures from pre-existing inter-individual measures in the original D-matrix, and never needed physically to fuse the original data records, which could be discarded or overwritten once the D-matrix had been calculated. This led Lance and Williams (1967) to investigate whether other strategies in common use could be calculated in the same way; they found that some could and some could not.

It is now necessary to outline the numerical model used by Lance and Williams. We assume the existence of two groups of individuals which we shall call (i) and (j), which are to be fused into a new composite group (k); a further 'outside' group (h) is not involved in this fusion. We shall already know the distance between (h) and (i); call this d_{hi}. Similarly we know the distance between (h) and (j), or d_{hj}, and between (i) and (j), or d_{ij}. What we want is d_{hk}, the distance between the outside group (h) and the composite group (k) formed by the fusion of (i) and (j). We set up the following linear model

$$d_{hk} = \alpha_1\, d_{hi} + \alpha_2\, d_{hj} + \beta\, d_{ij} + \gamma\left|d_{hi} - d_{hj}\right|$$

where α_1, α_2, β and γ define the strategy in question. Of these parameters, γ takes only the values of 0 or $\pm\frac{1}{2}$; the other three are normally functions of the numbers of individuals in the groups concerned. The details of these functions have been widely published in the literature and we need not repeat them here; it is, though, necessary to know which strategies are, and which are not, amenable to this type of solution.

Finally, it is important to note that the existence of combinatorial strategies provides a solution to a problem which was once regarded as quite intractable. This is the case in which the original data matrix is not available to the analyst, or even does not exist at all. This commonly arises because an experienced user wishes to calculate a special-purpose measure not catered for in standard programs; in such a case, the analyst receives only the matrix of measures. The classic example of the 'non-existent data' case is that of investigating the social structure of a small town by recording the number of times each pair of residents talk to each other over the telephone in a given period; then *only* the D-matrix exists. Both such cases can clearly be processed by combinatorial methods, which need only the D-matrix.

Individual strategies

We shall be concerned only with seven strategies, four of which are more important than the others. Their names, and their combinatorial situations, are as follows:

No combinatorial solution available or known

   (*a*) Information-statistic strategies

Both combinatorial and non-combinatorial solutions available

   (*b*) Centroid

   (*c*) Group-average

   (*d*) Incremental sum of squares

Only combinatorial solutions available

   (*e*) Nearest-neighbour

   (*f*) Furthest-neighbour

   (*g*) Flexible

We append brief notes on the nature of these strategies.

The *information-statistic strategies* are those which use information gain as a dissimilarity measure (see Chapter 6). In the *centroid* strategy, a group is thought of as in a space defined by the attributes as axes, and is considered as situated at its centroid; in practice, this implies that the value for the group of a numerical attribute is the group mean of that attribute, and the states of an ordinal or nominal attribute contain the proportion of the individuals in those states. In the *group-average* strategy, we take all the (say) n_1 individuals in one group and the n_2 individuals in the other and calculate the distances between all those in one group and all those in the other; there are $n_1 n_2$ such distances, and the measure is the arithmetic mean of all of them. The *incremental sum of squares* strategy is applicable only to a strictly Euclidean system; the distance between two groups is defined as the increase in the sum of squares of deviations from the group mean after fusion, as compared with the sum of the two within-group sums of squares before fusion. In *nearest-neighbour* we identify the two individuals, one in each group, which are nearest to each other; the distance between this pair of individuals defines the distance between the two groups. *Furthest-neighbour* is similar, except that the pair chosen is the pair whose members are furthest apart. Finally, the *flexible* strategy is defined purely algebraically from the basic Lance and Williams combinatorial model; it is the strategy defined by the quadruple constraint

$$[\alpha_1 = \alpha_2; \ \beta < 1; \ \alpha_1 + \alpha_2 + \beta = 1; \ \gamma = 0].$$

It is completely defined by β, the *cluster-intensity coefficient*. Before considering the uses of these strategies we must examine the criteria for choice of strategy.

Choice of agglomerative strategy

It must first be realized that a choice of dissimilarity measure may itself impose a choice of fusion strategy. If an information-statistic measure has been chosen, it would be absurd to proceed with one of the purely combinatorial strategies such as 'flexible'; this would be to discard the power inherent in information-statistic strategies. Similarly, if an *ad hoc* measure had been calculated for some special purpose, the 'incremental sum of squares' fusion strategy is not available, as this is only defined for a strictly Euclidean system. Three further criteria are important in practice: monotonicity, group-size dependence and computing time.

Monotonicity

An agglomerative classification produces a string of successive fusions; if there are n individuals, there will be $(n - 1)$ fusions in all. We may, and usually do, wish to display the fusions as a *hierarchy* or *dendrogram*. Now, the dendrogram is completely defined by the fusions; we could draw these out in any way we pleased on a sheet of rubber. We could twist, stretch, bend or otherwise distort the rubber sheet without altering the *order* of fusions; a dendrogram is, in fact, a topological entity and is invariant under distortion. However, in real life we wish to draw our dendrogram on a sheet of paper. As we pass upwards along any branch, we want each successive fusion to lie above the preceding ones, otherwise there will be 'reversals' which cause a later fusion to lie below earlier ones; the resulting dendrogram is extremely difficult to follow. We could ensure that no reversals occurred by the simple expedient of associating each fusion with the number of individuals produced by that fusion, and we could use this as an ordinate for plotting the dendrogram. However, this information is not particularly useful; we normally prefer to associate each fusion with the individual/ group or group/group measure at that moment in the process. The successive string of fusion measures must thus be *monotonic* if we are to have a tidy dendrogram.

Burr (1970) has shown that information gain is not necessarily monotonic; but the conditions under which monotonicity fails are so stringent that they are almost never encountered in real-life problems. For all practical purposes, therefore, information gain can be treated as if it were monotonic. Of the remaining strategies described above, all are monotonic except 'centroid'. In this strategy (as in Gower's original 'median' strategy) reversals are frequent and troublesome, and it is for this reason that both these strategies are virtually obsolete.

Group-size dependence
This is the most important single criterion and is best understood by considering a spatial model, with the individuals as points in a space defined by attribute axes. In some strategies we can think of the classification as proceeding by erecting walls within this space, so that groups of points become separated from each other, although their position in the original space remains unchanged. Such strategies are said to be *space-conserving*. The only completely space-conserving strategies are 'centroid' and 'flexible' in the case where $\beta = 0$; but 'group-average' is very nearly space-conserving and can be regarded as such for most practical purposes. In other strategies, as a group begins to grow in size by the addition of further individuals, the space behaves as though it becomes stretched in the neighbourhood of the group, and the group appears to recede from all other groups as it grows and it becomes progressively more difficult to join.

Such strategies are said to be *space-dilating*; their effect is to 'sharpen' any discontinuities in the system, so that the clustering is intensified. The data are thus distorted; but such strategies are particularly useful when it is known or suspected that no sharp discontinuities exist in the data, but it is wished to produce as 'clean' a classification as possible—to make, as it were, the best of a bad job. The dilatation may affect the individual/group and group/group measures in different ways. In either case the dilatation may be *indefinite* in that the measure continues to increase continuously as the group grows; or it may be *asymptotic*, in that the dilatation eventually 'levels off'. The 'flexible' strategy is space-dilating if β is negative, and the more negative the value of β, the more the dilatation and the more intense the clustering (hence the term 'cluster-intensity coefficient'); both individual/group and group/group dilatations are asymptotic. The 'incremental sum of squares' strategy has the individual/group dilatation asymptotic, the group/group dilatation indefinite; and in the information-statistic strategy, both dilatations are indefinite (for proofs, see Williams *et al.* 1971). As a result, an information-statistic classification clusters very intensely; moreover, if there are several large groups, each with one or two associated 'outlying' members, these members may no longer be able to join their most-like groups when their turn comes. As a result, the process may sweep up all such outliers into a 'nonconformist group' whose members share only the property that they are rather unlike everybody else, including each other. 'Furthest-neighbour' is also an intensely dilating strategy; but since the degree of dilatation is not under the user's control, it has tended to become superseded by 'flexible', where the dilatation is variable. In practice, 'flexible' with $\beta = -0\cdot25$ appears to meet most users' requirements.

It will cause no surprise to learn that there are also strategies such that the space appears to shrink as a group grows, the so-called *space-contracting* strategies; as a group grows it becomes progressively easier to join. Clustering is very weak or non-existent; the individuals tend to join a group one after another, a phenomenon known as *chaining*. The only space-contracting strategies among the seven here listed are 'nearest-neighbour' and 'flexible' with positive β. Although nearest-neighbour has theoretical mathematical advantages (Jardine and Sibson 1971), its 'chaining' tendencies make it undesirable for most biological applications.

Computing time

With the advent of computers as fast as the Cyber 76, computing time is less important than formerly; but for anybody using a second-generation computer such as the Control Data 3600, or the IBM 360/50, computing time can still be of great importance. We suppose that there are n individuals specified by s attributes; we can then define a quantity which is an approximate measure of the number of calculations to be undertaken, and we can define three cases. First, there are the information-statistic programs which begin by calculating the D-matrix, and for which there is no subsequent combinatorial solution, e.g. the Canberra program MULTBET; the number of calculations is then

$$s\,(n-1)^2\,.$$

Next are those programs which begin by calculating the initial D-matrix and then proceed combinatorially (e.g. the Canberra program MULCLAS); the number of calculations is

$$\tfrac{1}{2}\,s\,n\,(n-1) + \tfrac{1}{2}\,(n-1)\,(n-2).$$

Finally, there are the programs which accept a D-matrix from elsewhere and process it combinatorially, e.g. the Canberra program CLASS. The number of calculations is

$$\tfrac{1}{2}(n-1)(n-2).$$

It will be obvious that the slowest program is MULTBET, the fastest CLASS.

References

Burr, E. J. (1970). Cluster sorting with mixed character types. II. Fusion strategies. *Aust. Comput. J.* **2**, 98–103.

Jardine, N., and Sibson, R. (1971). 'Mathematical Taxonomy'. (Wiley: London.)

Lance, G. N., and Williams, W. T. (1967). A general theory of classificatory sorting strategies. I. Hierarchical systems. *Comput. J.* **9**, 373–80.

Williams, W. T., Clifford, H. T., and Lance, G. N. (1971). Group-size dependence: a rationale for choice between numerical classifications. *Comput. J.* **14**, 157–62.

<h1 style="text-align:center">12</h1>

Hierarchical Divisive Strategies

W. T. Williams

Monothetic Strategies

All-binary data

The first successful divisive monothetic program was that which implemented the 'association analysis' of Williams and Lambert (1960); it was originally intended for terrestrial ecological survey with presence-or-absence data for a number of plant species. It therefore handled only binary data and made no provision for missing values. It set up all possible 2×2 contingency tables between pairs of attributes (species) and calculated the inter-attribute correlation coefficients, storing only the sum of the absolute values associated with each species; that is to say, the quantity stored was a decision function, and took the form

$$\sum_{k \neq j} |r_{jk}| \, .$$

Division was on that attribute for which this was maximum. Although the method was widely used, it had two disadvantages. First, it proved rather sensitive to the presence of rare species or the absence of common ones; in statistical terminology, it was unduly sensitive to highly skewed data. Secondly, if the decision function itself was used as a 'level' for plotting the resulting dendrogram, reversals were frequent. The program has therefore been replaced by an information-statistic analogue (Canberra program DIVINF) which is largely free from these disadvantages. It still represents the most economical method of classifying data with a large number of individuals and binary attributes, providing a monothetic solution is acceptable.

Mixed data

Lance and Williams devised two mixed-data monothetic programs. One (MULASS) was a mixed-data analogue of the original 'association analysis' program; the other was a mixed-data analogue of DIVINF. Both made provision for missing values. Although perhaps at the time better than nothing, neither program proved particularly successful and both have now been withdrawn.

91

Polythetic Strategies

Early programs

The earliest programs of this type, which appeared almost simultaneously in different parts of the world, all worked on the 'dissimilarity' or 'deviant' principle. That pair of individuals which were most unlike each other was set aside; the remainder of the population was then allocated to one or other of these. Although the method has had some successes, it has two serious disadvantages. First, it requires the computation of all $\frac{1}{2} n (n - 1)$ inter-individual similarities, which precludes its use if n is large. Secondly, it has proved unduly data-dependent. With some sets of data, once a 'deviant' individual had been set aside, nothing else would join it. The result was a highly fragmented set of largely single-membered groups. Unless some means can be found of ascertaining in advance whether this is likely to happen, the strategy is unlikely to survive.

All-numerical programs

Only one Australian program is known to us, the Canberra program POLYDIV, intended for the case in which there is a large number of individuals defined by a fairly small number of continuous variables as attributes. The inter-attribute correlation matrix is calculated, the first principal component extracted, and the individual coordinates on this component obtained by the usual post-multiplication (see Chapter 7); these coordinates are then sorted into order, and the string dichotomized at the point at which the between-group sum of squares of deviations from the mean is maximum. The process is repeated as often as desired on the resulting subpopulations. Provision is made for missing values, and the program is fast and effective. A similar but somewhat more complex program exists in Britain (Lambert *et al.* 1973).

Mixed-data programs

The first successful mixed-data polythetic divisive program was that of Boulton and Wallace (1970), which uses a novel information-statistic measure. The original program was non-hierarchical, but a hierarchical version (Boulton and Wallace 1973) was later devised. Both programs accept a wide variety of attribute types, with provision for missing values. Nevertheless, they have two minor disadvantages which appear to have militated against their widespread use.

The first disadvantage arises out of the division procedure itself. Division begins by splitting the population into two equal groups at random; individuals are then moved across the division one at a time to increase the difference between the groups. After a number of cycles of such movements, the system will ultimately reach stability, when no further movements can improve the split. However, if the initial random split happens to be no-where near the final optimum split, convergence of this iterative procedure can be slow, particularly if there are a large number of individuals. As a result, the program may be expensive and time-consuming to use, and the extent to which this will be so cannot be known in advance. The second disadvantage arises out of the stopping rule (for a general consideration of stopping rules, see the following section). This is rigorously specified, and in some applications results in a degree of subdivision less fine than that required by the user.

A less mathematically rigorous strategy, which is free from these dis-advantages, is that of the Canberra program REMUL. This begins with the optimum monothetic split, which is inherently likely to be somewhere near the final optimum; the split actually used is that of the old program MULASS. Each individual is then tested in turn against the two groups as in the Boulton–Wallace program, but the iteration normally converges very rapidly. An allocation measure based on the Canberra metric is used, which makes for simple calculation. The program can take all types of attribute except linked numericals, and makes provision for missing values.

Terminal Procedures

Stopping rules

The whole point of a divisive process is that one can stop when one has obtained a sufficient number of divisions; it would obviously be absurd to continue a divisive process down to the point where one had only a final set of individuals. But how are we to define a 'sufficient' number of divisions? In the early programs, a statistical test was suggested; since the correlation coefficient and an information gain (or fall) can both be associated with a χ^2 statistic, division was terminated when this statistic fell below significance. However, the estimates of the statistic so obtained are conservative, not random, estimates; and there is reason to believe (Bottomley 1971) that, with one exception, all such quasi-statistical tests are invalid. The exception is that of the Boulton–Wallace program, which, as mentioned above, occasionally proves over-restrictive.

In most modern programs (such as DIVINF, POLYDIV and REMUL) a cruder procedure is consequently used: the user is asked to specify in advance how many 'final' groups he wants. Although this procedure sounds appallingly arbitrary it works remarkably well in practice; most users have a rough idea of how many groups they expect or would be able to handle. So long as the analyst always adds a few extra (to accommodate, for example, outlying individuals which demand groups for themselves), users appear to be satisfied.

However, the mere declaration of the number of final groups required does not completely define the system. As soon as there is more than one group in the system, i.e. after the first division, a decision has to be taken as to which group to divide next. In most divisive strategies two measures associated with any group are available or can be calculated. The first is some measure of heterogeneity of the group; the second is the amount of reduction in heterogeneity which will be obtained if the group is divided. The situation is analogous to that of a doctor with two patients but only enough medicine for one of them. A is more ill than B but B will show a greater response to treatment, so who should have the medicine? If we decide to divide next that group with the largest individual heterogeneity, we are said to be using a *Type 2 stopping rule*; if we divide that group which suffers the greatest reduction in heterogeneity, we are said to be using a *Type 1 stopping rule*. It will be obvious that the Type 1 rule requires the greater amount of calculation (since every group has to be notionally divided to find out what the reduction will be); in practice, therefore, the Type 2 rule is the more common in production programs.

Terminal reallocation

The program DIVINFRE (CSIRO Division of Land Use Research) uses what was, at the time it was written, a novel procedure. It first implements the monothetic program DIVINF for binary data to the required number of final groups. All these groups are then examined collectively, and every individual has a final opportunity to move into a new group. As with all reallocation procedures, the process is iterative, and in DIVINFRE it tends to be slow. Nevertheless, the result is essentially to simulate a polythetic division, since the monothetic character will usually be lost as a result of reallocation. What has really been done is to superimpose a non-hierarchical strategy at the termination of a hierarchical division; the results are often extremely satisfactory, though the hierarchical character has been lost. A similar procedure is available as an option in REMUL.

Whether such a procedure is useful rests on the interests of the user. If, as is usually the case in purely taxonomic investigations, it is the constitution of the final groups that is of primary interest, terminal reallocation almost invariably produces a marked improvement in group homogeneity. If, however, as is often the case in ecological survey, the user is primarily interested in the hierarchical route and the successive discontinuities it defines, terminal reallocation is undesirable.

References

Bottomley, J. (1971). Some statistical problems arising from the use of the information statistic in numerical classification. *J. Ecol.* **59**, 339–42.

Boulton, D. M., and Wallace, C. S. (1970). A program for numerical classification. *Comput. J.* **13**, 63–9.

Boulton, D. M., and Wallace, C. S. (1973). An information measure for hierarchic classification. *Comput. J.* **16**, 254–61.

Lambert, J. M., Meacock, S. E., Barrs, J., and Smartt, P. F. M. (1973). AXOR and MONIT: two new polythetic-divisive strategies for hierarchical classification. *Taxon* **22**, 173–6.

Williams, W. T., and Lambert, J. M. (1960). Multivariate methods in plant ecology. II. The use of an electronic digital computer for association-analysis. *J. Ecol.* **48**, 689–710.

<h1 style="text-align:center">13</h1>

Dendrograms and their Interpretation

H. T. Clifford

We have seen in the previous chapters that an agglomerative classification results in a *hierarchy* or *dendrogram*, a visual representation of the successive fusions as we pass from the individuals to the complete population. We have also noted that it is usual, though not essential, to take the individual/individual, individual/group or group/group measure at the moment of fusion as defining a *level* of that fusion. From early work, when often only a single measure and a single fusion strategy were available, it is difficult to avoid the impression that the dendrogram was sometimes regarded as a unique property of the population under study. We are now well aware that a change in dissimilarity measure or in the fusion strategy can radically change the order of fusions and can therefore change the dendrogram itself. Nevertheless, the interpretation of dendrograms is not completely straightforward; there are still problems that can act as traps for the unwary beginner. It is the purpose of this chapter to explore some of these.

A Simple Dendrogram

We begin by constructing a simple dendrogram. Table 13.1 represents the values of a single numeric attribute for seven individuals.

Table 13.1. Seven entities with their individual scores for a single attribute

	Entity						
	a	*b*	*c*	*d*	*e*	*f*	*g*
Attribute score	2	6	9	14	20	21	23

As dissimilarity measure we shall use the unstandardized Manhattan metric $|x_1 - x_2|$ (Ch. 6); as fusion strategy we shall use nearest-neighbour (Ch. 11). The D-matrix, or matrix of dissimilarities, is shown in Table 13.2.

Table 13.2. D-matrix corresponding to Table 13.1

	a	*b*	*c*	*d*	*e*	*f*	*g*
a	0	4	7	12	18	19	21
b	4	0	3	8	14	15	17
c	7	3	0	5	11	12	14
d	12	8	5	0	6	7	9
e	18	14	11	6	0	1	3
f	19	15	12	7	1	0	2
g	21	17	14	9	3	2	0

We see that the individuals *e* and *f* are the least dissimilar, and so are the first to be united; the dissimilarity is 1, so that the 'level' of the fusion is taken as 1 unit up the ordinate. Next come the two pairs (*b*,*c*) and (*e*,*g*), each with a dissimilarity of 3 units. We can therefore fuse *b* with *c* and fuse *g* with the group that already contains *e* and *f*. Proceeding in this way we obtain the dendrogram shown in Fig. 13.1. There are a number of

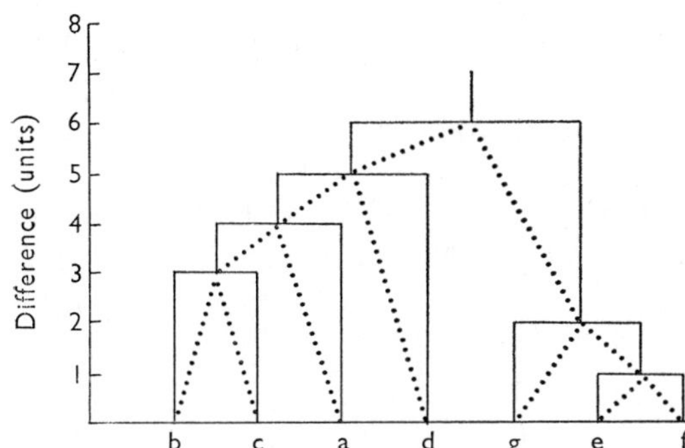

Fig. 13.1. A typical dendrogram, showing alternative methods of display.

conventions implicit in this diagram. For example it is usual, as here, to present the dendrogram with the individuals at the bottom of the diagram and the root (the whole population) at the top. It would be equally correct to draw the dendrogram the other way up; the scale would then need to be inverted, and it would be usual to work with similarities rather than dissimilarities in such a case. Next we note that the successive nodes can simply be joined by oblique lines (dotted in the figure); or, as in the continuous lines in the figure, the joins can be displayed as vertical lines joined by horizontal bars at the nodes. This is purely a matter of personal preference.

We have yet to consider how the seven individuals are to be ordered and spaced along the bottom of the diagram. The beginner, seeing a dendrogram such as Fig. 13.1 for the first time, can rather easily jump to the conclusion that b and f are the most unlike, simply because they are at opposite ends of the string. He forgets that a node simply represents a fusion between two equivalent entities, and any one of the nodes can be rotated without altering the sequence of fusions. For example, if we rotated the node joining d to (b,c,a) and the node joining g to (e,f) we should obtain the dendrogram of Fig. 13.2. It is important to realize that this is the *same* dendrogram.

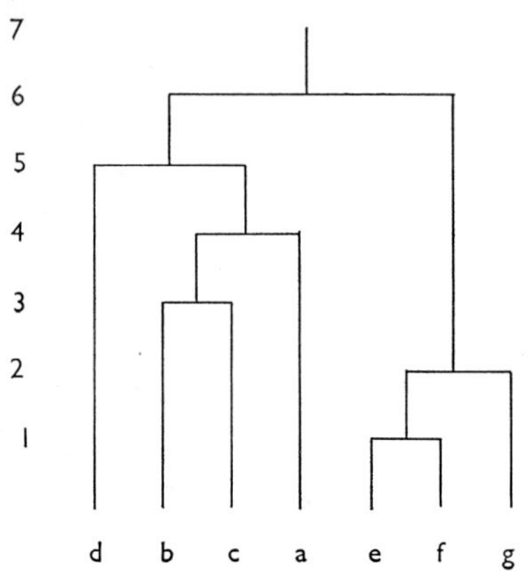

Fig. 13. 2. The same dendrogram as Fig. 13.1; the original dendrogram has been rotated round two of its nodes.

For seven individuals there will be six successive fusions or nodes; so there are in all 2^6, or 64, different forms in which this simple dendrogram could be presented. *Any* order will not do; it is clear from the figures that no rotation can place d between b and c or between a and c. This still leaves plenty of possibilities to choose from; the worker is at liberty, within these possibilities, to arrange the individuals in any order he pleases. It is usual to space the individuals equally along the abscissa, but even this is not essential. Dendrograms printed by computers imply no authority for the order printed; this is selected for computational convenience in plotting routines.

It does not follow that inter-group relationships cannot be elucidated; it is merely that such relationships cannot be obtained from the dendrogram. The correct procedure in such a case is to calculate the group/group measures between all pairs of terminal groups and then ordinate these by means of principal coordinate analysis (Ch. 8). Facilities for obtaining the inter-group measures for mixed data using a variety of measures are provided in the Canberra program GROUPER.

Group-size Dependence

This phenomenon, which has been discussed in outline in Chapter 11, is the cause of most errors in interpretation of dendrograms. It will be convenient first to demonstrate the process in action in its most extreme case, that produced by the use of information statistics. Table 13.3 represents three individuals specified by four binary attributes. Individuals A and C differ in a single attribute; B and C differ in three attributes; and

Table 13.3. Three entities specified by four binary attributes

	Attribute			
	1	2	3	4
A	1	1	0	0
B	0	0	1	1
C	1	0	0	0

A and B differ in all four attributes. A and C are obviously most alike, whereas A and B have nothing in common. Using the table in Chapter 6 you should be able to show that the information gains on fusion would be

$$
\begin{aligned}
AC \quad & 1 \cdot 3863, \\
BC \quad & 4 \cdot 1589, \\
AB \quad & 5 \cdot 5452.
\end{aligned}
$$

Suppose that these individuals are quadrats in an ecological survey in an area with only four species; and now suppose we collect more and more quadrats, all of which turn out to be exactly like C. If we recalculate the information gains as we add each new C-type quadrat, that for AB will not change; but both AC and BC will continue to rise. We need only consider AC. When we have acquired 94 C-type quadrats in all, the AC measure will have risen to 5·5486 units—higher than that of AB. At this stage, therefore, A will fuse with B, with which it has nothing in common, rather than join the big group C. AB has become, in fact, a 'nonconformist group'. If we completed the dendrogram by now adding C to (A+B) we should obtain the diagram of Fig. 13.3.

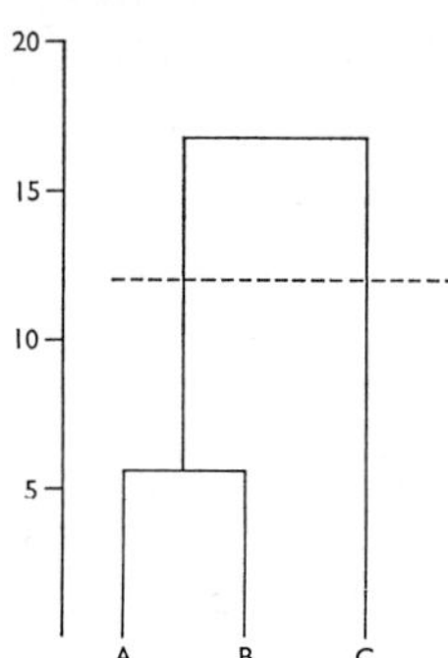

Fig. 13.3. The concept of 'hierarchical level'.

Users are now beset by a temptation. This is to select some level of information gain and to draw a line across the dendrogram at this level, such as the dotted line in Fig. 13.3. They then suggest that this is the 'level of heterogeneity' at which they propose to work. It will be clear from our simple example that it is nothing of the kind. Our group C is completely homogeneous, because it is composed of 94 identical quadrats; the other group is completely heterogeneous, because it consists of two individuals with nothing in common. Exactly the same situation can arise with real-life data. Clifford and Williams (1973) took a set of 45 grass genera and classified them into groups; one of these groups contained the 'panicoid' grasses, genera closely allied to *Panicum* itself. They added, in succession, 9, 18 and 27 sets of information identical with that for *Panicum*. Fig. 13.4

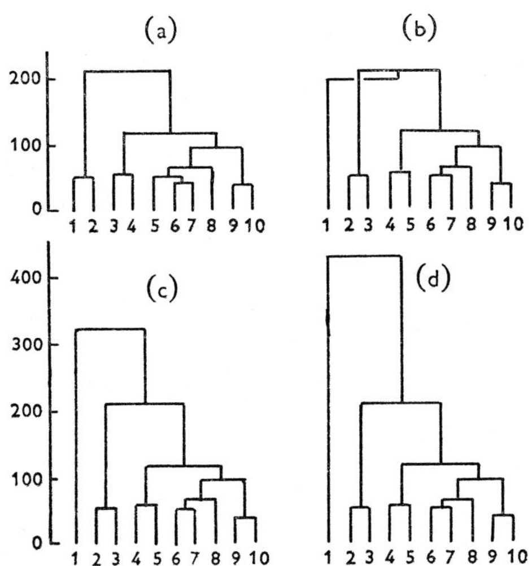

Fig. 13.4. The effect of increasing the size of a group by the addition of identical members.

shows the result; *Panicum* rapidly detaches itself from the rest of the system and, although it has close relatives elsewhere, now appears as a completely isolated group. They repeated the process using 9, 18 and 27 genuine additional genera, all fairly closely allied to *Panicum*. The result was, of course, more complicated, but the general principle—the isolation of the group simply because it was large—remained. It follows that a line purporting to represent a level of heterogeneity must *never* be drawn across a dendrogram unless the fusion strategy in use is strictly space-conserving. Since the only completely conserving strategy is centroid, which is almost never used, this virtually amounts to a complete ban on the procedure.

Comparison of Dendrograms

Often, in order to explore the relationships of a set of entities, they are classified in several ways, each of which is summarized in a dendrogram. The various classifications may result from employing different sets of attributes or from using different clustering strategies. In either event it may be required that the dendrograms be compared. However, before such comparisons can be undertaken a decision must be made as to the basis of comparison required.

To appreciate why this is so, consider two dendrograms each truncated at the 10-group level. It would be possible for each of the groups in each dendrogram to have the same composition with respect to entities and yet for the groups themselves to be quite differently associated in the two dendrograms. Such a situation would be unusual but the illustration is helpful in drawing attention to two fundamental and quite separate properties of dendrograms: their structure and their group compositions. Methods for comparing these two properties have been discussed by Williams and Clifford (1971).

Other methods based upon estimates of correlation coefficients and truncated dendrograms are neglected here because they are more sensitive to group-size dependency than the methods described below.

If the interest is primarily in group compositions a measure of dendrogram dissimilarity may be achieved as follows. Note the number of pairs of entities associated in the same group in dendrogram 1 but not dendrogram 2 and vice versa; sum these two numbers. The smaller the result the more similar are the dendrograms in group composition.

If, however, the primary interest is in the relationship between groups, some measure of dendrogram structure is required. Such a measure may be achieved in several ways, one of the simplest being to determine the number of nodes by which given pairs of entities are separated in the two dendrograms. In so far as the number of nodes separating pairs of entities is the same in each dendrogram they are regarded as similar, and in so far as the number of nodes separating pairs of entities is different in the two dendrograms they are regarded as dissimilar. A measure of dissimilarity is readily achieved; for each comparison subtract the larger from the smaller number and sum over all pairs of entities.

References

Clifford, H. T., and Williams, W. T. (1973). Classificatory dendrograms and their interpretation. *Aust. J. Bot.* **21**, 151–62.
Williams, W. T., and Clifford, H. T. (1971). On the comparison of two classifications of the same set of elements. *Taxon* **20**, 519–22.

14

Classification: Recent Advances and Future Possibilities

M. B. Dale

In order to review recent trends in classification studies it is necessary to return to the foundations of the methods and to provide some formalism within which the methods can be compared.

As with ordination, the possibilities of development of new classification methods involve either a greater specialization to include subject-specific constraints, or a relaxation of assumptions to provide more generality. Any classification method involves two parts: a definition of the nature of the classes being sought and a definition of the methods of search to be used in the finding. The latter emphasizes the heuristic search procedures involved, since it is rare that any optimal solution is known and still more rare that it is computationally feasible for large data sets. The former distinguishes two objectives, one concerned with representing relationships between items, the other with the distribution of values for descriptors within a class. For convenience, it is useful to separate extrinsic methods from intrinsic methods (Ch. 10), the former being concerned with evaluating structure in one set of variables through the mediation of another set, while the latter is solely concerned with the internal relationships of a single set of variables. It is also useful to consider the special problems of very large data sets, in effect infinite in size, where alternative methods may be necessary.

Models for Similarities

Most discussions of new classification methods concentrate on the method of search. This is somewhat surprising since it would seem useful to know you are looking for a needle before starting to destroy the haystack. Where attention has been given to the objective of the search it has usually been in terms of notions of similarity and dissimilarity. Indeed most classification methods have been regarded as methods of finding patterns in similarities and not in the raw data from which they are derived. Jardine and Sibson (1968) define classification as a mapping of similarities to distances on a

102

hierarchy and would reject any method that does not conform to their assumptions. Williams *et al.* (1972) adopt a similar view but do not regard the initially calculated similarity measures as inviolate, but only as initial estimates to be improved during the course of the analysis. Methods such as Wishart's (1969) mode analysis and the unimodal fuzzy sets of Gitman and Levine (1970) also identify patterns in the similarities.

If indeed the patterns are to be found in the similarities, then a variety of forms are possible. Jardine and Sibson (1968) argue that only nearest-neighbour sorting is acceptable if their axioms are to be met. This is, in effect, saying that each group shall have a short minimal spanning tree, which represents the hierarchical structure, and the effectiveness of the analysis can be measured by the discrepancy between observed similarity and that deduced from the tree representation. However, there is no reason to stop with a tree, since any well-defined graph structure could be used. Thus the structure sought might be maximal complete subgraphs, as in furthest-neighbour sorting, minimal unbranched spanning graphs, shortest loops and so on. The choice between these possibilities cannot be made on numerical grounds, but must be derived from the subject matter being analysed. It may be true that taxonomists, for example, are interested in minimal spanning trees, Wagner trees, Hu trees (Farris *et al.* 1970) or any other of the forest of possibilities, for good theoretical reasons based on the genetic mechanisms producing variation in living organisms. This does not show that ecologists, psychologists or anyone else need be interested in exactly and only the same range of possibilities (cf. Gower 1974; Wallace and Boulton 1968), nor that the numerical analyst is correct in imposing such a range of solutions on these users. It is, of course, to be hoped that a limited number of models will prove useful but this is likely to depend as much on human abilities to cope with models as on the 'real' structure of data.

Models for Data

In many cases the calculation of similarities imposes constraints on the data themselves. This is most obvious in the so-called (ij, k) measures of similarity introduced by Williams and Lambert (1966). With these a model is fitted to the data and if the fit is unacceptably bad (which might be probabilistically assessed) the data are subdivided into groups so as maximally to improve the fit. Thus the similarity between any set of items and any other set is measured by the change in fit of the model, that is by the difference in fit between the sets combined and the sum of the fits of the two taken independently. Now most similarity measures in fact impose one model, which Wishart (1969) termed the *minimal variance model*. In a 'normal' analysis, which groups individuals, a perfect group on this model would have all variables invariant within groups. This sounds plausible until an 'inverse' analysis is examined. In this case each item in a group must have constant abundance over all variables, which is much less plausible. Further, the normal and inverse models are fundamentally incompatible at this level, so that combinations of normal and inverse analyses are seldom profitable, except with binary data.

The first suggestion that minimal variance models were not compulsory is due to Macnaughton-Smith (1965) when he suggested his two-parameter model. If x_{ij} is the abundance of variable j in individual i, then define a parameter a_i for the individual and b_j for the variable such that

$$x_{ij} = a_i b_j / (1 + a_i b_j).$$

The parameters must be estimated from the data in order to assess the fit of this model but this is not difficult. Minimal variance models in effect either set all a_i values equal or set all b_j values equal.

The methods of Hartigan (1972) and Tharu and Williams (1966) adopt this latter approach rather obviously, but it is implicit in all minimal variance methods. Thus the two-parameter method includes minimal variance methods as special cases.

The original description of the method suggests why such a model might be useful. If people answer a series of questions, the number of correct answers depends both on the difficulty of the questions and on the ability of the people. In the ecological case we read 'sites' for people and 'species' for questions and the two-parameter model would suggest that the abundance of a species in a site is a function of both the difficulty of the site and the ability of the species, which is more reasonable than an assumption that all sites are identical or that all species are identical.

The parameters obtained during the course of fitting can also provide useful information (Dale and Anderson 1973) but it is the symmetry of the model which has the most surprising consequences.

In Table 14.1 several examples of the kind of groups sought are demonstrated and it is clear that these can be defined monothetically (a search procedure used in the present program). However, the search can involve either grouping items, in normal analysis, or variables, in inverse analysis. As a consequence of the symmetry of the model the results of trials of both normal and inverse divisions can be compared, and the better of the two chosen. Dale and Webb (1975) have argued that such a choice is equivalent to a choice between ordination and classification and can be used to determine specially significant groupings.

The two-parameter method is not the only one using a symmetric model. Noy-Meir (1973) has used another based on component analysis and there

Table 14.1

(a) Two-parameter

| | Sites | | | | |
	a_i parameter decreasing				
Species	6	4	2	1	1
b_j parameter	5	3	1	0	0
decreasing	4	2	0	0	0
	3	1	0	0	0
	2	0	0	0	0

(b) Minimal variance normal model

| | Sites | | | | |
	a_i parameter fixed				
Species	6	6	6	6	6
b_j parameter	4	4	4	4	4
decreasing	3	3	3	3	3
	2	2	2	2	2
	0	0	0	0	0

(c) Minimal variance inverse model

| | Sites | | | |
	a_i parameter decreasing			
Species	4	3	1	0
b_j parameter	4	3	1	0
fixed	4	3	1	0
	4	3	1	0
	4	3	1	0

---

are still others. These can, of course, be examined to determine their properties; but the important point is that the two-parameter method marks a significant change away from similarity-based models towards data-based models.

Extrinsic and Pragmatic Classification

It is interesting to note that while numerical classification has developed primarily methods for intrinsic analysis, classical statistics has been much more concerned with the problems of extrinsic analysis, of the effects of some operations on some other variables. There are some numerical classification methods which supplement the commonly used methods of multiple regression, especially when the population under study is not homogeneous.

Once the idea that classification involves fitting a model to data is accepted, it is not difficult to define the model for one set of variables while basing the criteria for search on another set. Macnaughton-Smith (1963) seems to have devised the first such method, although a similar method reappears, independently invented, in Hunt *et al.* (1966) and in Sonquist *et al.* (1972). All these methods are primarily concerned with predicting a single variable value, using a multivariate set of predictors. Mantel and Valand (1970) have suggested an ingenious method for extending this approach to multivariate prediction although they adopt a probabilistic approach and might almost be considered statisticians rather than pattern analysts.

Macnaughton-Smith (1965) also developed his multiple predictive analysis for more complex situations. Here the objective is to predict the effectiveness of treatments in producing certain results although the population to which they are applied is itself heterogeneous. Beeston and Dale (1975) have shown how this approach can be useful for preliminary evaluations of experimental programs.

In examining data there are three kinds of information which might be sought, kinds that Dale (1971) refers to as syntactic, semantic and pragmatic. Syntactic information is intrinsic in nature for it refers to the interrelationships within a set of variables. Semantic information is extrinsic since it involves the interpretation of variation in one set of variables in terms of at least one other set. In pragmatic information, the semantics must be valorized to suit the particular practical problem faced, and such valorization is the province of economics, in the wide sense. Belis and Guiascu (1968) have shown how cost and value information can be combined with predictability information (semantics). By coupling the use of their measure with the search procedures of multiple predictive analysis a pragmatic analysis should be possible, although our present studies show that some practical problems still remain. There are in fact several methods of combining value and probability (notably games theory) and even if the appropriate representation is known there still remains the question of whose values to use.

Large-scale Search and the Significance of the User

While a fundamental understanding of classification does not involve computer programs, any practical application will find the methods useful only if large amounts of data can be analysed. Most of the search methods at present in use, such as the polythetic agglomerative methods, become uneconomic with more than a few hundred items to classify, and almost all methods at present in use would fail with 1000 items or more. How then to cope with the very large sets which automatic data acquisition can provide, and *a fortiori* with the essentially infinite sets which continuous learning involves?

One possibility (see notes on program REMUL in Chapter 16) is to classify a small subset and then invoke allocation rules to place the remainder. It is fairly easy to see that any full development of this sort of approach must allow the new information to modify the classification both by redefining the groups and by initiating the fusion of some groups and the partition of others. Uttley (1970) has carried this to a logical conclusion since he starts with no small subset but develops all groups during the classification process; we hope to start a similar investigation in the near future.

In the opening paragraphs, one line of development suggested was the introduction of problem-specific knowledge. However, it does not seem impossible to let the computer itself observe its own 'behaviour' and develop some of this problem-specific knowledge for itself. This approach has been attempted by Buchanan *et al.* (1972). Their program is first of all suggesting structural formulae for chemical compounds based on mass spectrograms and nuclear magnetic spectra. However, they have a second level of program which attempts to correlate the derived structure with further features of the spectrum itself so that new rules of inference are formed. Once data are organized sequentially, it is unlikely that the user will be satisfied with allocating items to classes and a search will commence for establishing patterns in time (Notley 1970), for comparing patterns on several sets of variables (Holland 1971) and for modifying the inputs to the procedure in order more efficiently to separate the groups by modifying the descriptors. In this stage numerical classification has joined the wider field of artificial intelligence studies. Classification methods will certainly be used in solving these problems, but that is for the future.

While the search for adaptive methods continues, there is another approach which can be considered, that of introducing the user into the classification procedure. Dale and Quadraccia (1973) have investigated one means of accomplishing this, using interactive programming rather than batch processing. The problems here are those of providing adequate information to the user without overloading him, and providing control operations which will enable him to choose the next avenue of investigation. The whole question of presentation of results from classification and ordination methods has received little attention, and the logical points at which to introduce user interaction and control are but ill identified. How can a seven-dimensional ordination be effectively displayed, and what method can show the relationships between groups in a space-distorting classification? At present the necessary equipment for examining these problems is not available.

Conclusions

It should be clear from what has been said that there are still significant problems in classification and that a universally applicable classification method is not available. I hope also that there is some encouragement for the user to examine the structures which are really being sought rather than leaving it to the unfortunate analyst. The newly wed can be oblivious to surroundings, but a stable marriage will require some mutual interaction.

References

Beeston, G., and Dale, M. B. (1975). Multiple predictive analysis: a management tool. *Proc. Ecol. Soc. Aust.* **9**, 172–81.

Belis, M., and Guiascu, S. (1968). A quantitative–qualitative measure of information in cybernetic systems. *I.E.E.E. Trans. Inf. Theory* 593–4.

Buchanan, B. G., Faigenbaum, E. A., and Sridharan, N. S. (1972). Heuristic theory formation: data interpretation and rule formation. In 'Machine Intelligence'. (Eds B. Meltzer and D. Michie.) Vol. 6, pp. 267–90. (Edinburgh Univ. Press.)

Dale, M. B. (1971). Validity and utility of information theory in ecology. *Proc. Ecol. Soc. Aust.* **6**, 7–16.

Dale, M. B., and Anderson, D. J. (1973). Inosculate analysis of vegetation data. *Aust. J. Bot.* **21**, 253–76.

Dale, M. B., and Quadraccia, L. (1970). Computer-assisted tabular sorting of phytosociological data. *Vegetatio* **28**, 57–73.

Dale, M. B., and Webb, L. J. (1975). Numerical methods for the establishment of associations. *Vegetatio* **30**, 77–87.

Farris, J. S., Kluge, A. G., and Eckart, M. J. (1970). A numerical approach to phylogenetic systematics. *Syst. Zool.* **19**, 172–89.

Gitman, I., and Levine, M.D. (1970). An algorithm for detecting unimodal fuzzy sets and its application as a clustering technique. *I.E.E.E. Trans. Comput.* **C-19**, 583–93.

Gower, J. C. (1974). Maximal predictive classification. *Biometrics* **30**, 643–54.

Hartigan, J. A. (1972). Direct clustering of a data matrix. *J. Am. Stat. Assoc.* **67**, 123–9.

Holland, J. H. (1971). Processing and processors for schemata. In 'Associative Information Techniques'. (Ed. E. L. Jacks.) pp. 127–46. (Elsevier: London.)

Hunt, B., Marin, J., and Stone, P. J. (1966). 'Experiments in Induction'. (Academic Press: New York.)

Jardine, N., and Sibson, R. (1968). The construction of hierarchic and non-hierarchic classifications. *Comput. J.* **11**, 177–84.

Macnaughton-Smith, P. (1963). Predictive attribute analysis. *Biometrics* **19**, 364–6.

Macnaughton-Smith, P. (1965). Some statistical and other numerical methods for classifying individuals. Home Office Res. Rep. No. 6. (H.M.S.O.: London.)

Mantel, N., and Valand, R. S. (1970). A technique for nonparametric multivariate analysis. *Biometrics* **26**, 547–58.

Notley, M. G. (1970). The cumulative recurrence library. *Comput. J.* **13**, 14–19.

Noy-Meir, I. (1973). Divisive polythetic classification of vegetation data by optimized division on ordination components. *J. Ecol.* **61**, 753–60.

Sonquist, J. A., Baker, E. L., and Morgan, J. W. (1972). 'Searching for Structure'. (Revised ed.) (Inst. Social Res.: Univ. of Michigan.)

Tharu, J. T., and Williams, W. T. (1966).   Concentration of entries in binary arrays. *Nature (Lond.)* **210**, 549.

Uttley, A. M. (1970).   The INFORMON: a network for adaptive pattern recognition. *J. Theor. Biol.* **27**, 31–67.

Wallace, C. W., and Boulton, D. M. (1968). An information measure for classification. *Comput. J.* **11**, 185–94.

Williams, W. T., Clifford, H. T., and Lance, G. N. (1972).   Group-size dependence: a rationale for choice between numerical classifications. *Comput. J.* **14**, 157–62.

Williams, W. T., and Lambert, J. M. (1966).   Multivariate methods in plant ecology. V. Similarity analyses and information analysis.   *J. Ecol.* **54**, 427–45.

Wishart, D. (1969).   Mode analysis: a generalization of nearest neighbour which reduces chaining effects. In 'Numerical Taxonomy'. (Ed. A. J. Cole.) pp. 282–311. (Academic Press: London.)

15

Computational Considerations

G. N. Lance

The Problem of Machine Dependence

Except for what would now be regarded as extremely small problems, ordination and classification are out of reach of hand computation; they require too many calculations. The development of pattern-analysis methods has therefore had to await the development of modern high-speed computers. However, workers in different parts of the world, or even of the same country, tend to have access to different computers; and they naturally write their programs with their own computer in mind. It would be a happy situation if a program written for one computer could, without modification, be run on another; but this is seldom the case, and it is necessary to consider briefly why this is so. One obvious reason is size, since a problem cannot be run on a machine that is not large enough to store both the program and the data it is intended to process. Another obvious reason is speed, since a large problem may require, on a slow-running machine, more time than can be allocated to meet the demands of a single user. There are, however, more fundamental reasons, which we can divide into *hardware* (or physical) and *software* (or linguistic) reasons.

Hardware considerations

The unit in which a computer stores its information is the *word*, so called because it can be used for storing instructions as well as numbers. A single number, i.e. a single value of continuous variable, always occupies a single word. It follows that, so long as we are concerned only with continuous variables (what we called in Chapter 5 numeric attributes) the internal structure of the word itself need not concern us. This is the situation with ordination programs, and as a result such programs can often be transferred from one machine to another with a minimum of modification. The situation with classificatory programs is very different, since the majority of them need to deal with large numbers of nominal attributes; it would be impossibly wasteful of computer storage space to use a whole computer word to store a single 1 or 0. The word itself is made up of a string of binary digits, or *bits*, such that at any moment each can be either 1 or 0. Since this is also the situation with a binary attribute, it follows that only one bit is needed to store the condition of one attribute; and if the word contains 60 bits, the states of 60 attributes for an individual can be stored in a single word. Similarly, a 5-state nominal attribute needs only 5 bits, so that 12 such attributes can be stored in a 60-bit word. It is perhaps unfortunate that the number of bits in a word varies from one machine to another. The IBM 370 series has a 32-bit word; the Control Data 3600, with which many users will be familiar, has a 48-bit word; the Cyber 76 has a 60-bit word. The business of 'packing' data so that several attributes occupy a single word requires some fairly sophisticated programming, which is not easy to modify for a word of different length. This word-length dependence is probably the greatest single obstacle in the way of transferring a classificatory program from one computer to another.

Software considerations

There are a number of alternative languages in which users can write programs, the best-known being FORTRAN, ALGOL and COBOL. Of these, COBOL is primarily intended for commercial use and will not concern us. The ALGOL/FORTRAN dichotomy tends to be national; Europe commonly uses ALGOL, the U.S.A. and Australia mostly use FORTRAN.

The Canberra pattern-analysis programs are written almost entirely in FORTRAN. These so-called 'high-level' languages are ideal for writing mathematics, but the computers themselves cannot execute FORTRAN instructions directly. Every computer therefore has a FORTRAN *compiler*, which translates the FORTRAN instructions into a more basic code which is intelligible to the computer. These basic codes, or *machine languages*, differ greatly from one type of machine to another, so that each type of machine needs its own compiler. A single FORTRAN program translated by different compilers will obviously produce two different sets of basic code; but if both compilers are error-free—which all too frequently they are not—the results ultimately obtained from the two machines will be equivalent.

FORTRAN is essentially a human language, and human languages are never completely stable. The English spoken by Americans differs from that spoken by Australians, and both differ from that spoken in England. In the same way, FORTRAN has begun to develop its own dialects; a user may discover that an instruction which is accepted by his own compiler may not be understood by another. For example, a user writing for the Cyber 76 may wish to set two quantities, M and N, to zero. He can write $M = N = 0$; but such an instruction will not be understood by the PDP 10 compiler, which must have two separate statements, $M = 0$ and $N = 0$, which must be on separate cards.

Such FORTRAN discrepancies, though they can be exasperating, are easy to rectify. A more serious problem is posed by the existence of *assembly languages*. These are intermediate-level languages, but closer to the computer's own basic code than they are to high-level languages such as FORTRAN, and therefore extremely machine-dependent. They are extremely inconvenient for writing mathematics, but there are two reasons why they are sometimes used. The first arises because a computer may have a desirable hardware facility which FORTRAN does not know about, and therefore cannot utilize. For example, in classification programs concerned with large numbers of binary attributes, it is necessary to be able to count the number of bits in the '1' condition in a word. This can be done in FORTRAN, but it is time-consuming; the Cyber 76, however, has a hardware instruction for doing it extremely rapidly, which FORTRAN cannot access. The second reason is that FORTRAN compilers are not always perfectly efficient; they can sometimes produce machine code which, though it does the job assigned to it, does it in a somewhat roundabout way, with a consequent waste of computing time. Some subroutines in classificatory programs are entered an enormous number of times in the course of a single exercise; in such a case even a very small increase in computing time for a single entry can be cumulatively serious. As a result of both these considerations, the Canberra programs incorporate a small number of subroutines written in COMPASS, the Cyber 76 and 3600 assembly language; these will be completely incomprehensible to a compiler for any other computer.

It follows from all the foregoing that any group setting out to write a battery of pattern-analysis programs must first decide on the precise computer for which they are to be written. In our case this is the Cyber 76; a brief description of this machine is therefore necessary.

The Cyber 76 System

The computer

The Control Data Cyber 76 is currently one of the most powerful computers in the world; it can perform approximately 3 000 000 arithmetic operations per second in the central processing unit, or CPU. In fact, several arithmetic operations can be performed simultaneously, but this is of no concern to the ordinary user. The instructions to be obeyed, and the data on which they are to operate, must be stored somewhere in the system, and the Cyber 76 possesses four distinct 'levels' of storage. First, and most efficient, is the 'small core memory' (SCM), made up, as its name implies, of magnetic cores. This is the type of store from which numbers and instructions can be obtained most rapidly; it is also by far the most expensive, and the size of such a store is invariably limited by economic considerations. The Cyber SCM consists of 65 000 words. Next, and somewhat slower of access, is the 'large core memory' (LCM) of 256 000 words; then follow nine disc storage units, each capable of holding about 1 500 000 words. The last, and slowest, storage medium is magnetic tape; each tape can hold up to 750 000 words.

The monitor

During the last 20 years or so computers have become more and more complicated; but it has become necessary to make it possible for less and less experienced people to use them. In the earliest days a user would wait until the machine was idle, then read in his job; the machine would process this job, and this job only, produce the results in due course, then wait until the next user came along. With a machine of the power of the Cyber it is uneconomic to have any part of the machine idle. For example, it takes longer to transfer numbers from the disc to the SCM than it does to process the numbers once they are available; and the CPU cannot be left to wait until the numbers are available, but should be getting on with another job.

In fact, in the Cyber there can be over 64 jobs at some stage of execution at any given moment. No human operator could organize such a system at the Cyber's speed; and so all modern computers possess extremely sophisticated *monitor* or *control* systems which carry out the intricate multi-programming automatically. Their aim is to optimize the use of all elements in the system, so as to carry out as much work as possible in unit time subject only to such priority ratings as have been specified. The monitor on the old 3600 was called DAD; that on the Cyber is called SCOPE. Even a monitor of the complexity of SCOPE is not omniscient; it needs to be told whether the program and data are to be read from cards or magnetic tape, or whether they are already stored as 'files' on discs. It also needs to know which of its many associated line-printers is to receive the output. As a result, there also needs to be a *job control language* which gives SCOPE, or in some cases the operators, the necessary instructions. Unless a user's instructions are unusually complicated, however, there is little for him to learn in this regard; for most purposes less than 10 job control cards are needed.

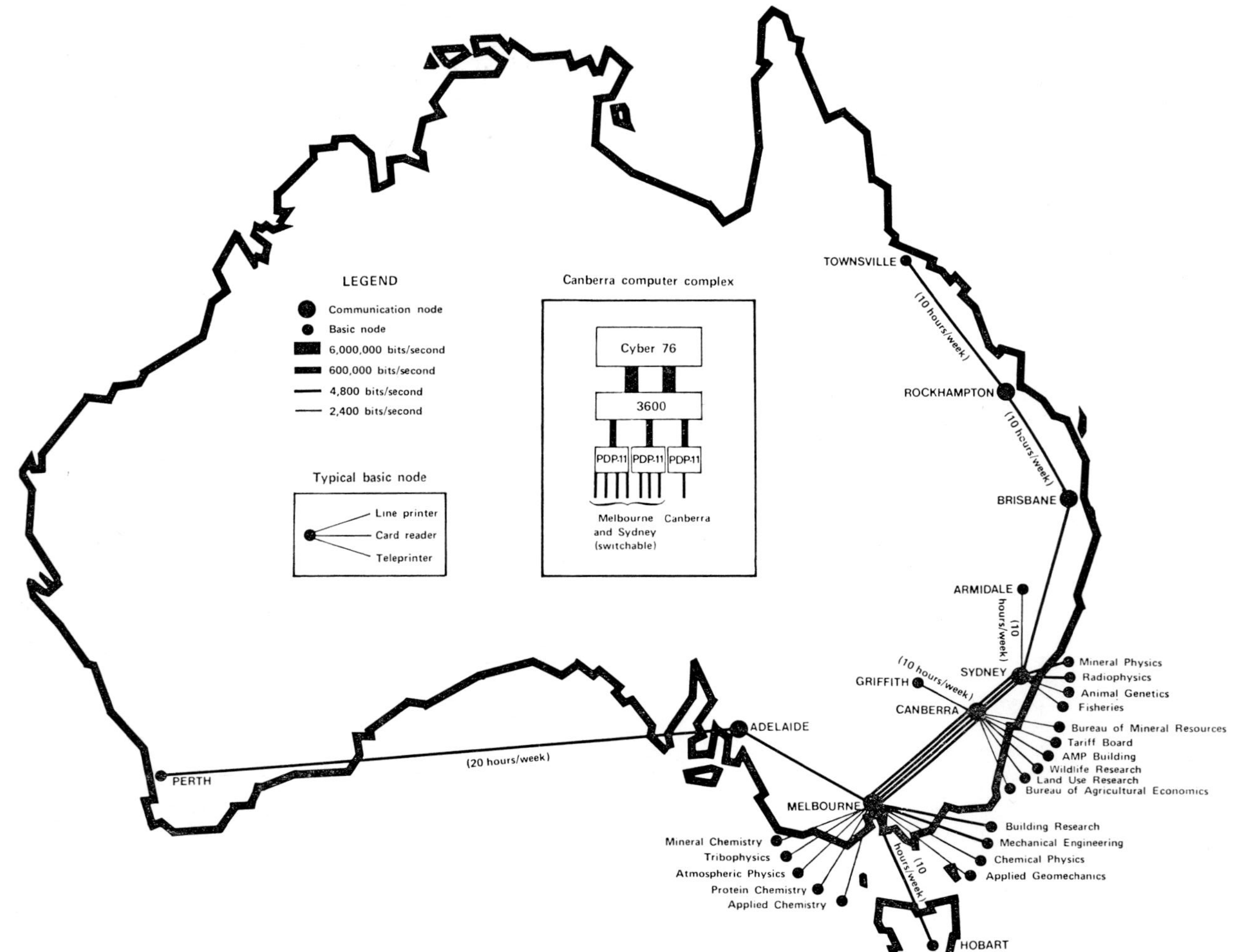

Fig. 15.1. The state of CSIRONET at 1 May 1974.

The network

CSIRONET is that part of the CSIRO computing system which enables users outside Canberra to access an efficient remote-job entry system, and to use Teletypes and other forms of console for 'interactive' computing. Fig. 15.1 shows the extent of the network, which can be viewed as a grandiose message-switching system. Users at a remote centre such as Townsville can use their own equipment to develop and test programs, which can then be stored centrally and made available to users elsewhere in the network. Results from executed programs can be sent to any or all of the 30 or so line-printers. The Canberra pattern-analysis programs are available on a library called TAXON, which can be accessed from any location where there is a CSIRONET node. An outline of these programs and of the method of accessing them is given in the next chapter.

16

The Canberra Programs and their Accession

P. W. Milne

The CSIRO Division of Computing Research in Canberra holds what is probably the most extensive set of pattern-analysis programs in the world. All but the most recent of these were originally written for the Control Data 3600; but all those which have proved of enduring value have now been converted to run on the Cyber 76, and nearly all of these are held on a permanent library file named TAXON, which can be called down from any CSIRONET terminal. The purpose of this chapter is to provide the potential user with an outline account of what programs are available and how to obtain access to them. Some of the programs provide additional options for special purposes, which will not be included in this list. Detailed specifications, which give a full account of all available options, together with instructions for preparation of data and the necessary control cards, can be obtained from P. W. Milne, CSIRO Division of Computing Research, Post Office Box 1800, Canberra City, A.C.T. 2601.

Synopsis of the Canberra Programs

Classification programs

Agglomerative polythetic

MULTBET

A non-combinatorial information-statistic program, accepting the widest variety of attribute types with provision for missing values. Intensely clustering, and apt to produce 'nonconformist groups'. Not recommended for use when the data are predominantly numeric. In the past has been regarded as the best general-purpose classificatory program, but present indications are that it may be superseded by the polythetic divisive program REMUL (see below).

MULCLAS

Implements the Euclidean strategy of Burr. Accepts nominal, linked binary, and numeric attributes with provision for missing values; calculates Euclidean distance and uses either the 'group-average' or 'incremental sum of squares' fusion strategies. Clusters fairly intensely, but not apt to produce 'nonconformist groups'. Indicated when the data are predominantly or entirely numeric.

At the time of writing MULCLAS has not been transferred to the TAXON library and special instructions are needed (see next section) to bring it down.

TRANMAT

A special-purpose information-statistic program. It accepts either transition matrices or state sequences from which transition matrices can be built up, and classifies these non-combinatorially. *Note*: MULTBET, MULCLAS and TRANMAT will, if requested, punch out the set of inter-individual dissimilarity measures for processing by an ordination program.

CLASS

An entirely combinatorial program, accepting only the upper triangle of a set of dissimilarity measures. Normally used either for 'group average' fusion (if a substantially space-conserving strategy is required) or for 'flexible' fusion. In the latter case, the cluster-intensity coefficient, β, must be specified; if there are no special requirements in this regard, it is usual to set $\beta = -0\cdot25$. The program is extremely fast.

CENPERC

This program was originally written to provide an information-statistic classification of percentage frequency data, such as arises in ecological survey when it is desired to accumulate quadrat records of presence-or-absence data on a per-paddock basis; it is still used for this purpose. It also makes provision for classifying the asymmetric 'diversity' measure (Ch. 6) and for dealing with the case when only spatially adjacent sites are permitted to fuse. However, it has recently been used methodologically, to investigate the efficacy of a number of information-statistic measures which have been suggested. As a result, the program is not yet completely finalized, and workers wishing to use it should take advice before doing so.

Divisive monothetic

DIVINF

Accepts only binary data, punched either in 'normal' or in 'inverse' form without missing values; a special compressed format is used to input the data. The decision-function is information-fall on division, and the user needs to specify the number of final groups that he requires. The program is extremely fast unless the number of attributes is very large.

Divisive polythetic

POLYDIV

Accepts only numeric data with provision for missing values. Its use is indicated when the number of individuals is large but the number of attributes fairly small.

REMUL

Accepts as wide a variety of attribute types as does MULTBET, with provision for missing values. It is substantially space-conserving; it makes provision for final overall reallocation of individuals to groups, in order to optimize group homogeneity. It also makes provision for allocation of new members to groups; this is to deal with the case when the data set is too large for primary classification, so that a modest subset of the data can be classified and the remainder of the data allocated. In such a case provision is made for the starting of new groups. The program is still undergoing minor improvements and so at the time of writing is not available on the TAXON library; it can, however, be made available to users on request.

Diagnostic

GROUPER

Expects the primary data, the specification of a set of groups produced by a classificatory program, and the dendrogram sequence to be simulated. At each of the specified fusions it will calculate, using the specified dissimilarity measure, the contribution of each attribute; the set of contributions is then sorted into descending order of importance.

Ordination programs

The basic programs

GOWER

Accepts the upper triangle of a matrix of similarity or dissimilarity measures, in free format with a comma as separator and an asterisk as terminator. The principal diagonal can be specified as everywhere 1 or 0, or a specific diagonal can be read in; various transformations of the off-diagonal elements can also be specified. Undertakes either a principal component or principal coordinate analysis; the output is limited to seven roots and associated vectors.

CANONGO

Accepts two sets of Gower output vectors and carries out a canonical analysis between them.

Diagnostic

GOWECOR

Accepts the basic data and a set of output vectors from a GOWER principal coordinate analysis. For each vector in turn it correlates the vector against each of the original numeric attributes, and each state of the original nominal attributes. The correlation coefficients are then sorted into descending order of absolute magnitude.

Accession of the Canberra Programs

This is essentially a matter of learning the relevant instructions in the job control language. We shall consider five cases.

(1) The simple FORTRAN program; this is the case when you wish to make some calculations—perhaps calculate some similarity measures—but you wish to examine them before deciding what to do with them. It therefore does not involve calling one of the Canberra programs, and it is included here to familiarize you with the basic language.

(2) You have a set of data cards, preceded by appropriate MULTBET control cards, and you wish to carry out a classification using MULTBET.

(3) You have a set of data, from which you propose to calculate a triangle of dissimilarity measures which you then wish to classify by means of CLASS.

(4) You have a set of data which you wish to classify by means of MULCLAS, and you also want the MULCLAS run followed by GROUPER, to provide a diagnosis of how the classification comes to be as it is, then followed by GOWER to provide an ordination of the same data, then lastly followed by GOWECOR to obtain a comparable diagnosis of the ordination.

(5) You have a set of binary data and you wish to obtain both 'normal' and 'inverse' runs using DIVINF.

You will have been allocated a *charge-code*, so that the computing costs can be charged to you; this is a string of eight characters, which we shall here denote by 'c.c.' It will also be necessary for you to give the job a name, which can be any string of up to seven alphanumeric characters; we shall denote this by IDENT. An *end of section* is a card with 7, 8 and 9 punched in column 1; an *end of information* is a card with 6, 7, 8 and 9 punched in column 1.

The job control language of the Cyber 76 is still undergoing slight modifications as the system is extended; the language given here is correct at the time of writing, but there may have been changes by the time this book appears in print. Users wishing to access these programs for the first time should therefore first check the accuracy of their job control statements with the nearest representative of the Division of Computing Research.

You will find, if you examine an individual program specification, that you are required to specify a parameter NLUN. This is the 'logical unit number'—now, on the Cyber, a file tape number, where the program expects to find the data. If NLUN is set equal to zero, the program will expect to read the data from cards; as will become clear from what follows, this is not always possible.

Case 1: *the simple* FORTRAN *program*

*CY, c. c., IDENT	This sends the program direct to the CYBER. You may prefer to replace the *CY by *DOC; this causes a copy of your document to be held (for approximately 24 hours) on the 3600. You can cause it to be run at any time by giving the order *CY, c. c., IDENT via a teletype or console; the advantage of this is that if you have made a mistake it can be corrected via the editing program TED, without reading in the cards again.
IDENT (T30)	This states that the estimated running time is 30 octal seconds (100 octal seconds is approximately 1 minute). If you wish to specify a particular priority, the request goes on the same card.
FTN (SL, R = 1, OPT = 0)	This will give you a listing of your program, any diagnostics concerning errors, and a reference map; compilation will be at the lowest (normal) level of optimization. There are many other possibilities which your local representative can explain.
LGO.	Tells SCOPE that after compilation an execution is to be performed; if you were only testing a new program you would omit this until you were sure that there were no errors in your program.
End of section	A signal to SCOPE that the form of cards to follow is different.

    FORTRAN main program
    FORTRAN subroutines
    COMPASS subroutines
End of section
    Data cards
End of information

Case 2: *a* MULTBET *run from data cards*

*CY, c. c., IDENT	As before.
IDENT (T100)	As before.
DISPOSE (TAPE 1, *PL)	Sends the output from which your dendrogram will be plotted to the plotter.
ATTACH (TAXON, TAXON, ID = CCRXTX)	The second 'TAXON' is the name of the file sought; the first 'TAXON' is what you are going to call it in your program (if any); in our case there is no need for a different name. The ID is the charge code under which it has been stored.
LIBRARY (TAXON)	Informs SCOPE that the file TAXON is held as a permanent library file.
MULTBET.	The name of the program you are seeking.

End of section
   MULTBET control cards
   Data cards
End of information

Case 3: *to calculate measures and run them in* CLASS

*CY, c. c., IDENT
IDENT (T20)
DISPOSE (TAPE 1, *PL)

FTN (SL, R = 1, OPT = 0)	Your program for generating measures.
LGO.	Ensures that the measures are generated by executing your program.

ATTACH (TAXON, TAXON,
   ID = CCRXTX)
LIBRARY (TAXON)
CLASS.
End of section

Your FORTRAN program.	This must write the measures away in a particular FORMAT on to file TAPE 12. When this has been completed you must REWIND TAPE 12 before you exit from your program.

End of section
  Data cards for your FORTRAN program to read.
End of section
   CLASS control cards, with NLUN = 12.
End of information

Case 4: *to read in cards, and run them successively in* MULCLAS, GROUPER, GOWER *and* GOWECOR

(*Note*: If MULCLAS is to write away the necessary information for GROUPER and the other programs to deal with, it *must* find the data on file TAPE 12; our first job will therefore be to copy the given cards on to TAPE 12 and rewind. In this case we shall carry out the rewinding in the job control language.)

```
*CY, c. c., IDENT
IDENT (T30)
DISPOSE (TAPE 1, *PL)
COPYS (INPUT,  TAPE 12)
REWIND (TAPE 12)
ATTACH (MULCLAS, MULCLAS, ID = CCRXTX)
MULCLAS.
ATTACH (TAXON, TAXON, ID = CCRXTX)
LIBRARY (TAXON)
GROUPER.
GOWER.
GOWECOR.
End of section
   Data cards
End of section
   MULCLAS control cards with NLUN = 12
End of section
   GROUPER control cards with NLUN = 13
End of section
   GOWER control cards with NLUN = 14
End of section
   GOWECOR control cards with NLUN = 13
End of information
```

Case 5: *to carry out both normal and inverse runs with* DIVINF

```
*CY, c. c., IDENT
IDENT (T40)
DISPOSE (TAPE 1, *PL)
COPYS (INPUT, TAPE 12)
REWIND (TAPE 12)
ATTACH (TAXON, TAXON, ID = CCRXTX)
LIBRARY (TAXON)
DIVINF.
```

REWIND (TAPE 12)
DIVINF.
End of section
  Data deck
End of section
  DIVINF control cards for normal run, with NLUN = 12
End of section
  DIVINF control cards for inverse run, with NLUN = 12
End of information

Postscript

It is vitally important that data cards be prepared with great care and thoroughly checked: even verification is not 100% secure, and for a complex deck intended for MULTBET or MULCLAS, actual proof-reading is safest. The computer will detect obvious errors; for example, if you have specified that a given nominal attribute has four states and the program encounters an apparent entry in the fifth state, it will report the error and will not attempt to execute the classification. However, if you were to punch a numeric integral attribute one column to the left of where it should have been, the number will appear to be 10 times what you intended and the computer has no means of knowing that this was an error.

It is equally important to ensure that control cards are accurate. All the programs are geared to processing more than one set of data, and need to be told when there is no more to come; this is signalled by a blank card at the end of the control cards, and a common error is to leave this out. Furthermore, MULCLAS and MULTBET require to be told whether any attributes are to be 'masked out' and ignored in the analysis; if all are to be used, so that there is no masking, another blank card is needed. For example, in Case 4 above, the complete set of control cards for MULCLAS would consist of (i) the two control cards which specify the type of analysis and the particulars of the data, (ii) a title-card, (iii) a blank card to inform MULCLAS that no masking is needed and (iv) a blank card to inform MULCLAS that there are no more data to follow. Mistakes in the data submitted to classificatory programs, particularly non-combinatorial programs such as MULTBET, can prove expensive to the user.

17

The Meaning of Pattern

W. T. Williams

In the previous chapters we have been concerned with the *techniques* of pattern analysis: given a complex set of data we have outlined some of the things that might be done with it, and the methods by means of which we might set about doing them. We have not yet considered the more fundamental question of what it all means when we have done it. We need to know more precisely what we mean when we speak of a 'pattern', and what mathematical status, if any, we can confer on it. Moreover, it is obviously desirable to know how these techniques relate to the more familiar ones of conventional statistics. These, then, are the subjects of the two final chapters in this Part.

Terminology

The term 'pattern analysis' was first used by Greig-Smith (1961) for a particular form of analysis which he had devised for use in plant ecology. It was first used in an agricultural context by Williams and Gillard (1971) who, having forgotten its earlier meaning, used it in the wider sense to cover *any* technique of classification or ordination. This extension of a formerly precise term is perhaps unfortunate, though it is unlikely to cause ambiguity; but, in Australia at least, the wider use is now so common that it would be difficult to introduce some completely new term. Nor is it easy to find anything better. In his now classic monograph, Macnaughton-Smith (1965) wrote of 'Some statistical and other numerical techniques for classifying individuals'. However, 'other numerical techniques' will not serve as a title on its own, and 'numerical analysis' includes many mathematical operations that have nothing to do with pattern. Nevertheless, the use of the term 'pattern analysis' implies that *some* mathematical meaning can be associated with the everyday word 'pattern'. To investigate the nature of this meaning is the purpose of this chapter.

It is first important to distinguish between the *recognition* and the *extraction* of pattern. The subject of 'pattern recognition' has long been respectable; it is concerned with questions like, 'What properties must a given squiggle possess to make it recognizable as the number 2 or the letter E?' But pattern extraction is concerned, given a set of data, to define within it some meaningful pattern of whose nature nothing is known in advance. The task sounds impossible; but it is not.

The Problem of Objectivity

Any set of entities, be they objects or numbers, possesses a pattern. Indeed, it is doubtful whether we can ascribe any meaning to the concept of a set of entities without pattern; this was, of course, the argument underlying Hume's rejection of Paley's 'Evidences'. The difficulty is that different people, looking at the same set of objects, will imagine different patterns in them. In fact, there is no reason to suppose, except in trivial or artificially constructed cases, that a set of data can ever be said to possess an objective, inherent pattern. As MacKay (1969) has pointed out, pattern is always *pattern-for-an-agent*, an interaction of the properties of the data set being examined and of the mind examining it. A pattern is therefore never correct or incorrect, right or wrong, true or false; it can only be *profitable* or *unprofitable*, and the agent, or type of agent, for whom it is or is not profitable must be specified, or the concept becomes meaningless. It follows that in pattern analysis *sensu stricto* there is no hypothesis to test, except the very weak one 'There exists in these data some pattern which might interest me'— weak, because it can always be fulfilled. The function of the extraction of pattern is not to test, but to *generate*, a hypothesis—which may, of course, be susceptible of later test by more conventional procedures.

The concept of hypothesis generation underlies all scientific research. True discovery is the generation of a new hypothesis, not based overtly on pre-existing knowledge, but grasped as a new pattern-for-agent by an unusually penetrating mind from an unpromising array of facts or numbers. The characteristic of an innovatory mind is that the patterns-for-agent that it extracts are more profitable than those extracted by less gifted workers.

Nevertheless, the longing for objectivity dies hard, and nowhere is this more clearly seen than in the attempts to grapple with the concept of a *cluster*. The formidable linguistic difficulties which surround this word have been examined exhaustively by Gasking (1960), from whose work it is clear that any definition based on the internal properties of a cluster will be so weak and so permissive in its exception, as to make it useless as a basis for numerical work. The first difficulty is one of context. Suppose there is an extensive monospecific and even-aged forest, and select a tree inside it at random. Consider now the 20 trees closest to this one. We do not normally think of these as a cluster; but we should immediately do so if they were smaller, or found to be of a different variety from the surrounding trees, or if the trees for some distance around were clear-felled. Our use of the word, in fact, appears to require not only that the members of a cluster be linked by some sort of internal relationship; they must be separated by some comparable negative relationship from all other elements that might be potential members of the cluster.

There is an even more fundamental difficulty in numerical work. Here, the extraction of a cluster rests on three things: the elements within which the cluster is to be sought, the attributes by which these elements are currently defined, and the numerical model to be used in extracting clusters. The importance of the two latter is too often underestimated. If we had a set of animals, and had been so misguided as to score only those attributes possessed by all mammals, we should find that cats, kangaroos, men and whales would not separate as distinct clusters; they would appear to be a completely homogeneous set—as indeed they would be by reference to the attributes chosen. This is an extreme example, but subtler versions of the phenomenon are easily perpetrated. The numerical model is even more important. We have taken 10 artificially constructed individuals, each specified by 5 continuous variables whose values have been extracted from a table of random numbers. This set has been classified by two—admittedly extreme—strategies, and the two dendrograms are shown in Fig. 17.1. Fig. 17.1(a) suggests that the

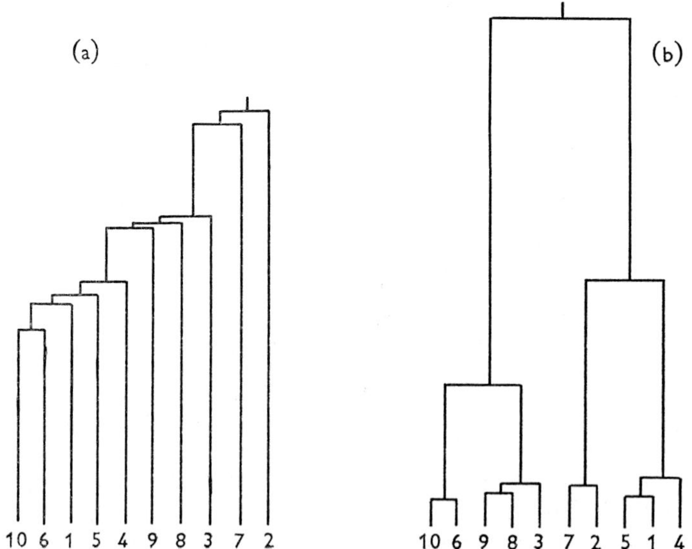

Fig. 17.1. Alternative classifications of the same set of data.

system contains no internal clusters whatsoever, whereas Fig. 17.1(*b*) suggests that it consists of two sharply distinct clusters, each of which contains two well-marked subclusters. Clusters do not, in fact, have any objective existence in the data. Like other patterns, they are clusters-for-an-agent, who can always extract clusters as intense as he wishes, or fail to extract them if that is what he would prefer.

Ensuing Controversies

The inherent subjectivity of pattern-analysis procedures does not fit easily into conventional mathematical thinking; and it is not surprising that this has engendered occasional controversies between mathematicians and biologists. The immediate cause has been a failure in communication, which in its turn largely arises from the schism which has arisen in schools between the teaching of biology and mathematics, so that almost nobody is ever taught both. The biologist has an intuitive idea of what he wishes to do, but is unable to translate this into terms meaningful to the mathematician; as a result the mathematician, anxious to help, has on occasion provided a rigorous and elegant answer to the wrong question.

Between 1930 and 1940 terrestrial ecologists in Britain were anxious to find means of extracting useful information from the mass of data they were collecting; at that time neither the computers nor the computer programs that would now be used were in existence. They consulted the mathematicians, who invoked conventional statistical procedures and fitted various forms of contagious Poisson distribution, such as the Neyman type A, to the ecologists' data. There is a considerable body of such work, which is elegant and interesting, but which in the event proved quite sterile; the models could be fitted but, once fitted, there was nothing that the ecologist could do with them. They were, in fact, convincing answers to a valid question, but not to the question the ecologist really wanted to ask. In many cases the correct question would have been, 'If there are major discontinuities in this set of data, where are they to be sought?' That particular question could not then have been answered, easy though it is now.

More recently, Cambridge mathematicians have been worried by the admitted instability of the intensely clustering procedures that biologists normally use. As a result, Jardine and Sibson (1968) devised a set of rigid mathematical criteria to which they considered all numerical classification should conform. Unfortunately, the resulting classifications were usually not of the type that a biologist wants; whatever the criteria might be that a biologist intuitively uses, they were different from these.

The Basic Criteria

It is obviously very important to try to find out what it is that the biologist really wants, what sort of criteria he is intuitively employing, what it is that he is trying to do. This is where the concept of pattern-for-an-agent becomes vitally important; what the biologist really wants is to use the computer as a stand-in for an agent. In other words, the programs that will satisfy him are those which will in some sense simulate the mathematical properties of the human mind. The idea that the human mind may exhibit general properties which are capable of mathematical expression is not new. For example, it has long been known of the classification of living organisms that if the number of genera containing n species is plotted against n, or the number of families with n genera is plotted against n, the result closely approximates the logarithmic sequence of general form $ax, ax^2/2, ax^3/3, ax^4/4, \ldots$. At one time it was usual to explain this phenomenon in terms of the frequency of mutations, or similar processes which result in the evolution of a new species; but it was Zipf (1949) who pointed out that anything which had been classified by the human mind, whether it was the functions of tools or the advertisements in the back of a telephone directory, obeyed the same law. In other words, the logarithmic sequence is not a function of the material being classified, but of the human mind which classifies it. In pattern analysis, then, it is properties of this type which we must seek.

Here we have been assisted by a fortunate accident. The mixed-data information-statistic program MULTBET was written simply to explore whether information-statistic classifications would prove useful; but it has proved spectacularly successful in angiosperm taxonomy. To take a single example, Leslie Watson (now at the Australian National University) did his M.Sc. work on the interrelationships of the genera of the Ericales (the two heath families Ericaceae and Epacridaceae). It took him about a year to collect his data, and almost as long to brood over it and to produce what he was satisfied was an improved classification. Exactly the same data were given to MULTBET, which in a few minutes produced a classification almost identical with Watson's; such few discrepancies as existed were, on detailed examination, seen to be minor improvements. The algorithms of MULTBET had, in fact, simulated the taxonomist's procedures with remarkable accuracy. Now these algorithms are such as to produce intense clustering; they will, to quote Chapter 11, 'sharpen up' the system, render distinct that which is indistinct, make the best of a bad job. But this is what a taxonomist has to do; faced with an intricate web of living organisms, he must divide them into groups to which he can give names which he hopes will remain stable, and by means of which he can communicate with taxonomists elsewhere. Moreover, MULTBET implements a minimum-variance strategy, in that it aims to reduce the inter-attribute correlations within groups to as near indeterminacy as is possible. Such a strategy, if it also clusters intensely, has a remarkable power of overriding and detecting minor errors of observation in the data. This then is the type of strategy that a biologist usually wants. Usually, but not always; but, knowing this, we are in a relatively strong position to ask the biologist the right questions, and to advise him on what program he should use. And there is a bonus; by a study of the failures and successes of our various programs, we are learning more about the functioning of the human mind.

References

Gasking, D. (1960). Clusters. *Australas. J. Philos.* **38**, 1–36.

Greig-Smith, P. (1961). Data on pattern within plant communities. I. The analysis of pattern. *J. Ecol.* **49**, 695–702.

Jardine, N., and Sibson, R. (1968). The construction of hierarchic and non-hierarchic classifications. *Comput. J.* **11**, 177–84.

MacKay, D. M. (1969). Recognition and action. In 'Methodologies of Pattern Recognition'. (Ed. S. Watanabe.) pp. 409–16. (Academic Press: London.)

Macnaughton-Smith, P. (1965). Some statistical and other numerical techniques for classifying individuals. Home Office Res. Rep. No. 6. (H.M.S.O.: London.)

Williams, W. T., and Gillard, P. (1970). Pattern analysis of a grazing experiment. *Aust. J. Agric. Res.* **22**, 245–60.

Zipf, G. K. (1949). 'Human Behaviour'. (Addison–Wesley Press: Cambridge, Mass.)

18

Pattern Analysis and Statistics

W. T. Williams

The Nature of Pattern Analysis

Pattern analysis is invariably aimed at the simplification of data; or, perhaps more accurately, at the efficient ordering of data. It is not, or should not be, concerned at all with probability. Unfortunately, earlier workers appear to have felt that the imposition of a significance test would bring their methods closer to those of conventional statistics, and would confer an air of respectability on methods which were at one time viewed with suspicion. There is reason to believe that almost all such tests—and there are many in the literature—are invalid (e.g. Bottomley 1971). But even if some could be marginally validated, they are certainly irrelevant. A numerical classification does not claim to demonstrate that there *are* discontinuities in the system under study; it demonstrates only that *if* discontinuities are present in these data when constrained to this numerical model, then this is where they are to be sought. A similar situation applies to ordination. In fact, its non-probabilistic nature is one of the strengths of pattern analysis, in that it is completely non-parametric. This is not to say that distributions do not matter; but when they matter, it is for quite different reasons from those of statistics. For example, the Canberra metric is insensitive to out-lying values, but has a singularity at zero; the Euclidean metric has no zero-problems, but is unduly sensitive to outliers. Information statistics are free from both difficulties, but are weak in their treatment of numerical attributes. Such considerations affect the choice of program to be used but they cannot invalidate results.

Confusion sometimes arises because certain techniques can be used in different contexts. Principal component analysis can be used solely as a pattern-finding technique, in which case it is completely distribution-free; but it can also be used for conventional statistical purposes, in which case the data should be substantially multivariate normal.

130

Confusion can also result from historical accidents. Multiple predictive analysis invariably appears in a pattern-analysis context, simply because it originally rose out of an extension of extrinsic classification; but it is a pure statistical technique, with a meaningful null hypothesis and a robust significance test—it is in fact the analogue, for discontinuous data, of the analysis of covariance. It is therefore important to be very clear as to when one should use pattern analysis and when statistics; it is even more important to consider those occasions when it might be appropriate to use both.

The Uses of Pattern Analysis

We shall distinguish five situations in which pattern analysis may be used, alone or in conjunction with standard statistical techniques. (There is, of course, a sixth case, that in which a conventional statistical analysis is self-sufficient and no other analysis is required, but that is not the concern of this book.)

Case 1: *The non-statistical case.* The most familiar example of this is the classification of living organisms, when the elements for classification are already-named taxa—species, genera, tribes or even families. In this situation there is no meaningful null hypothesis to test; the elements are already known to be different, and the information required concerns only the pattern and extent of the differences. Moreover, the population under study may well be finite; it may consist of all species of a given genus or family, or all genera of a given family or tribe. A statistician may well be puzzled in such a case as to the precise physical nature of the elements to be classified. If genera are to be classified, it is likely that some at least of the genera will contain a number of species; each genus is therefore associated with a range of variation. The standard practice in such cases is to select a single organism, animal or plant, which belongs to a given genus, and to record the attributes of this organism. The element is then not a sample in the statistical sense, but a *representative*; it stands in the same relation to its genus as a neotype does to its species. The method of selecting the representative may, however, itself present problems, some of which are briefly discussed by Clifford and Williams (1973).

A similar situation arises in terrestrial plant ecology, where a floristic survey is carried out to investigate the possibility of environmental discontinuities. Quadrats are set out over the area (in a grid, not at random) and the presence or absence, or sometimes the quantity, of all vascular plant species is recorded. Again, although the quadrats bear some resemblance to statistical samples, they are not; they are being used simply as representatives of the area around them. Cases such as these are no longer controversial; it is recognized that probability, and therefore statistics, do not apply.

Case 2: *The pseudo-statistical case and the prior search.* The most familiar example of this case arises from sward trials of plant introduction material. In a particular example (Burt *et al.* 1974) 27 accessions of *Stylosanthes* were grown at 8 climatically distinct sites over a period of 3 years; the swards were replicated and randomized in the usual manner. At first sight, this appears to be a conventional $27 \times 8 \times 3$ factorial experiment with replication, probably using dry-matter yield as a variate. There is, of course, no difficulty in carrying out an analysis of variance on such results; but the statistical validity of such a procedure is doubtful. The accessions are mostly known to be different, since they often include several different species; and the sites have been deliberately chosen to be as climatically diverse as possible. But an analysis of variance is in effect testing the null hypothesis that every accession–site result could have been obtained as a random sample from a single parent population. This is tantamount to estimating the probability of the truth of a hypothesis known in advance to be false. Again, all that is really required is a concise summary of the pattern and extent of such yield differences as exist, and for this purpose pattern analysis is indicated. The purpose of the replicates in this situation is not to provide an estimate of stochastic variation, but to act as a safeguard against the possibility that accession performance will be confounded with micro-site differences; the pattern analysis (usually a classification) would then be carried out on the means over all replicates.

It is the next phase of the analysis that rather easily becomes contentious. The pattern analysis will almost invariably show that only a very small number of the possible comparisons are of agronomic interest; and since it is at this stage that accessions may be considered for pre-release, it is naturally desirable to test these comparisons as rigorously as possible. There is, of course, no statistical objection to designing a further experiment to test these particular comparisons; but then at least one year, and probably three years, will be lost. It is therefore extremely tempting to regard the pattern analysis as a prior search procedure, to recreate the separate replicate results, and to carry out statistical analyses on the comparisons indicated by the search. If a variance ratio obtained in this way is regarded simply as a descriptive measure of the relative magnitudes of accession and stochastic variation, no harm is done; the difficulty arises when it is desired to assign a level of significance to the ratio, since the mere success of the search procedure ensures that any resulting estimate is biased. As a result, statisticians dislike this procedure, and there is no doubt that if it is adopted, the results must be interpreted with considerable caution. We may suspect, however, that most agronomists would prefer to sacrifice some degree of statistical rigour in exchange for a considerable saving in time.

Some comfort may be obtained from the fact that all existing one-parameter classificatory models are minimum-variance models. It follows that if two species which it is desired to compare came from the same classificatory group, the difference between the dry-matter means is a conservative, not a random, estimate; a difference which nevertheless achieves significance is therefore reasonably reliable. In contrast, if the two species came from different classificatory groups, the difference will be biased upwards, and it is failure to achieve significance which would have meaning. There is undoubtedly need for research in this field to devise a means of estimating the probability corresponding to a given variance ratio, given the prior information that a particular classificatory structure has been imposed on the variates.

Case 3: *Pattern analysis as a prior transformation.* We first consider the multivariate case. If an experiment is expensive to set up it is natural to measure as many outcomes as may be possible. In such a case one is confronted with a set of variates, which may be highly inter-correlated. Provided all the variables are continuous and not excessively abnormal in their distributions, this case presents no difficulty in conventional statistics. If the system is not over-defined, a multivariate analysis of variance can be undertaken; and in any case the system can be reduced by principal component analysis and this followed by a set of univariate analyses on the resulting components. However, the existence of mixed-data pattern analysis permits an extension of this technique. Even if some of the attributes recorded are nominal, and even if a proportion of observations is missing, the individuals whose analysis is desired can always be mapped into a reduced Euclidean space by principal coordinate analysis, and the resulting Gower vectors submitted to analyses of variance. It is true that in such a case the distribution of the entries on the coordinates will not be known, so that the ensuing significance tests must be interpreted with caution; but at least *some* analysis is now possible for what was formerly a completely intractable case.

A more important use of pattern analysis as a prior transformation is the *sequential* case. Many agricultural experiments produce results in the form of sequences of, for example, cumulative liveweight gains, and the analysis of such sequences has always been a difficult problem. In earlier work all that could be done was to analyse totals at intervals—say annually, or at points of particular interest on the growth curves. The first true sequential analysis known to us was that due to K. P. Haydock, in Yates *et al.* (1964), who fitted quartic regressions to the sequences and then carried out analyses on the successive regression coefficients. Despite its success, this strategy has attracted few imitators; this may be because high-order coefficients may be difficult to associate with any specific biological interpretation or because their calculation can rather easily be ill conditioned. Dale *et al.* (1970) have recently suggested an alternative and simpler solution. The sequences are 'chopped' into a number of discrete states; each sequence of such states now serves to define a one-step transition matrix; each such matrix can be associated with an information content, and each pair of matrices with an information gain. These information gains can, if that is what is desired, be classified. More relevant in the present context is that they can also be mapped into Euclidean space by principal coordinate analysis, and the resulting Gower vectors analysed as before. The strategy is often surprisingly effective; and one of its most interesting properties is that the original main effects tend to separate on different vectors. The algebraic reason for this is unknown, but it presents a further advantage. Despite an initial soil survey, an experimental site may be later found to be unexpectedly heterogeneous, and this heterogeneity may be strongly patterned. This in its turn may produce an uncomfortably large error sum of squares, which may swamp the main effects. However, such a pattern is often partitioned out by the sequence-mapping process, and then occupies a vector of its own. In such a case its nature can often be quickly identified, and the main effects compared against error terms from which this pattern has been removed.

The procedure is not limited to sequences of a single variable. Pasture composition, in terms of the quantities of a number of volunteer and sown species, may have been recorded at, say, monthly intervals. If there are q paddocks, in which s species have been recorded on t occasions, the qt paddock–occasions are classified with respect to the s species. This will provide a small number of classificatory groups, which are used as successive states for the generation of transition matrices; the process then continues as before.

Case 4 : *The posterior search*. In some types of experiment—mineral nutrition is a notorious example—an apparently innocent analysis of variance may sometimes demonstrate the existence of a number of significant high-order interactions. These are always difficult to interpret; and the difficulty is increased if more than one variate has been analysed and the interaction pattern is different for the different variates. Standard statistical procedures can be used to elucidate such a situation by means of a careful examination of all effects and residuals; but this is laborious and time-consuming, and may still be relatively unprofitable if the system is multivariate and complex. In such a case a pattern analysis of all means over all variates can be extremely helpful, in that it may immediately expose the overall configuration of the data and disclose those areas which are primarily responsible for the interactions. Here pattern analysis is being used to elucidate a complex statistical analysis, a descriptive technique for displaying the statistical results more efficiently. An example is given by Gates *et al.* (1971).

Case 5: *The rescue operation.* Most biometricians at some time find themselves confronted by the well-intentioned experiment in which almost everything has gone wrong. Some animals have died, some treatments have 'crashed', some paddocks may have been flooded, some unsuspected soil discontinuity may have imposed a strong but alien pattern on the results; the design may now have become wildly non-orthogonal, with some levels of some treatments missing altogether, and the results may manifestly bear little relation to the imposed treatments. Statistical analysis is then usually unprofitable, since the unwanted variation will be absorbed in the error sum of squares, which will probably dominate the analysis. It does not follow that the results should be discarded. The pattern of the unwanted variation may itself be meaningful and useful, in that it may assist in avoiding similar disasters in the future. In fortunate cases it may even be possible, by the use of principal coordinate analysis, to partition out the unwanted variation and to demonstrate that some element of the treatment effects sought is in fact present.

The Administrative Problem

In its early days, statistics was certainly concerned with the efficient ordering of data, and could without difficulty have included pattern-analysis techniques within its boundaries. As statistics developed, however, it became increasingly concerned with distribution theory and the estimation of probabilities; and it was not long before all non-probabilistic techniques, from curve-fitting to classification, were excluded from the formal study of statistics. We may regret this; but the distinction is probably a useful one and it is in any case too late to reverse the process now. It has, however, had the unfortunate result that pattern-analysis expertise is now almost confined to biologists. This chapter has been concerned to suggest that the two sets of techniques can be used to complement each other; but it is difficult to arrange a marriage between two techniques when their respective practitioners are isolated from each other in their training.

Training itself presents a difficulty, since there are as yet no suitable general pattern-analysis text-books and virtually no formal courses. Still, half a century ago, statistics was in a very similar situation, in that many statisticians learned their trade by apprenticeship rather than by formal training. Pattern analysis appears to be developing along similar lines.

References

Bottomley, J. (1971). Some statistical problems arising from the use of the information statistic in numerical classification. *J. Ecol.* **59**, 339–42.

Burt, R. L., Edye, L. A., Williams, W. T., Gillard, P., Grof, B., Page, M., Shaw, N. H., Williams, R. J., and Wilson, G. P. M. (1974). Small-sward testing of *Stylosanthes* in northern Australia: preliminary considerations. *Aust. J. Agric. Res.* **25**, 559–76.

Clifford, H. T., and Williams, W. T. (1973). Classificatory dendrograms and their interpretation. *Aust. J. Bot.* **21**, 151–62.

Dale, M. B., Macnaughton-Smith, P., Williams, W. T., and Lance, G. N. (1970). Numerical classification of sequences. *Aust. Comput. J.* **2**, 9–13.

Gates, C. T., Williams, W. T., and Court, R. D. (1971). Effect of droughting and chilling on maturation and chemical composition of Townsville stylo (*Stylosanthes humilis*). *Aust. J. Agric. Res.* **22**, 369–82.

Yates, J. J., Edye, L. A., Davies, J. G., and Haydock, K. P. (1964). Animal production from a *Sorghum almum* pasture in south-east Queensland. *Aust. J. Exp. Agric. Anim. Husb.* **4**, 326–35.

PART 3

Applications of Pattern Analysis

Case 3.1. Description of Introduced *Stylosanthes* Material using Morphological and Performance Attributes

In the search for better pasture legumes for northern Australia the genus *Stylosanthes* merits special attention. Not only does it contain two species of known pastoral value (*S. humilis* and *S. guyanensis*) but both these species have yielded cultivars that have given the species a wide geographical range. Because of this potential a large collection of *Stylosanthes* material, comprising some 154 accessions, had been amassed and was available for study. Several species other than those already mentioned were present in the collection. Some of these were held to have value in the countries from which they were collected, but little or nothing was known of the growth characteristics of others. *Stylosanthes* is, in fact, a 'difficult' genus taxonomically, and it has only recently been possible to identify some of the accessions in the collection. The initial work was aimed at describing the material in such a way that the descriptions would be useful for plant introduction and evaluation purposes, and for persons working in related disciplines. As usual in this type of work, seed supply was limiting; a spaced-plant experiment was all that was possible. Details of the experiment are given in Burt *et al.* (1971).

Selection of attributes

The material to be evaluated was collected from a wide range of climatic types and geographical locations; it could be of potential value over an equally wide range. The decision was therefore taken to evaluate at three climatically diverse sites: South Johnstone (wet tropics), 'Lansdown' (dry tropics) and Samford (subtropics). After discussion between the persons involved, a list of what we have called 'performance attributes' was drawn up. This list, consisting entirely of numeric attributes, reflected the experience and intuition of those involved in deciding which characteristics were likely to be of value; it must be recognized that at this early stage the list was inevitably based mainly on knowledge of *S. humilis* and *S. guyanensis*.

Another early decision was to record, mainly from pressed specimens, the type of information that agronomists often use to distinguish individual genotypes: stem hairiness, viscidity, length of stipules, greenness of leaf etc. These were called 'morphological attributes'. As we had no *a priori* information as to which characteristics were likely to be of value a wide range of data was recorded (Table 3.1.1). Many of these characteristics were gradings (e.g. length of beak on the pod) and so were ordered multistates. Others were disordered multistates, e.g. leaf shape. Yet others were qualitative, for instance leaf colour green or blue/green.

Table 3.1.1. Data recorded at South Johnstone and Lansdown (agronomic) and Lansdown alone (morphological)

Data recorded from: H, herbarium specimen; L, live specimen

Character	Observation
MORPHOLOGICAL	
1. *Floral Characters*	
(H) Spike shape (Q.4)[A]	(0) Extended, (1) compact
(H) No. of spikes per axil (Q.5)	(0) Single, (1) multiple
(H) Upper articulation indument (M.7)	(1) Absent, (2) short, (3) long
(H) Upper articulation colour, brown or grey (M.6)	(1) Pale, (2) medium, (3) dark
(H) Beak length relative to upper articulation length = 1 (M.8)	(1) $<\frac{1}{2}$, (2) $>\frac{1}{2}$ <1, (3) $= 1$, (4) >1
(H) Beak thickness (M.9)	(1) Slender, (2) medium, (3) thick
(H) Beak curl (M.10)	(1) Coiled, (2) medium, (3) straight
(H) Seed colour (M.11)	(1) Pale, (2) yellow to dark brown, (3) black
(H) Seed length (M.12)	(1) <2 mm, (2) $>2-<3$ mm, (3) $>3-<4$ mm
(H) Seed shape (M.1)	(1) Elongate, (2) normal, (3) square
2. *Stem and Leaf Characters*	
(H) Viscosity of stems determined by presence/absence of exudate at tip of bristles (M.14)	(1) Viscid, (2) slightly viscid, (3) non-viscid
(L) Stem colour (Q.2)	(0) Dark, (1) pale
(L) Stem thickness at 5 cm above ground level (M.13)	(1) <3 mm, (2) $>3-<6$ mm, (3) $>6-<9$ mm, (4) >9 mm
(H) Stem hairs (Q.7)	(0) Absent, (1) present
(H) Stem hairs, if present (Q.8)	(0) Straight, (1) curled
(H) Stem hair distribution (M.4)	(1) All over, (2) one side, (3) widely scattered
(H) Stem hair density (M.15)	(1) Very sparse, (2) sparse, (3) medium, (4) dense
(H) Stem hair length (M.16)	(1) <1 mm, (2) $>1-<2$ mm, (3) $>2-<3$ mm
(H) Stem bristles (Q.6)	(0) Absent, (1) present
(H) Stem bristles, distribution (M.5)	(1) All over, (2) one side, (3) just below the node, (4) top $\frac{1}{2}-\frac{3}{4}$ of internode, (5) present on young but absent on old stems, (6) widely scattered, (7) present on old but absent on young stems
(H) Stem bristles, density (M.17)	(1) Very sparse, (2) sparse, (3) medium, (4) dense
(H) Stem bristles, length (M.18)	(1) <1 mm, (2) $>1-<2$ mm, (3) $>2-<3$ mm, (4) $>3-<4$ mm
(H) Stipule sheath, actual length (mm) (N.12)	

Table 3.1.1 (*Continued*)

Character	Observation
MORPHOLOGICAL (*Continued*)	
(H) Stipule horns, actual length (mm) from the centre top of the stipule sheath (N.11)	
(H) Distance between base of centre leaflet to junction point of the two outer leaflets (M.19)	(1) $<$1 mm, (2) $>$1$-$$<$2 mm, (3) $>$2$-$$<$3 mm, (4) $>$3$-$$<$4 mm, (5) $>$4$-$$<$5 mm
(H) Centre leaflet length (M.20)	(1) $<$1 cm, (2) $>$1$-$$<$2 cm, (3) $>$2$-$$<$3 cm, (4) $>$3$-$$<$4 cm, (5) $>$4$-$$<$5 cm
(L) Leaf colour (Q.3)	(0) Blue-green, (1) green
(H) Leaf shape as defined by Clapham *et al.* (1952) (M.3)	(1) Linear, (2) ensiform, (3) lanceolate, apex acute, (4) oblong, apex mucronate, (5) elliptic
AGRONOMIC	
1. *Branching Habits*	
(L) Mean no. of branches originating from the main stem at or below 5 cm (N.7), June 1967	
(L) Mean crown branch density (N.8)[B] and (M.21), May 1968	(1) Very poor, (2) poor, (3) fair, (4) good, (5) very good, (6) excellent
(L) Mean growth habit (M.2)	(1) Erect, (2) erect ascending, (3) semi-erect, (4) prostrate, (5) procumbent, (6) bush
2. *Flowering Habits*	
(L) Flowering onset, mean day length[C] in minutes (N.1)	
(L) Flowering onset, range in minutes between the first and last plant to flower (N.2)	
(L) Flowering onset, range in days between the first and last plant to flower (N.3)	
(L) Flowering onset, mean no. of days from Dec. 1, 1966 to flowering (N.10)	
3. *Performance as Spaced Plants*	
(L) Mean dry matter yield (g) from one plant in June 1967 (N.4)	
(L) Mean dry matter yield (g) from all surviving plants in May 1968 (N.5)	
(L) Mean seed yield (g) from one plant 8–12 weeks after onset of flowering (N.6)	

Table 3.1.1 (*Continued*)

Character	Observation
AGRONOMIC (*Continued*)	
(L) Mean survival (%) of plants on Dec. 31, 1967 after being cut once in June 1967 for dry matter yield (N.9)	
(L) Perenniality (Q.1)	(0) Annual, (1) perennial

[A] Q, qualitative attributes numbered 1 to 8. N, numerical attributes numbered 1 to 12. M, multistate attributes numbered 1 to 21.

[B] Included as both a numerical (N.8) and multistate (M.21) attribute.

[C] Day length (sunrise to sunset) in which plants commenced flowering.

Although many of the morphological characteristics used by taxonomists were not measured (for example one of the key characters in *Stylosanthes* taxonomy, the presence or absence of an axis rudiment, was not recorded), some taxonomic bias was inevitable. This was desirable in that taxonomic units are a meaningful method of recognition and communication. Taxonomy also has a useful predictive element; for instance, some species tend to be poisonous. Taxonomy alone, however, was insufficient for our purpose for several reasons, viz:

(1) Taxonomic classifications are often based on minute floral characteristics that are difficult to measure, particularly in the field.

(2) Many taxonomic classifications employ dichotomous keys. Plants similar over a wide range of characteristics may then be widely separated in the classification because of some earlier separation, possibly on the basis of a single minute and agronomically unimportant characteristic. Once again, presence or absence of an axis rudiment in *Stylosanthes* may be quoted.

(3) Species and performance limits seldom coincide. First, we have instances of 'wide' species which contain many distinct forms. In *S. guyanensis* such forms differ in climatic adaptation, resistance to frost, resistance to flooding, day-length adaptation, rhizobial response etc. It seems possible that some of these forms would be more similar agronomically to other species than they would be to other forms of *S. guyanensis*. Secondly, some species may be so closely defined that we may fail to recognize their similarity to very similar species. Clayton (1965), for instance, states '*Digitaria decumbens* . . . is probably no more than a single strain selection

from *D. pentzeii*. Its widespread acceptance as a species disguises the fact that even better forage strains might be available in the parent material'. Susceptibility of *D. decumbens* (pangola grass) to a now common virus disease has since emphasized the dangers of this type of neglect.

We thus needed to describe the collection on the basis of performance plus morphological characteristics. We had three prerequisites. First, we needed to be able to talk about a 'type' or 'form' of plant, i.e. we wanted to be able sharply to define groups of similar plants with obvious overall similarities. Secondly, within each plant type we needed to be aware of variation in characters or groups of characters; for instance, variations in flowering times. Finally, we needed to know something of the overall similarities and differences between the various forms.

Analysis of data

In the early work (Burt *et al.* 1970) 'Andersonian plots' were used. Accessions were intuitively separated into morphological 'types' within a species, each 'type' being represented by a given colour. Yield and flowering time were then plotted against each other and various agronomic data superimposed (Fig. 3.1.1). This method provided reasonable clustering but the amount of data that could be used was very limited, the position of various nonconformist individuals could not be resolved and the intuitive choices were debatable. It was therefore decided to have recourse to numerical methods of classification.

Four sets of data were available: the morphological set and the three sets of performance data. Although it would have been possible to analyse all the information together, this approach was rejected as it was felt it would heavily bias the results towards the performance characteristics. We decided initially on a combination of the morphological and 'Lansdown' performance data, the latter being the most complete of the three performance sets.

At the time at which this work was carried out there was some uncertainty as to which of the mixed-data classificatory procedures would be the most suitable. Three analyses were therefore carried out: these used the information-statistic (I.S.) program MULTBET, the Euclidean 'incremental sum of squares' (I.S.S.) and the 'mean squared distance' (M.S.D.) strategies (this is a group-average fusion of Euclidean distance measures). The last two used the program MULCLAS. The first two strategies are space-dilating, the M.S.D. strategy is nearly space-conserving.

The M.S.D. analysis produced marked 'chaining', the 'nonconformist' accessions being rapidly delineated and placed in small groups (Fig. 3.1.2). The first group, for example, contained a single introduction which failed to flower although otherwise similar to other accessions.

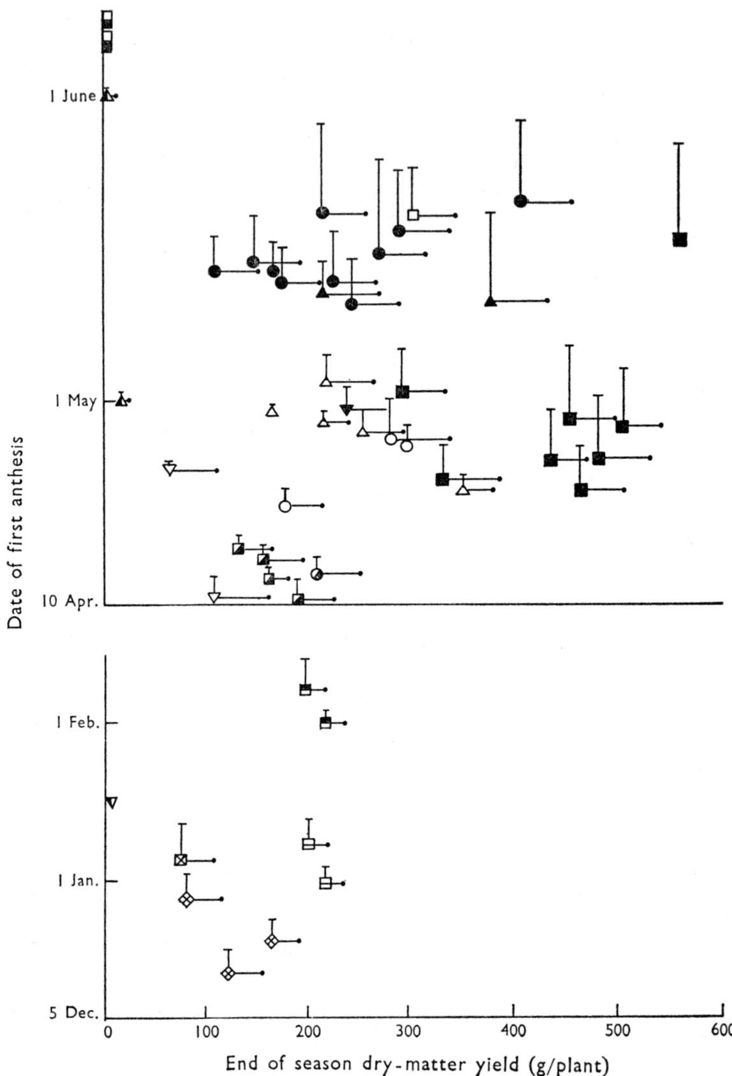

Fig. 3.1.1. The relationship between time of first anthesis, end of season yield (g/plant), seed yield and crown type for selected genotypes of *S. guyanensis*. Seed yield represented by vertical bars; crown type represented by horizontal bars, the length of the line increasing with increasing denseness of crown. Symbols refer to various plant types, e.g. solid triangles refer to plants similar to the commercial stylos (closed circles) but with a denser crown.

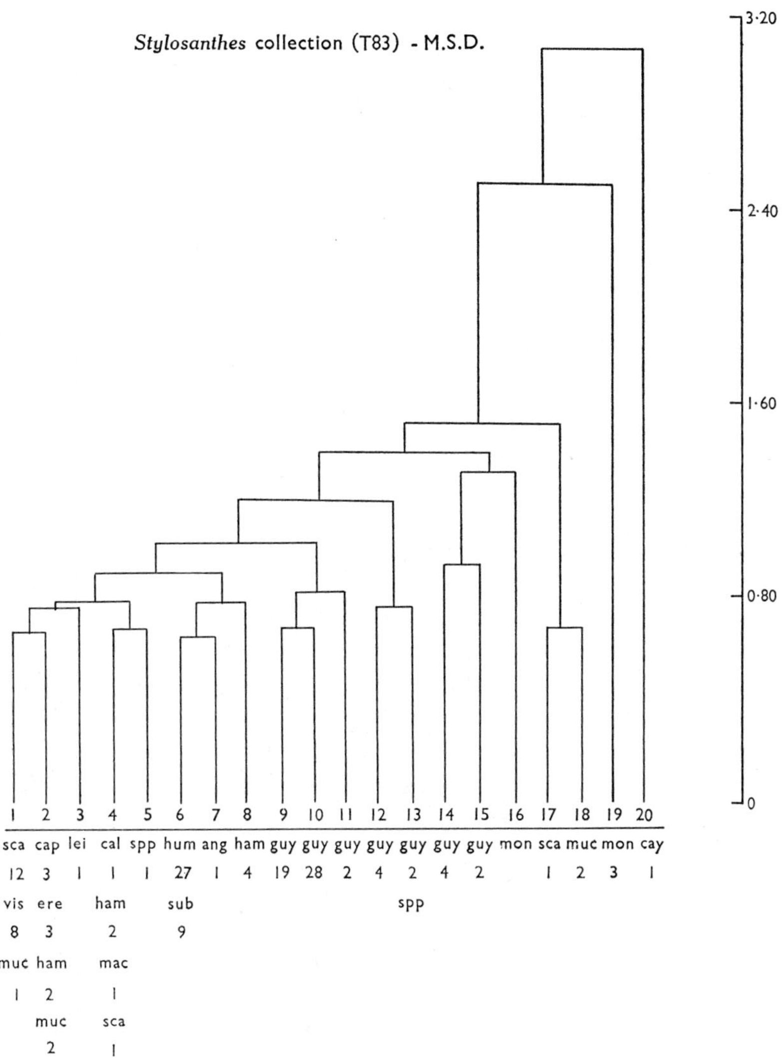

Fig. 3.1.2. Mean-squared distance classification.

The larger groups contained unacceptable mixtures of species and species forms. The I.S.S. analysis produced groupings somewhat similar to the I.S.; the hierarchy, however, dichotomized at lower levels and distinctions between groupings were less well defined.

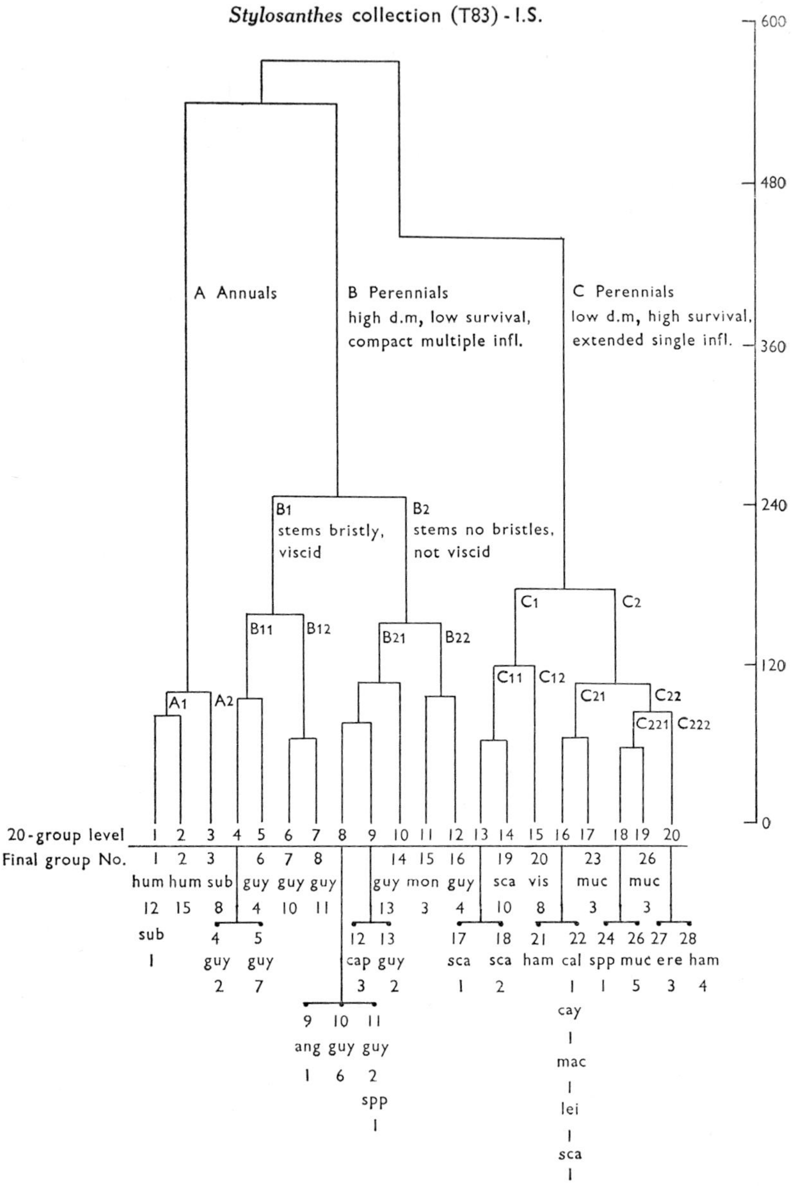

Fig. 3.1.3. Information-statistic classification.

The procedure we adopted, therefore, was to accept the I.S. hierarchy down to the 20-group level, at which rapid fragmentation occurred (Fig. 3.1.3), and to reallocate and split groups on the basis of the other two analyses.

Reallocation and splitting

The results of the various classifications were compared at the 20-group level, using two-way tables. In some instances, there was evidence for misclassification. The I.S.S. and M.S.D. analyses, for instance, placed accession 133 with the accessions found in group 15 (Table 3.1.2); I.S.

Table 3.1.2. Constitution of groups before reallocation

Group No.[A]	Treatment numbers	Species
1	68[B]	*S. subsericea*
	69, 70, 77, 78, 79, 80, 93, 94, 95, 96, 97, 98	*S. humilis*
2	71, 72, 73, 74, 75, 76, 84, 85, 87, 88, 89, 90, 91, 92, 145	*S. humilis*
3	81, 122, 123, 124, 125, 127, 129	*S. subsericea*
4[C]	18, 37, 38, 42, 43, 52, 59, 62, 149	*S. guyanensis*
5	19, 20, 21, 100	*S. guyanensis*
6	10, 13, 17, 22, 24, 30, 32, 33, 142, 143	*S. guyanensis*
7	12, 26, 27, 28, 34, 46, 54, 55, 56, 57, 58, 148[B]	*S. guyanensis*
8[C]	1	*S. angustifolia*
	11, 14, 16, 23, 25, 45, 51, 61	*S. guyanensis*
9[C]	3, 4, 5	*S. capitata*
	39, 146	*S. guyanensis*
10	15, 29, 31, 35, 36, 40, 41, 44, 47, 50, 60, 144	*S. guyanensis*
11	101, 102, 103	*S. montevidensis*
	104[B]	*S. guyanensis*
12	48, 49, 53, 147	*S. guyanensis*
13[C]	140, 141, 151	*S. scabra*
14	116, 117, 118, 119, 120, 121, 130, 131, 138, 139	*S. scabra*
15	132, 134, 135, 136, 137, 152, 154	*S. viscosa*
16[C]	2	*S. calcicola*
	6	*S. cayennensis*
	99	*S. macrocarpa*
	150	*S. leiocarpa*
	153	*S. scabra*
	64, 65, 66, 67	*S. hamata*
17	111, 112, 113	*S. mucronata*
	133[B]	*S. viscosa*
18[C]	106, 107, 108, 109	*S. mucronata*
	126	*S.* species
19	110, 114, 115	*S. mucronata*
20[C]	7, 8, 9	*S. erecta*
	63, 82, 83, 86	*S. hamata*
	105[B]	*S. mucronata*
	128[B]	*S. subsericea*

[A] See Fig. 3.1.1.
[B] Treatments subsequently reallocated.
[C] Initial groups subsequently split (see Fig. 3.1.1).

placed it in group 17. Accession 133 (*S. viscosa*) obviously did not belong in group 17, all members of which are *S. fruticosa*, and was more correctly placed with other *S. viscosa* accessions in group 15; it was so placed. Less obvious reasons for reallocation also occurred. Both the I.S.S. and M.S.D. analyses placed treatment 104 with group 8 (*S. guyanensis*) accessions rather than in group 11 (*S. montevidensis*); in many characteristics groups 8 and 11 were similar and accession 104 somewhat intermediate. Examination of herbarium specimens and climatic background information showed that accession 104 is taxonomically *S. guyanensis* from a rather extreme altitude/latitude situation. It was thus reallocated to group 8.

Splitting of heterogeneous groups was also carried out. Group 9, for instance, fragmented into *S. capitata* and *S. guyanensis* and the legality of this dichotomy was recognized. Where divisions were less secure, as in group 20 (*S. viscosa*), subgroups were recognized and designated by serial letters (A, B etc.). In other cases, particularly where single accessions were involved, nonconformity was recognized and delineated but splitting was not stipulated; accession 82 (*S. hamata*) was obviously different from other members of group 28 and was so described. The introduction of similar material could, of course, justify group delimitation in these instances. Finally, where a group was clearly delineated from other groups but was heterogeneous and could not sensibly be split, heterogeneity was simply recognized. *S. viscosa*, groups 20A and 20B, is of this type.

External information, i.e. information not used in the analyses, was frequently used in reallocation and splitting procedures. This could be climatic, as in one of the examples already mentioned, or morphological or agronomic. A few cases arose in which an accession appeared to possess the morphological characteristics of one group but the agronomic characteristics of another. Conflict of purpose is always likely to arise in these final stages of group delineation. To the agronomist, a promising group might warrant subdivision on the basis of even individual characteristics, such as flowering time. Further introductions might, however, show that there was continuous variation within the original group and that subgrouping would have been purely arbitrary. It has to be accepted that agglomerative hierarchical strategies, in which the position of an individual is fixed by an early and unredeemable fusion, usually give rise to the need for reallocation. Later experience has shown that the need for reallocation is greatly reduced if a divisive program with continuous reallocation (as in REMUL) is used

Attribute contributions

It was of particular interest to determine the extent to which performance attributes had contributed to the classification. The program GROUPER was used to do this.

Without going into detail it can be stated that performance attributes were of importance at all levels in the classification, reinforcing morphological data. At the highest dichotomy, for instance, groups B and C (Fig. 3.1.3) differed in both inflorescence structure and dry-matter yield. At the lowest levels, we find that leaflet length, stem thickness, crown branch density and flowering time were used to distinguish groups 10A and 10B.

Further analyses

Two further analyses were carried out. In the first of these we analysed morphological data only, the aim being to explore further the implications of the use of performance data. In the second the 'Lansdown' performance data were replaced with those of South Johnstone; this helped us to assess the extent to which M–A classifications are sensitive to location.

Comparison of the *morphological-only* and M–A classifications revealed several differences. The morphology-only classification improved on the M–A in that *S. angustifolia* and *S. capitata* left the *S. guyanensis* complex and joined *S. humilis* and *S. fruticosa* respectively; this was due to close similarities in flower and pod characteristics. Individual accessions were, however, often badly placed; for instance accession 151 (*S. scabra*) was transferred to an *S. fruticosa* group. Distinctions of high agronomic importance were also lost; groups 6 and 7, for instance, were amalgamated.

For the purposes mentioned initially the inclusion of performance data is desirable.

Replacement of the 'Lansdown' performance data set with that from South Johnstone produced relatively few changes in the accession groupings; these have been detailed previously (Burt *et al*. 1971). The values for the individual attributes, however, changed markedly. Provided that the accessions tested grow reasonably well, M–A groupings are not very sensitive to location.

Advantages and disadvantages of the classification

Some immediate advantages are obvious to those using the techniques described; paradoxically they are often related to the lack of reliable significance tests and the then absence of reallocation systems. This necessitates a detailed, often soul-searching exploration of the basis of the dichotomies and the structure of the resultant groups. Nonconformist individual and heterogeneous groups come under close surveillance, frequently with recourse to preserved or living specimens. In this way attention is focused on slight differences in performance and morphology. Similarly unexpected combinations of characters may be brought to light; differences in first-year/second-year yield were of importance in separating forms of *S. fruticosa*. The reality of such combinations must be evaluated, leading to the so-called 'hypothesis generation' function of numerical analysis.

For plant introduction purposes the delineation of the various species forms confers numerous benefits; it aids with the selection of accessions for future work, conveniently summarizes the large data mass and provides a framework to which data for later accessions can be added.

For other disciplines the classification provides a useful means of considering the whole genus, as represented by the collection. Selection of material for subsequent agronomic testing, rhizobial or mineral nutrition studies is facilitated. The data so collected can then be used as an outside source of information to clarify further the structure of the collection.

The main danger in this approach is that potential users may fail to recognize that the various groups are not rigidly defined entities. Each group is heterogeneous to some extent and some attributes, such as yield, may be continuous variables. In selecting accessions for further study this must be recognized.

R. L. BURT

References

Burt, R. L., Edye, L. A., Grof, B., and Williams, R. J. (1970). Assessing the agronomic potential of the genus *Stylosanthes* in Australia. Proc. XI Int. Grassl. Congr., Surfers Paradise, Australia, pp. 219–23.

Burt, R. L., Edye, L. A., Williams, W. T., Grof, B., and Nicholson, C. H. L. (1971). Numerical analysis of variation patterns in the genus *Stylosanthes* as an aid to plant introduction and assessment. *Aust. J. Agric. Res.* **22**, 737–57.

Clayton, W. D. (1965). Some aspects of grass taxonomy. Proc. IX Int. Grassl. Congr., Sao Paulo, Brazil, pp. 1331–3.

Case 3.2. Sward Performance of *Stylosanthes* Accessions over a Range of Climates

The previous case (3.1) reported an experiment in which 154 *Stylosanthes* accessions were grown under spaced-plant conditions in three different climates. Several accessions appeared to be promising under these conditions, but it was obviously desirable to ascertain how they would perform under sward conditions and under a wider range of climates. Twenty-seven genotypes were therefore selected for a further experiment; these comprised seven species and were selected, as far as possible, to represent the range of forms and species available (a list of the species concerned is included in Table 3.2.4). Each was sown as a pure sward with three replicate blocks. Dry-matter harvests were taken at 6-week intervals during the growing season, samples being separated into sown and volunteer species. Germination and perennial plant counts were also taken annually. The experiment was carried out at seven locations. Each of these is pastorally important and they represent a wide range of climatic types (Table 3.2.1). Attempts to

Table 3.2.1. The location and generalized climates of the sites used in the experiment

Sites	Lat. (°S.)	Long. (°E.)	Moisture index[A]		Mean rainfall (mm/yr)	Jan. temp. (°F)[B]		July temp. (°F) [B]	
			Summer	Winter		Mean max.	Mean min.	Mean max.	Mean min.
1. Darwin	12 .5°	131 .2°	0 .8	0 .0	1488	90	77	87	68
2. South Johnstone	18 .0°	146 .0°	0 .8	0 .6–0 .8	3259	88	72	75	58
3. Kangaroo Hills	18 .7°	145 .7°	0 .6–0 .8	0 .2–0 .4	640	92	70	79	49
4, 8. 'Lansdown'	19 .5°	146 .8°	0 .6–0 .8	0 .2–0 .4	860	88	76	76	60
5. Rodds Bay	24 .0°	151 .3°	0 .8	0 .4–0 .6	800	86	72	72	52
6. Samford	27 .5°	153 .0°	0 .8	0 .8	1050	86	69	69	49
7. Grafton	27 .9°	152 .9°	0 .8	0 .8	882	89	67	70	44

[A] After Fitzpatrick and Nix (1970).   Moisture index defined as the ratio of estimated actual to potential evapo-transpiration.
[B] Data presented are those recorded for the nearest centre (Bureau of Meteorology 1956).

standardize procedures were generally successful although in some cases it was not possible to run the experiment for the prescribed 3-year period. At one site ('Lansdown') the experiment was repeated, supplementary irrigation being used to ensure good establishment in the second sowing. The experiment has been reported in full by Burt *et al.* (1974).

Approach to data analysis

A very large volume of data was collected and this could be approached in several ways. First, it would have been possible to analyse the information for each of the sites independently; this would, however, make inter-site comparisons difficult. Secondly, it would have been possible to compare only the information from the final year of the testing. This was also rejected, not only because the vagaries of the environment might affect the final outcome, but also because we thought that yield trends with time could be of importance. The third possibility was to omit much of the detail and concentrate on defining which species or suites of species were most likely to be adapted to the various areas; a more detailed examination of the data for these genotypes could follow. This was the approach adopted.

Choice of attributes for analysis

For the purpose of this exercise we decided to use two basic attributes: the annual cumulative yields of the sown and of the invasive species. This information by itself, however, was not sufficient for our purpose; it may, for instance, fail to highlight the ability of a genotype to perform relatively well in 'bad' years when all yields are low. Rankings of both legumes and grasses, from 1 to 27 in decreasing order of yield, were therefore used. In this case, where a given rank could be associated with very different absolute yields at the different sites or in different years, there would be no necessary correlation between ranking and absolute yield. They could therefore be treated as independent variables.

Choice of strategy for analysis

Even with the reduced data input the complete data set was still large: 27 accessions × 8 sites × 3 years × 4 attributes × 3 replicates, or 7776 pieces of information. It was still difficult to decide what should be compared. As we were working with material for which we had no 'standard' genotypes, and which differed markedly in their production characteristics, we could not make comparisons of the type 'genotype A is better than standard B in environment C'.

The immediate problem was thus one of finding some pattern, or patterns, which would allow us mentally to encompass the results. We therefore needed a suitable classificatory procedure which would provide us with an objective summary of the major patterns of variation.

For this analysis we chose to work on replicate means; this reduced the size of the matrix and would make interpretation easier. As the matrix still contained a considerable amount of missing data it was decided to use the LINKED algorithm of Lance and Williams (1967) which is known to be appropriate for such information. In this, for each attribute, a measure (in this case the Canberra metric) is calculated over each year separately for which the necessary information is available; the measures are then meaned over those years for which they exist. The classifications were agglomerative using the flexible-fusion strategy in the program CLASS. Three analyses were carried out.

Analysis 1

In this analysis the system was regarded as (27 accessions $\times$ 8 sites), that is 216 site–accessions, defined by 4 performance attributes linked over 3 years. This establishes the performance of a given accession at a given site.

The dendrogram was truncated at the 25-group level for preliminary examination. We found that several of the resulting groups differed from their nearest neighbour by factors that we deemed to be minor; these were bulked leaving a total of 16 groups. Data within these groups were then examined and found to be sufficiently homogeneous for the present purpose.

The constitution of these groups, defined by sites and accessions, is shown in Table 3.2.2. The group means of the four attributes for each of the years are shown in Table 3.2.3. The order in which the groups appeared in

Table 3.2.2. Constitution of performance groups by accessions and sites
Entries represent site numbers

Species	Accession no.	1	2	3	4	5	6	7	8	9	10	11	12	13	14	15	16
S. guyanensis	1		2								1, 5, 6, 8		3	4, 7			
	2		2								5	6, 8		1, 4	3, 7		
	3		2						7	1	5, 6, 8		3		4		
	4			6							5, 8		3	4	1, 7		
	5				4					1	8, 5				3, 7		6
	6										5, 8	6	3		1, 4, 7		
	7		2								5, 8	6	3	7		1, 4	
	8		2	6	1						8		3		4, 5	7	
	9		2						1		6, 8		3		7	4	
	10		2	6							8		3		1, 4, 5, 7		
	11		2								5, 6, 8		3	7	1	4	
	12		2						4	1	6				3, 7		
	13										5, 8		3		1	4, 7	6
	14	6	2						7		5, 8		3	4	1		
	15		2								5, 8				1, 4		
S. hamata	16	5, 8	2				3, 4				6					1, 7	
S. humilis	17					5					8				1, 7	4	6
	18			1	3		5					8		4		7	6
	19					3					1, 8	5		4		7	6
	20				3						1, 6, 8	5		4, 7			
S. mucronata	21				3				4	8	5				7	1	6
	22			1	3						5, 6, 8			4		7	
S. scabra	23				3		1		4	8	5, 6					7	
	24			4	3					8	5	6		1, 7			
S. viscosa	25			1	3				4	8				7			
	26			1	3						5, 8	6		4, 7			
S. subsericea	27	5, 8					3				1			4, 7			6

Table 3.2.3. Mean attribute values of performance groups

Attribute	Year	Performance group															
		1	2	3	4	5	6	7	8	9	10	11	12	13	14	15	16
Legume yield	1	5268	1599	7587	119	66	576	807	21	442	1178	4530	125	76	51	130	277
(kg/ha)	2	7327	12241	—	1933	2996	3994	4512	599	2323	876	1724	1128	485	414	118	—
(1)	3	—	5528	—	2449	6583	6888	9787	947	1076	225	—	2898	297	507	259	—
Legume rating	1	1·6	6·5	3·0	16·4	17·7	2·0	2·7	18·9	20·8	14·4	6·2	11·9	9·9	19·3	5·6	22·6
(2)	2	1·5	6·5	—	6·6	8·0	1·0	3·0	7·3	6·8	16·7	11·0	18·7	8·1	16·5	16·4	—
	3	—	6·5	—	7·8	8·4	5·7	3·0	4·9	5·0	14·3	—	17·2	8·3	14·5	12·8	—
Weed yield	1	1675	0	5580	1013	164	272	782	607	1892	3989	3349	276	392	453	492	9329
(kg/ha)	2	1387	222	—	1425	2407	941	2804	2978	2766	3067	3096	3605	2523	3035	3435	—
(3)	3	—	1725	—	3013	9439	2273	10006	4488	418	371	—	10205	4575	4603	5269	—
Weed rating	1	25·4	—	17·3	4·2	17·7	18·3	16·3	5·9	18·4	10·8	22·4	9·3	15·8	19·0	10·2	4·6
(4)	2	22·8	6·5	—	19·6	19·7	25·0	15·0	4·1	16·3	13·1	15·3	6·8	17·4	12·8	8·2	—
	3	—	6·5	—	15·6	13·7	26·0	9·5	17·4	20·8	11·0	—	10·3	16·6	13·8	7·4	—

the initial computer output is without any practical significance; we have therefore arranged them in an order that we consider to be more convenient for our purposes.

Inspection of the group means revealed that we had four yield trends:

(*a*) Legume yield high in all years (Groups 1–3). Initial yields were high and tended to increase. Weed rankings were also high, indicating that the legume tended to suppress weed invasion. Legume rankings were low, the accessions included being among the highest-yielding of those tested. Total dry-matter production, legume and weed, was high; this type of performance was found in the higher-yielding environments. Reference to Table 3.2.2 shows that this type of performance was provided by *S. hamata* and *S. subsericea* at sites 5 and 8 (Rodds Bay and 'Lansdown' with irrigation) and by *S. guyanensis* at South Johnstone.

(*b*) Yield of legume initially low but increasing with time (Groups 4–8). This increase, as shown by the reduced rankings, was relatively greater than that of most of the accessions tested. Weed yield also increased but so did weed ranking, i.e. relatively the position of the legume had improved.

The position of group 9 poses some problems; legume yields decline in the third year although rankings improve (i.e. decrease). In Table 3.2.2 we find that this group contains information from only two sites, 1 and 8. There was no third-year information for site 8 and at site 1 the season was unusually harsh. We believe that, had the data for site 8 been available, the yield pattern would have been maintained. We have therefore tentatively placed group 9 with the 'improvers', 4–8.

It is noticeable that these groups relate mainly to tropical areas with relatively short growing seasons and virtually exclude reference to *S. guyanensis*. The build-up in yield may be attributed to a thickening up of the pasture or, in some instances, to increased yields per plant. The latter characteristic was noted to occur in several of the species found in these groups.

(*c*) Performance initially high but decreasing with time (groups 10 and 11). These may be interpreted as plants that establish well but are not adapted to the environment and die out. Group 12 is somewhat anomalous—legume yields increase but rankings fall. Reference to Table 3.2.2 shows that this group describes the performance of *S. guyanensis* at one site, Kangaroo Hills. Growing conditions improved throughout the successive years of the experiment, the final year being extraordinarily favourable. This explains this peculiar performance pattern and group 16 has been placed tentatively with 10 and 11.

(*d*) Performance uniformly poor (groups 13–16). Noticeably these groups contain a large number of entries from 4, 6 and 7, i.e. two subtropical sites and one dry tropical site that experienced abnormally harsh seasons. Failures obviously occurred with most species.

Preliminary discussion. Although discussion of this analysis will be largely deferred until a later section, comments on the use of these tables are probably in order. If we wish to look at the performance of a given accession we simply look along the relevant row in Table 3.2.2 and then refer to Table 3.2.3; *S. hamata*, for instance, is outstanding at sites 5 and 8, good at sites 3 and 4 (although taking longer to reach high yields) and poor at sites 1, 6 and 7. At the sites at which it is adapted it suppresses weed invasion. Alternatively we may wish to look at a given type of performance; we find, for instance, that *S. subsericea* performs in much the same way as *S. hamata* at sites 5 and 8 (column 1, Table 3.2.2) whereas *S scabra* tends to perform the same at site 1 as *S. hamata* at sites 3 and 4 (column 6, Table 3.2.2).

Analysis 2

Here we regarded the system as comprising 8 individuals (sites) defined by (27×4) attributes linked over 3 years. We were attempting to classify the various environments on the basis of the performance of the suites of accessions tested.

Although the classification was agglomerative we shall adopt the usual practice of reading the dendrogram downwards. GROUPER was used to examine the attributes responsible for each of the splits.

The first split (Fig. 3.2.1) separates site 2, South Johnstone, on the basis of high legume yields and initially low weed yields. This is perfectly reasonable, South Johnstone being the only 'wet tropical' site. The next split separates 5, 6 and 8 from 1, 3, 4 and 7. This was mainly on the basis of weed yields, which were high initially at 5, 6 and 8 but fell with time. The opposite was true at 1, 3, 4 and 7; the differential performance of two *S. humilis* accessions was also of importance. Site 6 was then separated from 5 and 8; performance of *S. hamata*, *S. subsericea* and some *S. humilis* accessions, which was poor at site 6 but good at 5 and 8, was the major factor involved. Finally, differential performance of some *S. humilis* accessions, together with the weed rankings associated with some *S. guyanensis* accessions, separated 5 and 8. The other side of the dichotomy showed splits based on similar considerations.

Preliminary considerations. From a climatic point of view the dendrogram exhibited some unexpected groupings. The similarity of sites 5 and 8 (Rodds Bay and 'Lansdown' + irrigation) is plausible in retrospect; both have short, hot, summer growing seasons and the difference in soil moisture regime in this period would be reduced by the use of irrigation. Noticeably the major differences between these sites centred around two day-length-sensitive species, *S. humilis* and *S. guyanensis*. This is again to be expected in view of their difference in latitude.

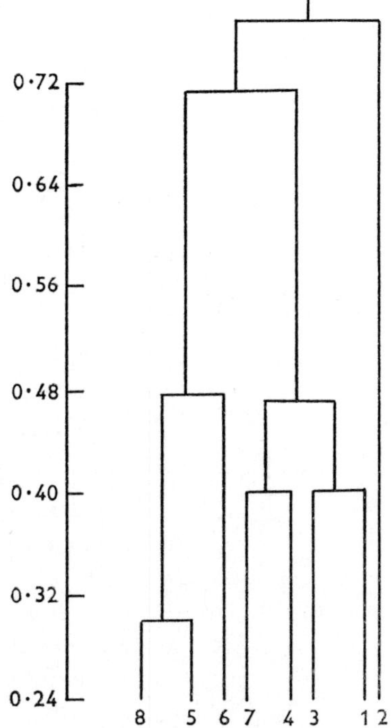

Fig. 3.2.1. Classification of sites.

Similarities between 4 ('Lansdown') and 7 (Grafton) were unexpected and are, in fact, unreal. The similarity was based on high failure rates at these centres; at Grafton this was almost certainly due to its subtropical position, together with a severe nutritional problem, whereas at 'Lansdown' failure is attributable to very severe drought conditions.

Similarities between 1 (Darwin) and 3 (Kangaroo Hills) were also unexpected. This was almost certainly due to the abnormally good years at Kangaroo Hills.

In addition to the very general use that we have made of this classification, to be presented in a later section, this type of analysis may have other uses. It could, for instance, be used to examine the importance of soil type in a given climatic region or to equate performance on different soils in different climates. Certainly it has, in the present instance, stimulated thought on the siting of testing centres and the possibility of manipulating one environment to simulate another.

Analysis 3

In this analysis we regarded the system as 27 individuals (accessions) defined by (8 × 4) attributes linked over 3 years. It provides a summary of accession performance over all sites.

Examination of the dendrogram showed that the nine-group level would be a convenient working level. At this level four groups consisted solely of *S. guyanensis*, one uni-membered group of *S. hamata* and two of *S. humilis* (to one of which the single accession of *S. subsericea* had been added). The remaining two groups consisted of *S. scabra* and *S. viscosa* or *S. fruticosa* and *S. viscosa*. These groupings, and the number of groups required to accommodate the various species, actually correspond relatively well with the results obtained from the spaced-plant experiment.

Delineation of the groups was based on a complex series of genotype × environment interrelationships. Even the most isolated of the groups, that containing the single accession of *S. hamata*, was distinguished by the data from at least five sites using all four attributes. In the context of this paper lengthy discussions of this analysis are not warranted and have therefore been omitted.

Consolidation of the analyses

The overall results of the experiment can be summarized very briefly in one table. To do this we constructed a two-way table (Table 3.2.4), the rows of which are defined by the accession groups of analysis 3 (i.e. by the overall similarities of the accessions) and the columns by the site groups of analysis 2, i.e. the overall similarities of the environments. Two criteria are then adopted to differentiate between 'good' and 'poor' performance: (1) the attainment of a yield of at least 1500 kg/ha by the third year and (2) inclusion in a performance group (analysis 1) with a legume ranking of less than 8. Those accessions which satisfied the first criterion were given a + sign, those which satisfied the second an ○ sign and those which did not, or were not planted, a full stop.

General discussion

For the purpose of this discussion we shall consider three aspects of the work that has been described. Firstly, what have we learnt about the ability of the various accessions or groups of accessions to perform satisfactorily in the different climates? Secondly, what have we learnt about the individual accessions *per se*? Finally, what about the methods employed?

Zones of adaptation. Considering firstly the subtropical areas, as represented by sites 6 and 7, we find that most of the species failed; *S. guyanensis* occasionally showed some promise. It is difficult, however, to be objective because of the lack of data from these sites.

CASE 3.2

Table 3.2.4. Summary of accession-in-site performance
(For symbols, see text)

Accession no. and species	Sites								M–A group[A]
	2	1	3	8	5	4	7	6	
17 *S. humilis*	.	.	.	.	+O	.	.	.	1 (A)
18 *S. humilis*	.	+O	+	.	+O	.	.	.	1 (A)
19 *S. humilis*	.	.	+O	.	.	.	.	.	1 (A)
20 *S. humilis*	.	.	+	.	.	.	.	.	1 (A)
27 *S. subsericea*	.	.	+O	+O	+O	.	.	.	3 (A)
21 *S. fruticosa*	.	.	+	+O	.	O	.	.	25 (C)
22 *S. fruticosa*	.	+O	+	.	.	.	.	.	26 (C)
25 *S. viscosa*	.	+O	+	+O	.	O	.	.	20 (C)
16 *S. hamata*	+O	.	+O	+O	+O	+O	.	.	28 (C)
23 *S. scabra*	.	+O	+O	+O	.	O	.	.	19 (C)
24 *S. scabra*	.	.	+	+O	.	+O	.	.	18 (C)
26 *S. viscosa*	.	+O	+	+O	+O	.	.	.	20 (C)
3 *S. guyanensis*	+O	+O	+	.	.	.	O	.	10 (B)
12 *S. guyanensis*	+O	+O	.	.	.	(O)	.	.	4 (B)
13 *S. guyanensis*	.	.	+	.	.	.	.	.	16 (B)
4 *S. guyanensis*	.	.	+	.	.	.	.	+O	5 (B)
6 *S. guyanensis*	.	.	+	.	.	.	.	.	8 (B)
7 *S. guyanensis*	+O	.	+	.	.	.	.	.	8 (B)
8 *S. guyanensis*	+O	+O	+	.	.	.	.	+O	14 (B)
10 *S. guyanensis*	+O	.	+	.	.	.	.	+O	8 (B)
14 *S. guyanensis*	+O	.	+	.	.	.	O	+O	8 (B)
15 *S. guyanensis*	+O	.	.	.	.	.	.	.	8 (B)
5 *S. guyanensis*	.	+O	.	.	.	(+O)	.	.	6 (B)
1 *S. guyanensis*	+O	.	+	.	.	.	.	.	7 (B)
2 *S. guyanensis*	+O	.	.	.	.	.	.	.	8 (B)
9 *S. guyanensis*	+O	O	+	.	.	.	.	.	8 (B)
11 *S. guyanensis*	+O	.	+	.	.	.	.	.	10 (B)

[A] See Burt *et al.* (1971), Fig. 1, the letter in parenthesis referring to the three major groupings.

159

In dry tropical regions, in a series of abnormally harsh years (site 4), two accessions, *S. hamata* and *S. scabra*, were outstanding, particularly the former. *S. fruticosa* and *S. viscosa* persisted. In better years (sites 4 and 8) these same species did well and *S. subsericea* showed some promise. Under these improved conditions *S. humilis* showed potential, *S. guyanensis* showing some merit at the 'intermediary site', Darwin. Finally, in the wetter areas (site 2) the potential of *S. guyanensis* was realized.

Some of these results, those for instance for *S. guyanensis*, were to be expected in the light of past experience. Others have been confirmed by subsequent testing. The method would seem, in fact, to be convenient and workable.

Performance of individual accessions. At the time at which this experiment was commenced we knew very little of the performance characteristics of the various plants. Working from the data presented, which are based only on annual yields, we shall examine the performance of one accession, that of *S. hamata*, to see what can be learned about this accession.

S. hamata has a wide range of adaptation (Table 3.2.4) but is basically adapted to summer growing areas that experience long dry seasons. Generally it is adapted to the same range of conditions as the species *S. humilis*, although much more generally adapted than individual accessions of that species. In such areas in abnormally 'bad' years it establishes, persists and gives a reasonable yield when *S. humilis* does not (site 4). In the higher-yielding environments (Table 3.2.2, column 1) it gives very high first-year yields, its yields increasing with time. It suppresses weed invasion in both high- and low-yielding environments. It differs markedly from other plants adapted to dry tropical regions (for instance Table 3.2.2, column 5) in which the legume yield is initially poor and which allow more weed ingress.

The approach and methods. The initial decisions, to concentrate on site comparisons and trends with time and to simplify the data input, seem to have been well founded. Rankings of performance proved to be of considerable value. It is noticeable, in Table 3.2.3, that ranking tended to give a much better understanding of species performance than yield; we note, for instance, that the apparent adaptability of *S. guyanensis* to site 3, in terms of final-year yield, is not justifiable.

The use of numerical methods has thus helped to elicit a pattern that has proved to be valid. Further, more detailed examination of yield patterns during a growing season, plant perennation counts and so forth can now be undertaken for accessions of interest.

R. L. BURT

CASE 3.2

References

Bureau of Meteorology (1956). Climatic averages, Australia: temperature, relative humidity, rainfall. (Melbourne.)

Burt, R. L., Edye, L. A., Williams, W. T., Grof, B., and Nicholson, C. H. L. (1971). Numerical analysis of variation patterns in the genus *Stylosanthes* as an aid to plant introduction and assessment. *Aust. J. Agric. Res.* **22**, 737–57.

Burt, R. L., Edye, L. A., Williams, W. T., Gillard, P., Grof, B., Page, M., Shaw, N. H., Williams, R. J., and Wilson, G. P. M. (1974). Small-sward testing of *Stylosanthes* in northern Australia: preliminary considerations. *Aust. J. Agric. Res.* **25**, 559–75.

Fitzpatrick, E. A., and Nix, H. A. (1970). The climatic factor in Australian grassland ecology. In 'Australian Grasslands'. (Ed. R. Milton Moore.) pp. 1–26. (Aust. Natn. Univ. Press: Canberra.)

Lance, G. N., and Williams, W. T. (1967). Note on the classification of multi-level data. *Aust. Comput. J.* **9**, 381–2.

Case 3.3. The Climatic Background of *Stylosanthes* Accessions

The problem

Plant evaluation is expensive in terms of time and effort; plant introduction, on the other hand, can import thousands of plants with relative ease. The problem is one of selecting which material should be imported and where it should be tested. Obviously the choice of material is dependent upon the requirements of the program; plant breeders, for example, may be quite specific, requiring accessions of a single species in the search for a single attribute such as frost resistance. In the example with which we are dealing here we are concerned with several species, or even genera, and the main 'selective sieve' is that they are adapted to the dry tropics.

Various methods are available to aid the selection of material for import. It is known, for instance, that certain species or genera are more likely to be of value than others. Ecological considerations can also be of value and indeed are currently being investigated. Geographical and climatological studies have, however, been most widely employed for this purpose. As such studies are described elsewhere in this volume it is probably sufficient to say that there are problems, first, in classifying climates in a meaningful way for plant introduction purposes and, secondly, in deciding whether or not successful pasture species are likely to come from similar homoclimes.

To explore this field we have attempted to integrate plant biogeographical studies with some of the *Stylosanthes* evaluation programs. In the first (1) we superimposed background climatic data onto the morphological–agronomic (M–A) classification described previously (Section 3.1) and so investigated the climatic background of the various species and species forms. Secondly (2), using the same background information, we classified the climates from which individual accessions were collected. Finally (3), principal coordinate analysis was used to examine the same data.

Climatic data

Detailed climatic information is not available for many of the areas in which the accessions were collected. The data used (Table 3.3.1) represent a compromise between that which we should like to have used and that which was available. Many of the climatic parameters commonly used by agronomists were included. Climatic data for 158 accessions were finally assembled.

1. Superimposition of the M–A classification on climatic data. In this and subsequent sections we have accepted the *Stylosanthes* classification as based on a mixture of morphological and agronomic data (Burt *et al.* 1971).

Table 3.3.1. Climatic characteristics used in the analysis

1   Latitude (°)[A]
2   Altitude (m)
3   Number of months giving 85–92% of total annual rainfall
4   Highest monthly maximum temperature during G^B (°C)
5   Lowest monthly minimum during G (°C)
6   Lowest monthly maximum in D^C (°C)
7   Lowest monthly minimum in D (°C)
8   Average annual rainfall (mm)

[A] Expressed as degrees or fractions of a degree, not distinguishing between N. and S.
[B] Months during which 85–92% of total annual precipitation falls.
[C] Months in which remainder of precipitation falls.

The existence of two additional groups, 29 and 30, has since been recognized (Edye *et al.* 1974). Accessions for which climatic data were unavailable were deleted from the hierarchy, some branches thus becoming redundant. The 158×8 climatic attributes were then subdivided to simulate the M–A hierarchy and tests of statistical significance were carried out across the divisions. This is reasonable because these characters were not used in the original classification.

Results. For the present purpose it is not necessary to consider the results *in toto* but only to examine the value of this approach. We begin by comparing data for two of the three main branches of the trichotomy and subsequently look in detail at some individual M–A groups.

In the first of the branches, that dealing mainly with the *S. guyanensis* complex, temperature differences between the groupings were of major importance (Fig. 3.3.1). Location factors, altitude and latitude, figure occasionally and rainfall attributes only rarely. For those interested in *Stylosanthes per se* it is interesting to note that rainfall was important in separating groups 4 + 5 + 6 (which contain the more 'subtropical' cultivar Oxley) from 7 + 8 (which contain the 'wet tropical' cultivars Schofield, Cook and Endeavour).

In the second branch of the trichotomy, dealing basically with species that have proved to be adapted to the dry tropics, rainfall figured much more prominently. We may tentatively conclude that different forms of *S. guyanensis*, as represented in the collection, come from environments which differ most markedly in temperature regime. With the 'dryland' species rainfall is much more important. Comparisons at the lower levels of the hierarchy are usually precluded by the lack of information.

One feature of this analysis which is not immediately obvious is that widely separated branches of the dichotomies may have similar climatic backgrounds. Groups 4 + 5 + 6 have climatic means very similar to those for groups 15 + 16 and very different from groups 7 + 8. Reference will be made to this in subsequent sections.

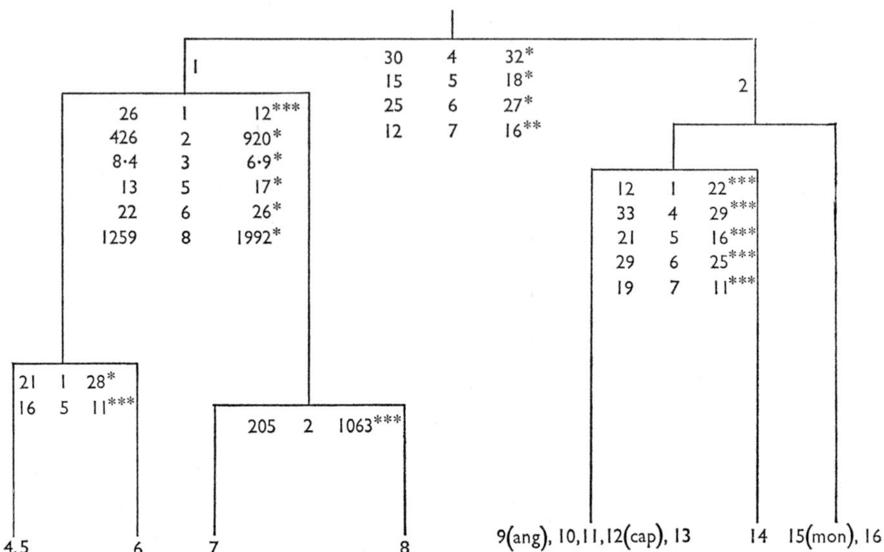

Fig. 3.3.1. Differences in the climatic background of the various forms of *S. guyanensis*, *S. angustifolia* (ang), *S. capitata* (cap) and *S. montevidensis* (mon).

The background climatic data for individual accessions within M–A groups was also examined. In some cases the group was collected only over a narrow geographical and climatological range. In Table 3.3.2 we find that most members of group 14 are from similar altitudes and latitudes; the member from the lowest latitude comes from the highest altitude so that environmental conditions are similar. Some heterogeneity was, however, particularly evident for one accession (No. 6) which had originated in a higher-altitude, higher-rainfall area. This member had been previously designated as a subgroup. In this way it is possible to substantiate previous decisions on subgrouping and, where these were not made, draw attention to members that are potentially typical.

This approach may also be of value in inter-disciplinary studies. In Table 3.3.2 symbiotic nitrogen fixation ability, measured against a 'standard' rhizobial strain, has been included. Two abnormally high values are noted (treatments 3 and 5). The accessions both originated in an area which experienced higher temperature regimes than other members of the group. With most other M–A groups such abnormally high values were associated with accessions from lower-rainfall areas. In addition to helping to examine group heterogeneity such observations may well be of value in rhizobial ecology.

Table 3.3.2. Climatic background of M–A group 14

M–A group	Accession[A]	1	2	3	4	5	6	7	8	Symbiotic efficiency
14A										
	1	22·9	663	7	29	16	24	11	1363	8
	2	21·2	562	7	30	16	26	12	1267	3
	3	21·2	629	6	31	18	27	11	1428	141
	4	19·9	915	6	28	17	24	13	1472	9
	5	21·2	621	6	31	17	27	11	1428	101
14B	6	20·3	1110	6	28	17	24	13	1784	7

[A] For the sake of brevity data for only 6 of the 12 members of this group are presented; other members conform to accessions 1, 2 and 4.

2. Classification of climatic data. For this purpose the data matrix was 158 individuals by 8 continuous variables. The program used was POLYDIV.

Two features of the resultant classification negated its use for plant introduction purposes. Firstly, environments known to support quite different suites of species were placed in the same groups. Secondly, species or species forms known to have relatively narrow distributional limits were scattered widely throughout the classification.

In general the classification appeared to be more meaningful for subtropical areas. This may possibly reflect the relative dearth of information for rainfall characteristics (two attributes only) in the analysis.

The overall result, however, was not unexpected. Problems of selecting the most important attributes for determining plant distribution have been encountered previously.

3. Principal coordinate analysis. In this section we have used principal coordinate analysis to explore the relationships between the climatic backgrounds of the various M–A groups and subgroups. The 158 accessions represented some 44 groups and subgroups. The number of individuals in each group varied from 1 to 12. For each group the means of the 8 climatic attributes were calculated giving a 44×8 matrix. The triangle of $\frac{1}{2} \times 44 \times 43$ distance measures was calculated using the Canberra metric and mapped into a reduced Euclidean space using the principal coordinate analysis of Gower.

Results—preliminary interpretation. The first three components accounted for 48, 20 and 19% respectively, i.e. 87% of the total variation. We are therefore concerned primarily with these. Inspection of the components (Table 3.3.3) shows that major discontinuities occur after values of about 0·6. Meaningful patterns can be extracted down to this value, which was therefore chosen as the level at which to work.

Table 3.3.3. Attribute values for background climatic information

Component 1		Component 2		Component 3	
Value	Attribute	Value	Attribute	Value	Attribute
0·63	1	0·77	8	0·69	1
0·61	2	0·74	3	0·39	3
0·38	3	0·43	7	0·32	4
0·02	8	0·13	5	0·11	8
−0·56	4	0·07	6	0·01	7
−0·89	7	0·08	2	−0·07	5
−0·92	5	−0·21	1	−0·27	6
−0·93	6	−0·65	4	−0·73	2

The first component is concerned with high values for altitude and latitude, being negatively associated with high minimum temperatures, i.e. low minimum temperatures are generally found at high altitudes and/or latitudes. Examination of the data set shows that high positive values and, to a considerable extent, low negative values are associated with *S. guyanensis*. The first component is, in other words, basically concerned with that species.

In the second component we find that high positive values are associated with high rainfall received over a long period of time. High negative values reflect high maximum temperatures during the growing season. This component thus represents a gradient from long reliable growing seasons to areas with short, hot, erratic growing seasons. It is basically concerned with such species as *S. hamata*, *S. fruticosa* and *S. humilis*.

The third component separates altitude (high negative values) and latitude (high positive values).

Interpretation. A graph with all 44 points is difficult to follow, and for ease of presentation the points have been displayed in two separate graphs: Fig. 3.3.2 shows the *S. guyanensis* complex, Fig. 3.3.3 shows *S. humilis*, *S. scabra* and *S. subsericea*. In both figures component 1 is used as the vertical axis and component 2 as the horizontal. Quadrant A is subtropical, low rainfall with relatively high temperatures during the growing season and cool dry-season temperatures. Quadrant B is tropical, semi-arid with high growing-season temperatures. Quadrant C is tropical, without extremes of high or low temperatures and with long growing seasons. Quadrant D represents the cooler, moister subtropics.

The results presented reveal some interesting features about the species and species forms represented in the collection. Considering all forms of *S. guyanensis*, there appears to be a general trend in which 'subtropical' forms come from low-rainfall and 'tropical' forms from humid areas. Examination of Fig. 3.3.2 shows that we actually have three groups of M–A types: 6A, 6B; 5A, 14A, 14B, 8A, 4, 10B; and 10A, 7A, 11. A general trend may exist but, once 6B is excluded, it is very slight. The value on the second component also clarifies the position of group 8B; it is from relatively high altitudes.

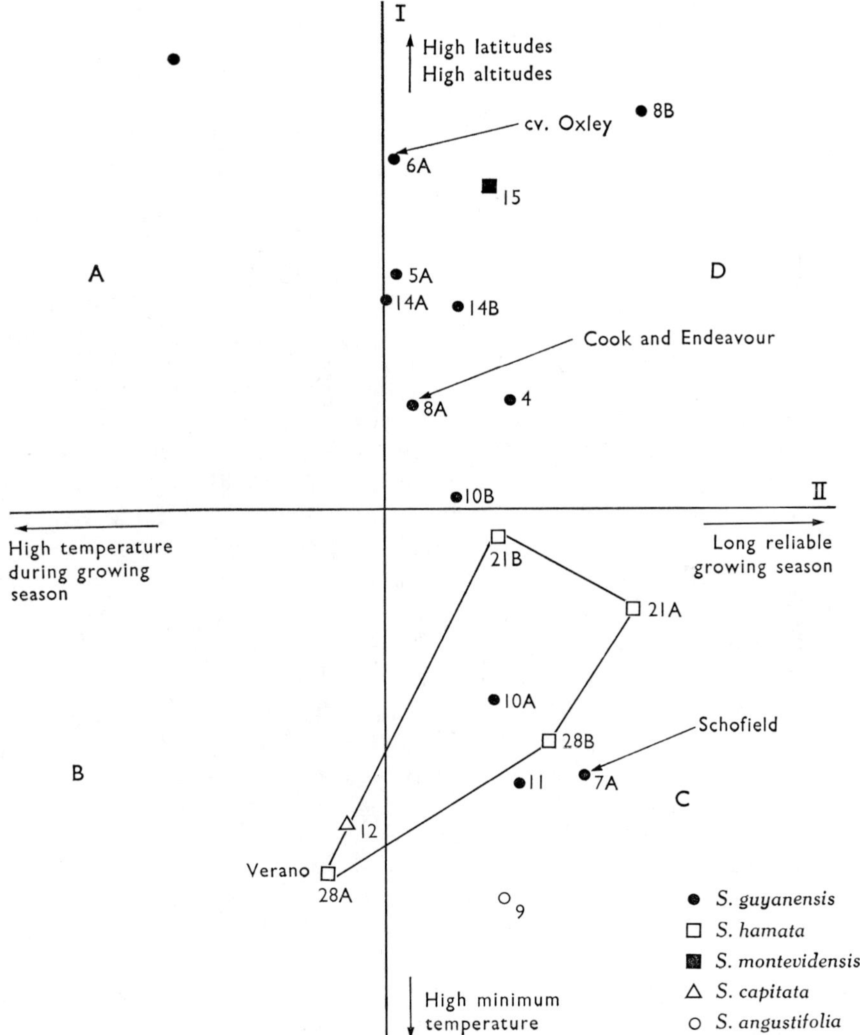

Fig. 3.3.2. Ordination of members of *S. guyanensis* complex with respect to climatic attributes.

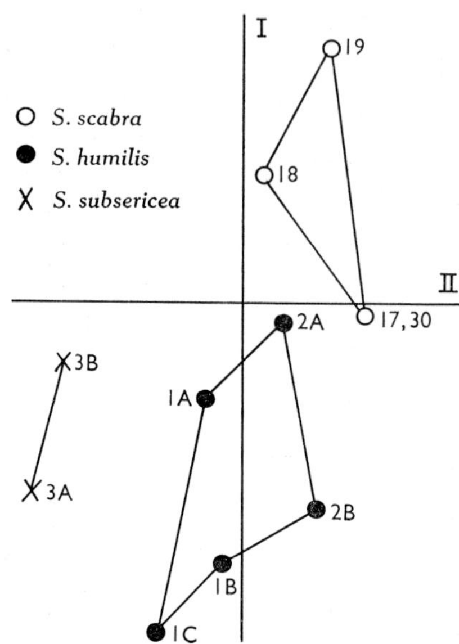

Fig. 3.3.3. Climatic ordination of other *Stylosanthes* species.

The position of species placed with *S. guyanensis* in the M–A classification becomes obvious in this analysis; *S. montevidensis* obviously 'belongs', whereas *S. capitata* and *S. angustifolia* probably do not.

In contrast to *S. guyanensis*, *S. hamata* is from low altitudes and latitudes but comes from markedly varying rainfall regimes. Part of this range apparently overlaps with *S. guyanensis* although such areas are in different phytogeographic regions. One form of *S. hamata* (28A) is found in semi-arid areas. *S. humilis* has a relatively limited climatic distribution.

S. guyanensis cv. Oxley, suitable for 'subtropical' areas, and cv. Schofield, suitable for the wetter tropics, come from equivalent climates. Cultivars Cook and Endeavour are more centrally placed. *S. hamata* cv. Verano and *S. humilis* cv. Paterson and Lawson are suitable for the dry tropics. *S. hamata* cv. Verano is from the semi-arid quadrant as are *S. humilis* accessions similar to the released varieties. Forms of *S. scabra* are currently showing high potential in the dry tropics. Without going into detail it seems that the forms most likely to be suitable for the dry, intermediate and wetter 'dry tropics' could have been forecast from these studies. Newly introduced forms, from areas located in the semi-arid quadrant, must obviously be viewed in this light.

These data suggest that studies of this type may be of value in predicting areas from which material should be introduced and in describing where material should be tested. If we examine what happened at 'Lansdown', in a sward-testing situation, we get some indication of the possibilities. In a bad establishment year the agronomically desirable forms come from groups 20B and 18 (considering only those forms of *S. scabra* and *S. viscosa* included in the experiment). The following year, with some irrigation, both of these forms did well and forms 20C and 19 showed promise, i.e. forms with the lowest values of component 1, those for component 2 being similar, were adapted to a 'bad' year in the dry tropics.

Superimposition or ordination?

Both systems of analysis have vices and virtues. Superimposition did not help to show the species and form overlap very clearly. It was, however, useful in looking at individual features of the environment and in looking at group heterogeneity. Coordinate analysis does the opposite; individual attributes can only be viewed by inference but distribution of forms can be viewed graphically. In the example examined I suggest that the two are complementary; in other situations it may be desirable to use one or the other.

Epilogue

In a recent publication Hyland (1970) states, 'The controlled use of "wild" plant germ plasm is largely dependent upon accurate description and taxonomic identification, followed by comprehensive evaluations by competent authorities'. Several other authors writing in the same volume refer to the importance of climatic studies in this field.

From the data presented in this and other papers in this volume, it appears that numerical methods are of value, not only in the initial descriptive stage of plant evaluation, but in later testing programs. By integrating biogeographical studies with the evaluation programs, further aids to plant introduction and evaluation are obtained. The further development of this approach could be useful in planning plant expedition missions and in the rational use of material contained in the 'gene banks'.

R. L. BURT

References

Burt, R. L., Edye, L. A., Williams, W. T., Grof, B., and Nicholson, C. H. L. (1971). Numerical analysis of variation patterns in the genus *Stylosanthes* as an aid to plant introduction and assessment. *Aust. J. Agric. Res.* **22**, 737–57.

Edye, L. A., Burt, R. L., Nicholson, C. H. L., Williams, R. J., and Williams, W. T. (1974). Classification of the *Stylosanthes* collection, 1928–69. CSIRO Aust. Div. Trop. Agron. Tech. Pap. No. 15.

Hyland, H. L. (1970). Description and evaluation of wild and primitive introduced plants. In 'Genetic Resources in Plants'. (Ed. O. H. Frankel and E. Bennet.) pp. 413–20. (Blackwell: Oxford.)

Case 3.4. Analysis of Agronomic Data from *Stylosanthes* Introductions

Numerical classificatory programs have, in the past, found their main application in formal taxonomy, commonly at species or genus level. More recently, however, there have been several attempts to explore their usefulness in the classification of cultivated species, even of a set of cultivars of the same species. The purpose of such investigations appears to be twofold. There is first an essentially taxonomic requirement, aimed at facilitating identification and at producing succinct summaries of information to ease problems of communication. The second requirement is primarily economic, aimed at elucidating the pattern of variation in those attributes that are of economic importance, irrespective of taxonomy. Unfortunately, the two requirements are not fully compatible, since they imply different sets of attributes. The taxonomic requirement calls for morphological attributes which can be easily and rapidly recorded; the economic requirement calls for attributes describing, for example, yield and quality. An uncertainty of purpose is often revealed by the use of both types of attribute in the same data matrix. The drawback of this procedure is that the more numerous set of attributes will tend to dominate the resulting classification. In the work of Burt *et al.* (1971) on 154 accessions representing 14 species of *Stylosanthes*, morphological attributes outnumbered economic by some three to one, while Akinola and Whiteman (1972) studied 95 accessions of the single species *Cajanus cajan* (L.) Millsp., and economic attributes predominated; no work is known in which economic attributes alone have been used.

There are three reasons why a classification based on economic attributes alone might be an invaluable supplement to a classification based on morphological or a mixture of morphological and economic attributes. First, two morphologically different species may exhibit almost identical economic performance. Secondly, two accessions of closely similar morphology may differ in performance, so that a group based largely, or even partly, on morphological characters may be heterogeneous for performance. Lastly, if a single accession is grown in a site climatically different from that in which its performance is familiar and satisfactory, it may remain morphologically invariant but its performance may change profoundly. These considerations suggest that (i) a pure morphological classification will be valuable for identification and description; (ii) a mixed morphologic–economic classification will be useful for description and may also provide some guide as to economic potentiality of complete groups; and (iii) given either of the preceding, an additional purely economic classification should provide a more reliable basis for the assessment of individual accessions.

Our work (Edye *et al.* 1973) provided a test of the last of these suggestions and comprised two major projects. First, we used the data for the original 154 *Stylosanthes* accessions (Burt *et al.* 1971) and classified them on their economic attributes alone (in this context they are referred to as agronomic attributes) in order to ascertain in what respects the resulting classification differed from the original mixed-type data classification. Secondly, the same accessions were grown at three climatically different sites, to investigate the stability of the classification under change of environment.

Agronomic attributes

The agronomic attributes which were recorded in spaced-plant experiments at South Johnstone (wet tropics), 'Lansdown' (dry tropics) and Samford (subtropics) are summarized in Table 3.4.1. The method of recording each attribute was standardized over all three sites.

Table 3.4.1. Attributes used in the agronomic classification of the *Stylosanthes* accessions

---

Flowering onset
1. Mean day-length in minutes for flowering
2. Range in day-length in minutes between the first and last plant to flower
3. Mean number of days from 1 December 1966 to flowering
4. Range in days between the first and last plant to flower

Yield as spaced plants
5. Mean dry matter yield (g) from one plant per plot in June 1967
6. Mean dry matter yield (g) from all surviving plants in May 1968
7. Mean seed + pod yield (g) from the other plants per plot, when first seed crop mature
8.[A] Mean seed + pod yield (g) from same plant as (7) when second seed crop mature

Survival
9. Mean survival percentage of plants in (5) on 31 December 1967 after cut once at a height of 5 cm
10. Mean survival percentage of plants in (5) on 19 May 1968 after cut at a height of 5 cm in June and December 1967

Branching habits
11. Mean number of branches originating from the main stem of all plants at or below 5 cm in June 1967
12. Mean crown branch density (ranked 1 min. to 6 max.) of all plants at or below 10 cm in May 1968

[A] At Samford only.

Analysis of data

The Canberra program POLYDIV which was designed to handle large numbers of individuals defined by a small number of continuous attributes was used to analyse the data. This program provides for missing values.

The dimensions of the data matrix are 154 (accessions) × 3 (sites) × 11 (agronomic attributes), and detailed exposition of the results from a data matrix of this size would be quite impracticable. Nevertheless, it is an essential part of the study to explore the possibility that an all-agronomic classification will permit a more detailed assessment of agronomic potentiality than will a mixed morphological–agronomic classification alone, so that at least some aspect must be selected for detailed study. When detailed appraisal is at issue, I shall therefore concentrate on the *S. guyanensis* group: this contains the greatest number of accessions, shows considerable promise and is agronomically extremely variable.

Agronomic groups for 'Lansdown'

The POLYDIV hierarchy was truncated at the 12-group level, since below this individual accessions began to fragment off; even at this level, there was one single-membered group (*S. cayennensis* No. 6, which failed to flower) which has been excluded from the discussion that follows. The course of the dendrogram itself is of no interest, and we may take the groups in any convenient order. The group means of the 11 attributes are not very revealing to those unfamiliar with *Stylosanthes* performance, and I have elected to present the groups in the form of an artificial key. This is given in Fig. 3.4.1 ('high' is taken arbitrarily as within the upper third of the population range), which also shows the constitution of the groups by species.

Certain familiar agronomic types are immediately obvious. Group 8, with high survival and good second-year yield, is the type of the desirable perennial for this climate (seed-set in this group, though not high, was also adequate). Group 10, not persistent, but with good first-year yield and good seed-set, is the type of desirable annual. Group 9 had poor survival, but good second-year yield if the plants survived; its members would probably be suited to a somewhat wetter climate than that of 'Lansdown'. Group 4 contained the most persistent accessions in the second year, and its members would be indicated for dry climates where persistence was of greater importance than high yield. The accessions in groups 1 and 2, if day-length-sensitive, would be unsuitable for the tropics. The accessions in group 6, with thoroughly unsatisfactory performance, were either beset by nodulation problems or were apparently quite unsuited to the 'Lansdown' climate.

There was good separation of annuals and perennials. Annuals were almost confined to groups 6, 10 and 11, though three *S. hamata* perennated

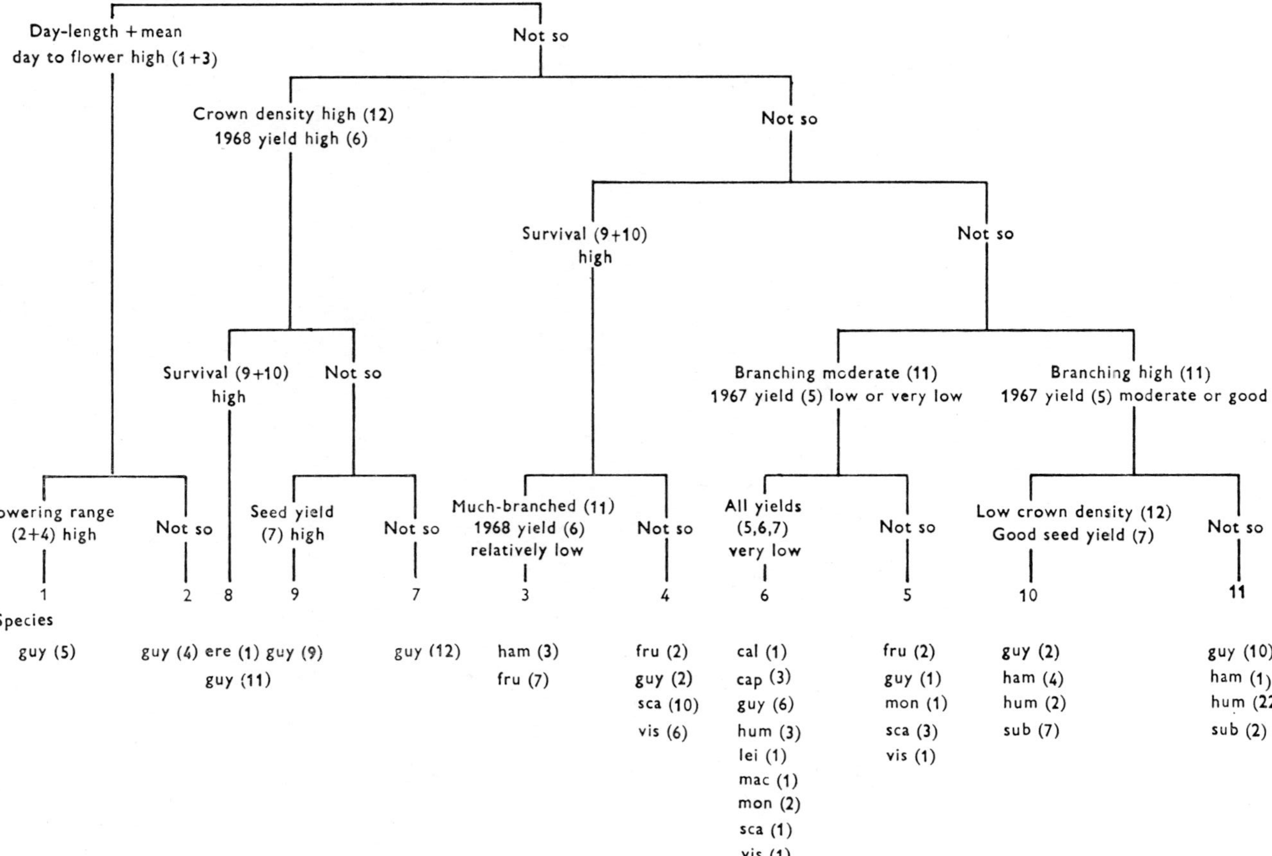

Fig. 3.4.1. Artificial key to 'Lansdown' agronomic groups. Attribute numbers concerned shown in brackets at each branch of key; species composition, and number of accessions of each species, given below group numbers. Species are denoted by the first three letters of their names.

and appeared in group 3. There was also surprisingly good species separation (e.g. group 4, dominated by *S. scabra* and *S. viscosa*). On the other hand, the extremely variable *S. guyanensis* complex appeared in every group except group 3.

Comparison with morphological–agronomic groups

On counting subgroups as separate groups, the population represents 36 of the original M–A groups (Burt *et al.* 1971; Edye *et al.* 1974*b*). A two-way table of M–A against agronomic groups, with the number of accessions as cell entries, is given in Table 3.4.2. Owing to the disparity in number between rows (M–A groups) and columns (agronomic groups), most columns may be expected to transgress several rows. It is perhaps of interest to note the similarities between *S. scabra* and *S. viscosa* (in agronomic group 4) and between *S. hamata*, *S. fruticosa* and *S. erecta* (in agronomic group 3). However, the most interesting aspect of the table is the light it throws on the variability within *S. guyanensis*. Of the M–A groups within the species, M–A groups 6, 11 and 16 are agronomically homogeneous, so that with perennials such as these any one of their members may be taken as typifying the group. In contrast, M–A groups 7, 8 and 14 are remarkably hetero-geneous. Since the best of the perennials (i.e. those in agronomic group 8) are largely taken from M–A groups 7 and 8, it is clear that these groups will need intensive study.

It will now be evident that a purely agronomic classification can be used to explore what we may call the 'agronomic fine structure' of an M–A classification. It was already known that some accessions in M–A groups 7 and 8 performed satisfactorily at 'Lansdown'. However, of the 10 accessions in M–A group 7, 5 appeared in group 8 of the agronomic classification; these 5 would be the natural choice for further selection and study. Similarly, of the 13 accessions in M–A group 8, 3 appeared in agronomic group 8, and again would be chosen for further work. Alone, a purely agronomic classification is too crude to permit identification, owing to the limited num-ber of attributes that can be devised for collection; but a pure morphological or mixed morphological–agronomic classification is liable to be insufficiently sensitive to agronomic heterogeneity.

Inter-site comparisons

Overall comparison. The three sets of data were combined into a single set, regarded as 462 (i.e. 3×154) individuals defined by the 11 attributes common to all 3 sites. The interest in classifying such a set lies in ascertaining whether the subdivisions will reflect primarily species or site groups. If the former, the M–A groups are agronomically stable under change of environment; if the latter, the M–A groups will differ markedly in agronomic

Table 3.4.2. Comparison of agronomic and M–A groups

M–A group	Species	Agronomic group										
		1	2	3	4	5	6	7	8	9	10	11
1	*S. humilis*	—	—	—	—	—	—	—	—	—	2	8
2A	*S. humilis*	—	—	—	—	—	3	—	—	—	—	—
B		—	—	—	—	—	—	—	—	—	—	2
C		—	—	—	—	—	—	—	—	—	—	2
D		—	—	—	—	—	—	—	—	—	—	10
3A	*S. subsericea*	—	—	—	—	—	—	—	—	—	2	2
B		—	—	—	—	—	—	—	—	—	5	—
5	*S. guyanensis*	—	—	—	—	—	2	—	2	—	—	2
6A	*S. guyanensis*	4	—	—	—	—	—	—	—	—	—	—
B		1	—	—	—	—	—	—	—	—	—	—
7	*S. guyanensis*	—	—	—	—	—	—	2	5	2	—	1
8	*S. guyanensis*	—	—	—	1	—	—	4	3	5	—	—
9	*S. angustifolia*	—	—	—	—	—	—	—	—	—	—	1
10A	*S. guyanensis*	—	—	—	—	—	—	1	—	1	—	2
B		—	—	—	1	—	—	—	1	—	—	—
11	*S. guyanensis*	—	—	—	—	—	3	—	—	—	—	—
12	*S. capitata*	—	—	—	—	—	3	—	—	—	—	—
13	*S. guyanensis*	—	—	—	—	—	1	—	—	—	—	1
14	*S. guyanensis*	—	—	—	—	1	—	5	—	1	2	4
15	*S. montevidensis*	—	—	—	—	1	2	—	—	—	—	—
16	*S. guyanensis*	—	4	—	—	—	—	—	—	—	—	—
17	*S. scabra*	—	—	—	—	1	—	—	—	—	—	—
18	*S. scabra*	—	—	—	1	1	—	—	—	—	—	—
19	*S. scabra*	—	—	—	9	1	—	—	—	—	—	—
20A	*S. viscosa*	—	—	—	3	1	—	—	—	—	—	—
B		—	—	—	3	—	—	—	—	—	—	—
C		—	—	—	—	—	1	—	—	—	—	—
21	*S. hamata*	—	—	3	—	—	—	—	—	—	1	1
22	*S. calcicola*											
	S. leiocarpa	—	—	—	—	—	3	—	—	—	—	—
	S. macrocarpa											
23	*S. fruticosa*	—	—	2	—	1	—	—	—	—	—	—
24	*S.* sp.	—	—	1	—	—	—	—	—	—	—	—
25	*S. fruticosa*	—	—	2	2	1	—	—	—	—	—	—
26	*S. fruticosa*	—	—	3	—	—	—	—	—	—	—	—
27	*S. erecta*	—	—	2	—	—	—	—	1	—	—	—
28	*S. hamata*	—	—	—	—	—	—	—	—	—	3	—
30	*S. scabra*	—	—	—	—	—	1	—	—	—	—	—

performance at different sites. It will suffice to note that the divisions over-
whelmingly represented site differences. The effect was already strongly
marked at the 5-group level, and was strengthened with every further subdivision. This is in striking contrast to the case when morphological attri-

butes were included; in a previous study (Burt *et al.* 1971), replacing the 'Lansdown' agronomic data by either of the other sets made little change in the resulting M–A groups. It again follows that, although M–A groups are stable under change of site so far as recognition and description are concerned, an agronomic-only classification is essential to assess details of performance.

South Johnstone. This set was now classified independently; it was taken to the 15-group level, after which fragmentation occurred. It would be unprofitable to discuss the groups in the same detail as for 'Lansdown', but the following points should be noted: group 1 had an extremely high yield; yields in 10 and 11 were also very high; yields in 14 and 15 were somewhat lower (though still high by 'Lansdown' standards), but the plants were more branched with a higher crown density. All these can be regarded as 'good' groups. The 'poor' group in this case was group 8, with low survival and low yield; with a few exceptions, it contained all the annuals. The most persistent and high-yielding perennials were the *S. guyanensis* accessions in agronomic groups 1 and 15. The other perennial species performed badly.

Since the main interest at South Johnstone resides in the *S. guyanensis* complex, we shall confine comparison with M–A groups to this complex; and we shall omit consideration of M–A groups 11, 13 and 16, since these performed poorly at all sites. Table 3.4.3 is a two-way table of agronomic groups ('Lansdown' × South Johnstone), with accession numbers and corresponding M–A groups as cell entries.

We note, for example, that at 'Lansdown' an all fine-stem stylo group (M–A 6A) occurred in agronomic group 1 together with M–A 6B. The group was characterized by flowering under long days (mean 776 minutes) and by a wide range in flowering time (132 days) because some plants flowered very early in the first growing season and others very early in the second growing season. They established well with an average yield in the first year but relatively poor yield in the second year. Survival during the first dry season was 100% which dropped to 50% at the end of the second growing season. They were poorly branched.

At South Johnstone accession 21 (M–A 6A) was separated from the other fine-stem stylos, which failed to survive during the first wet season and did not flower. *S. guyanensis* 21 occurred in agronomic group 3 with other accessions which flowered under long day-lengths (mean 726 min) and also had a wide range in flowering time. The accessions in agronomic group 3 lacked vigour but gave reasonably high seed yields. Although they survived well during the first dry season, they did not survive the second wet season (mean 11%). They were reasonably well branched.

Table 3.4.3. Agronomic classification of some _S. guyanensis_ at 'Lansdown' and South Johnstone

Upper number in italics is accession number; lower number is morphological–agronomic (M–A) group number

South Johnstone agronomic groups	1	4[A]	5	6	'Lansdown' agronomic groups 7[B]	8[C]	9	10	11
1[B]					_28_ 8	_12_ 8		_31_ 14	
2		_38_ 8			_44_ 14				_18, 24, 61_ 5 7 14
3	_21_ 6A		_47_ 14		_15, 35, 41, 60_ 14 14 14 14		_29_ 14	_36_ 14	_40, 50_ 14 14
8	_19, 20, 100_ 6A 6A 6A			_149_ 5					_148_ 14
10						_43, 54_ 5 8			
11						_143_ 7	_27, 34_ 8 8		
12		_45_ 10B		_37_ 5		_42_ 5	_142_ 7		_62_ 5
13	_52_ 6B								
14					_10, 55, 56, 16_ 7 8 8 10A	_59, 51_ 8 10B	_58_ 8		
15[AC]					_13, 46_ 7 8	_17, 22, 30, 32_ 7 7 7 7	_33, 26, 57, 144_ 7 8 8 10A		_11, 14_ 10A 10A

[A] Highest survival. [B] Highest yield. [C] Best combination.

The best overall accessions at 'Lansdown' fell into M–A groups 5, 7, 8 and 10B; the best at South Johnstone were in M–A groups 7, 8 and 10A. The versatility of groups 7 and 8 is remarkable, in that they performed equally well in the wet and dry tropics; these groups clearly merit further study.

Samford. The Samford data were also classified into 15 groups. The 'good' groups were 1, with good survival; 3, with highest yield; and 4, with the best combination of yield and survival. The 'poor' groups were 6–10, which contained all the annuals. Groups 6 and 7 with higher survival in the second year included *S. hamata* and *S. subsericea* which perennated. However, all accessions in these five groups had very low dry matter and seed yields.

Table 3.4.4 shows, for the *S. guyanensis* complex, the Samford × South Johnstone comparison. The fine-stem stylos (M–A 6A) occurred in the high-survival group 1. At Samford flowering data were missing for this group, but all are known to flower under long days and were the only accessions to set seed in the first year. M–A group 14, which was substantially homogeneous at South Johnstone, was much less so at Samford; its members appeared both in the 'best' group 4 and the 'poor' group 5. The remarkable versatility of M–A group 8 was once more in evidence, in that several of its members appeared in the high-yielding group 3.

General considerations

It will be obvious that, for a set of accessions with as wide a species range as that considered here, pure agronomic classifications would be quite unsuitable for descriptive purposes: they are of necessity based on too few attributes, and are extremely site-sensitive. On the other hand, it will be clear from the foregoing that many M–A groups are agronomically heterogeneous and that the nature of their heterogeneity varies from site to site. In the *S. guyanensis* complex, for example, M–A 6A is homogeneous at 'Lansdown' and Samford but not at South Johnstone; M–A group 13 is homogeneous at South Johnstone only, group 16 at 'Lansdown' only. The M–A groups have sufficient integrity to make it possible to state that certain groups are best at particular sites; but a supplementary agronomic classification is obviously essential for selecting the most suitable accessions within an M–A group.

Conclusions

Burt *et al.* (1971) showed that a morphological–agronomic classification would reduce the material to a relatively small number of easily typified groups; the present chapter has shown that such a classification can be further refined by exploring the pattern of the agronomic attributes alone.

Table 3.4.4. Agronomic classification of some _S. guyanensis_ at Samford and South Johnstone

Upper number in italics is accession number; lower number is morphological–agronomic (M–A) group number

South Johnstone agronomic groups	Samford agronomic groups										
	1[A]	2	3[B]	4[C]	5	6	9	12	13	14	15
1[B]			_12_							_31_	_28_	
			8							14	8	
2			_38_	_44_			_24_	_18_		_61_		
			8	14			7	5		14		
3	_21_			_15, 40, 41, 60_	_29, 36, 50_					_35, 47_		
	6A			14 14 14 14	14 14 14					14 14		
8	_19, 20, 100_	_149_			_148_							
	6A 6A 6A	5			14							
10			_43, 54_									
			5 8									
11			_34_		_27, 143_							
			8		8 7							
12			_45_			_42_				_62_	_142_	
			10B			5				5	7	
13	_52_											
	6B											
14			_59_	_16_							_10, 55, 56, 58_	_51_
			8	10A							7 8 8 8	10B
15[AC]			_22, 26, 46_	_11, 14_	_13, 17, 30, 32, 33_						_57, 144_	
			7 8 8	10A 10A	7 7 7 7 7						8 10A	

[A] Highest survival.   [B] Highest yield.   [C] Best combination.

The double classificatory method thus serves to define a systematic procedure for the preliminary assessment of a wide range of material. Other numerical methods may also be pressed into service for this purpose; for the largely continuous attributes characteristic of agronomic characters, it is inherently likely that methods of ordination, rather than classification, would be appropriate.

In all such work, however, two possible limitations must be borne in mind. With legumes, there is always the possibility that ineffective or only partly effective nodulation will affect the agronomic performance (Edye *et al.* 1974*a*). However, there were 10 strains of *Rhizobium* in the mixed inoculum that was used in the experiment that are capable of effective nodulation on a range of species ('t Mannetje 1969), and available soil nitrogen was probably adequate in the cultivated nurseries. Secondly, it must be acknowledged that success in a spaced-plant trial does not ensure success in a sward. It is essential that before selection and release as a cultivar, the most promising of the spaced-plant accessions should be further tested in sward trials. Agronomic data from sward experiments at a number of sites can also be used for an agronomic classification of accessions as an aid to the regional evaluation of a large collection of introduced pasture plants (Edye *et al.* 1975).

L. A. EDYE

References

Akinola, J. O., and Whiteman, P. C. (1972). A numerical classification of *Cajanus cajan* (L.) Millsp. accessions based on morphological and agronomic attributes. *Aust. J. Agric. Res.* **23**, 995–1005.

Burt, R. L., Edye, L. A., Williams, W. T., Grof, B., and Nicholson, C. H. L. (1971). Numerical analysis of variation patterns in the genus *Stylosanthes* as an aid to plant introduction and assessment. *Aust. J. Agric. Res.* **22**, 737–57.

Edye, L. A., Burt, R. L., Williams, W. T., Williams, R. J., and Grof, B. (1973). A preliminary agronomic evaluation of *Stylosanthes* species. *Aust. J. Agric. Res.* **24**, 511–25.

Edye, L. A., Burt, R. L., Norris, D. O., and Williams, W. T. (1974*a*). The symbiotic effectiveness and geographic origin of morphological–agronomic groups of *Stylosanthes* accessions. *Aust. J. Exp. Agric. Anim. Husb.* **14**, 349–57.

Edye, L. A., Burt, R. L., Nicholson, C. H. L., Williams, R. J., and Williams, W. T. (1974*b*). Classification of the *Stylosanthes* collection, 1928–69. CSIRO Aust. Div. Trop. Agron. Tech. Pap. No. 15.

Edye, L. A., Williams, W. T., Anning, P., Holm, A. McR., Miller, C. P., Page, M. C., and Winter, W. H. (1975). Sward tests of some morphological–agronomic groups of *Stylosanthes* accessions in dry tropical environments. *Aust. J. Agric. Res.* **26**, 481–96.

't Mannetje, L. (1969). *Rhizobium* affinities and phenetic relationships within the genus *Stylosanthes*. *Aust. J. Bot.* **17**, 553–64.

Case 3.5. Statistical and Pattern Analysis of a Small-sward Trial

Many *Stylosanthes* species have been introduced as potential pasture legumes for northern Australia. The accessions show considerable variation as spaced plants in morphological and agronomic attributes both within and between species. Pattern analyses have been used to classify the collection into morphological–agronomic (M–A) groups (Burt *et al.* 1971; Edye *et al.* 1974*b*) and to describe the symbiotic effectiveness and geographic origin of the groups (Edye *et al.* 1974*a*) and their agronomic performance as spaced plants (Edye *et al.* 1973; Williams *et al.* 1973).

Subsequently, classificatory methods of pattern analysis were used to summarize data from small-sward experiments when the matrices were considered too large for analysis by classical statistical procedures (Burt *et al.* 1974; Edye *et al.* 1975*b*). With the possible increasing use of pattern analysis to simplify complex data sets, it seems desirable to relate pattern analysis to the results obtained from the more familiar statistical procedures that are commonly used. This should give a measure of the confidence that can be placed in pattern analyses which have only recently been applied in agricultural research.

For this comparison, a simple data matrix was chosen that could be adequately analysed by either approach. The results that were used came from a small-sward comparison of 15 *Stylosanthes* accessions grown at 4 sites in the dry tropics of Queensland. The accessions were tested at Musgrave (M), 'Lansdown' with irrigation (I), Fanning River (F) and 'Lansdown' without irrigation (S) on a solodic soil. The experiment extended over three growing seasons at all sites except the last where it was terminated after the second season. The number of accessions of each species included in the experiment is shown in Table 3.5.1: for full details of the accessions and experimental methods see Edye *et al.* (1975*a*). Briefly, the accessions were laid out as three randomized complete blocks at each site and they were grown with the grass *Urochloa mosambicensis*.

Choice of attributes

Three sets of attributes have been analysed by both methods for this chapter: the annual dry matter yield of *Stylosanthes* accessions and associated *Urochloa;* the percentage of *Stylosanthes* in the total annual yield of the swards; and the yield of seed pods that could be recovered from the soil surface in August each year.

Numerical analyses

The preliminary statistical methods used were the analysis of variance, the calculation of least significant differences at $P = 0.05$ so that groups of accessions could be defined whose means did not differ significantly, and the principal component analysis. Missing plots were statistically calculated and the means corrected when necessary. For some attributes the original data were transformed, e.g. square root, arcsine and log $(x + 1)$ transformations, before the analysis of variance was undertaken. With simple data sets, more sophisticated statistical analyses are possible but they have not been undertaken in this comparison because we are mainly concerned with the preliminary evaluation of results. Two different methods of pattern analysis were used and they will be briefly mentioned below for each attribute. In all the classificatory work, the means of the three replicates were used and the analyses were carried out on the Control Data 3600 computer in the CSIRO Division of Computing Research, Canberra.

Dry matter yields

Statistical analyses. There were significant differences ($P < 0.05$) between the accessions in yield of dry matter at all sites in all the years that were statistically analysed (Table 3.5.1). *S. hamata* 8 gave the most consistent performance of all the accessions and it was always in the top three in all years at all sites except at Fanning River in 1969/70 when it was ranked fifth. *S. hamata* 8 was one of the few accessions that persisted during 1970/71 at S.

There were significant differences ($P < 0.001$) between the first-year *Urochloa* yields associated with each accession at all sites except S (Table 3.5.2). The significant differences ($P < 0.05$) were partly attributed to poor competition from ineffectively nodulated accessions 6, 7, 9, 10 and 12. This response was not significant in the second year at all sites even though significant differences ($P < 0.05$) occurred at F which cannot be explained. A different response occurred in 1971/72 at I where the highest *Urochloa* yields were associated with *S. hamata* 8 and *S. subsericea* 5 which were among the three highest-yielding legumes in 1969/70 and 1970/71.

Pattern analysis. We are primarily concerned with the analysis and interpretation of trends in the annual yield of the accessions and associated *Urochloa* over all sites. Because there are marked inter-site and seasonal differences in dry matter yields, it is advantageous to use the rank order of annual dry matter yields of the accessions and associated *Urochloa* in addition to the absolute annual yields. Thus, any given rank may be associated with very different yields and so both attributes have been treated as if they were independent variables. (See Burt *et al.* 1974 for further explanation.)

Table 3.5.1. Annual yield of *Stylosanthes* accessions at four sites

Stylosanthes accession		Annual yield of dry matter (tonnes/ha)										
		Musgrave			'Lansdown' irrigated			Fanning River			'Lansdown' solodic	
		1969/70	1970/71	1971/72	1969/70	1970/71	1971/72	1969/70	1970/71	1971/72	1969/70	1970/71
S. subsericea	1	0·29	3·16	1·49	4·91	0·38	1·70	0·82	0·64	4·27	0·16	0·00
	2	2·81	1·68	1·80	5·45	1·94	1·99	0·69	1·24	3·32	0·15	0·00
	3	0·96	0·49	0·17	6·46	1·10	1·79	1·04	2·01	4·23	0·54	0·15
	4	0·61	0·97	2·52	3·91	0·60	1·78	0·36	1·07	4·20	0·06	0·00
	5	1·88	1·92	3·37	6·09	1·54	1·16	0·84	1·39	3·34	0·20	0·05
S. hamata	6	0·10	2·00	0·00	1·39	0·79	1·15	0·30	1·60	4·28	0·12	0·00
	7	0·11	0·94	0·00	0·94	1·11	2·25	0·11	0·55	2·91	0·03	0·00
	8	3·78	6·06	5·34	5·77	2·44	3·85	0·76	5·86	4·71	0·26	0·09
	9	0·17	1·40	0·00	1·07	1·11	1·89	0·24	2·08	6·20	0·05	0·00
	10	0·33	1·17	0·00	0·62	0·14	1·27	0·48	2·44	5·76	0·12	0·00
S. guyanensis	11	0·18	2·97	0·97	1·93	0·21	2·29	0·07	1·47	2·28	0·01	0·00
Hybrid	12	1·13	3·57	1·96	1·28	0·76	1·53	0·66	0·08	1·65	0·14	0·00
Hybrid	13	4·28	8·02	4.27	4·04	0·26	1·58	0·30	0·00	1·55	0·01	0·00
S. humilis	14	3·10	3·63	2·81	4·19	0·62	1·70	0·70	1·16	3·59	0·04	0·02
	15	5·72	3·66	3·68	5·02	1·22	2·26	1·00	1·21	4·21	0·10	0·00
Significant at		***	***	***	***	*	***	***	***	**	***	n.a.
L.S.D. $P = 0·05$		1·05	1·83	2·50	0·91	1·26	0·88	0·28	1·87	1·98	0·14	—

* $P < 0·05$; ** $P < 0·01$; *** $P < 0·001$; n.a., not analysed.

Table 3.5.2. Annual yield of *Urochloa* in *Stylosanthes* mixtures at four sites

Stylosanthes accession		Musgrave			'Lansdown' irrigated			Fanning River			'Lansdown' solodic	
		1969/70	1970/71	1971/72	1969/70	1970/71	1971/72	1969/70	1970/71	1971/72	1969/70	1970/71
S. subsericea	1	1·56	0·96	0·63	3·79	7·52	4·03	0·52	3·74	3·39	1·14	5·07
	2	1·29	0·72	1·22	3·58	6·95	3·63	0·98	4·67	3·77	1·24	4·47
	3	0·36	1·56	0·49	2·20	6·17	3·61	0·41	4·70	3·93	1·38	5·04
	4	0·19	0·96	0·54	4·38	7·86	3·19	1·24	3·82	3·53	1·63	5·44
	5	1·05	0·69	0·84	3·36	6·97	5·11	0·57	4·76	3·54	1·20	5·82
S. hamata	6	1·88	0·51	1·13	5·69	6·78	3·67	1·43	3·80	3·73	1·39	4·63
	7	2·30	1·30	0·67	5·23	5·15	2·92	1·26	5·92	3·24	1·48	4·83
	8	0·70	0·14	0·11	3·96	8·47	4·12	0·92	2·06	3·29	1·21	4·65
	9	3·19	1·73	1·32	6·81	5·98	3·30	1·58	3·35	2·83	1·46	4·59
	10	1·99	1·34	0·85	6·00	6·14	3·34	1·18	3·75	3·44	1·67	5·17
S. guyanensis	11	1·62	0·37	1·09	4·46	5·73	2·81	1·60	2·29	2·96	1·46	3·83
Hybrid	12	0·33	0·31	0·11	4·58	5·23	2·97	0·89	4·27	2·58	1·57	5·84
Hybrid	13	1·48	0·41	0·95	4·29	7·02	3·52	1·32	3·02	3·02	1·41	5·50
S. humilis	14	0·21	0·16	0·34	3·71	7·01	3·67	0·78	4·47	2·98	1·63	5·01
	15	1·08	0·18	0·42	2·90	6·32	3·77	0·43	2·94	2·65	1·28	4·89
Significant at		***	n.s.	n.s.	***	n.s.	*	***	*	n.s.	n.s.	n.s.
L.S.D. $P = 0·05$		1·68	1·13	1·16	1·00	2·00	1·08	0·38	1·92	1·06	0·43	1·59

Annual yield of dry matter (tonnes/ha)

* $P < 0·05$; *** $P < 0·001$; n.s., not significant.

The dimensions of the complete data matrix are 15 accessions $\times$ 4 sites $\times$ 3 years (except at site S) $\times$ 4 attributes. Since there are missing items of data, the program LINKED was used (Lance and Williams 1967). The data were treated as 60 site–accessions defined by 4 attributes linked over 3 years. The classificatory hierarchy was truncated at the 9-group level because the groups were considered sufficiently homogeneous for data evaluation, and the group means were calculated by GROUPER. The constitution of the groups by accessions and sites is shown in Table 3.5.3. It is immediately clear that there has been greater segregation of sites than of accessions; the number of performance groups representing each site is 3 at all sites except I which has 2 groups. The group means of the 4 attributes for each of the 3 years are summarized in Table 3.5.4; this is the basic performance table and requires relatively detailed consideration in conjunction with Table 3.5.3.

The performance trend of the accessions in successive years (Table 3.5.4) could be theoretically one of 4 types and all 4 types are discernible within our 9 groups as shown below:

Type	Performance of accessions	Group No.	Sites represented
(a)	High in all years	*1, 4, 8*	M–F, I, S
(b)	Initially low but improving with time	*6*	F–S
(c)	Initially high (low rank) but diminishing with time	*7*	F
(d)	Poor in all years	*2, 3, 5, 9*	M, M, I, S

Performance group *8* was placed in type (a) because even though the dry matter yields were low, the ranks were also low (Table 3.5.4): this is indicative of the low rainfall and poor physical and moisture storage properties of the solodic soil at site S.

The high-performance accessions at each site can now be read from Table 3.5.3: at M, group *1* accessions 8, 12, 13, 14 and 15; at I, group *4* accessions 2, 3, 5, 8 and 15; at F, group *1* accession 8; and at S, group *8* accessions 3, 5 and 8. Accession 8 is in the high-performance group at all four sites, and accessions 3, 5 and 15 at two sites. Particular note should be made of the improving performance of the group *6* accessions at F, especially accessions 6, 9 and 10 which had very high dry matter yields in the third year (Table 3.5.1).

Urochloa yields were predominantly site-dependent and the within-site pattern was not marked except in the first year. The significant differences in *Urochloa* yield described in the previous section are discernible in Table 3.5.4.

Table 3.5.3. Constitution of performance groups by accessions and sites

(Entries represent sites[A])

Accession		Group No.								
		1	2	3	4	5	6	7	8	9
S. subsericea	1		M			I		F		S
	2		M		I			F		S
	3		M		I			F	S	
	4		M			I	F			S
	5		M		I			F	S	
S. hamata	6			M		I	F			S
	7			M		I	F			S
	8	M, F			I				S	
	9			M		I	F			S
	10			M		I	F			S
S. guyanensis	11		M			I	F			S
Hybrid	12	M				I	F			S
	13	M				I	F			S
S. humilis	14	M				I	S	F		
	15	M			I			F		S

[A] M, Musgrave; I, 'Lansdown' irrigated; F, Fanning River; S, 'Lansdown' solodic.

Table 3.5.4. Attribute means for performance groups

Attribute	Group No.								
	1	2	3	4	5	6	7	8	9
Legume rating									
Year 1	3·8	8·2	13·2	3·2	10·4	11·6	3·8	2·0	9·3
Year 2	2·7	10·0	11·2	3·0	10·5	8·1	8·3	2·0	10·0
Year 3	3·5	8·0	13·5	5·8	9·1	8·5	8·2	—	—
Legume yield[A]									
Year 1	3·14	1·17	0·18	5·78	2·46	0·28	0·85	0·33	0·09
Year 2	5·17	1·96	1·38	1·67	0·63	1·04	1·28	0·10	0
Year 3	3·79	1·70	0·0	2·23	1·76	3·58	3·81	—	—
Grass rating									
Year 1	10·7	9·2	2·5	12·6	5·7	4·6	12·2	12·3	7·3
Year 2	13·2	6·5	4·2	6·6	8·7	8·4	6·0	6·0	8·6
Year 3	11·2	7·0	4·8	4·4	9·8	9·1	6·5	—	—
Grass yield[A]									
Year 1	0·81	1·08	2·34	3·21	4·90	1·34	0·61	1·27	1·43
Year 2	0·54	0·86	1·22	6·99	6·47	3·87	4·18	5·17	4·91
Year 3	0·87	0·86	1·00	4·04	3·33	3·18	3·38	—	—

[A] Mean yield (tonnes/ha).

Table 3.5.5. Rank for percentage legume in total annual yield of *Stylosanthes/Urochloa* swards

Rank[A]	Musgrave			Lansdown irrigated			Fanning River			Lansdown solodic
	1969/70 Acc. Group[B]	1970/71 Acc. Group	1971/72 Acc. Group	1969/70 Acc. Group	1970/71 Acc. Group	1971/72 Acc. Group	1969/70 Acc. Group	1970/71 Acc. Group	1971/72 Acc. Group	1969/70 Acc. Group
1	14	8	8	3	8	8	3	8	9	3
2	15	15	14	5	2	11	15	10	15	8
3	8	14	15	15	5	7	1	11	10	5
4	4	13	12	2	15	15	5	9	8	2
5	3	12	4	8	9	4	14	3	1	1
6	12	11	13	1	3	9	8	15	4	12
7	13	6	5	14	12	2	2	6	14	6
8	2	1	2	13	7	12	12	5	6	15
9	5	5	3	4	6	3	10	14	3	10
10	1	2	1	11	14	14	4	4	5	4
11	10	4	11	12	4	1	13	2	2	9
12	11	3	6	6	1	13	6	1	7	14
13	6	10	7	7	11	10	9	7	11	7
14	9	9	9	9	13	6	7	12	12	13
15	7	7	10	10	10	5	11	13	13	11
Significant at[A]	***	*	***	***	*	*	***	***	*	***
Range (%)	94–5	98–33	98–0	73–9	23–2	48–19	72–4	73–1	68–32	28–1

[A] Based on arcsine transformations.
[B] Linked (]) accessions do not differ significantly ($P = 0.05$).
* $P < 0.05$;   *** $P < 0.001$.

A principal component analysis of the annual yields of *Stylosanthes* and *Urochloa* at each site gave results consistent with the performance groups of Table 3.5.3 using only the first vector. However, the group separation using principal component analysis was not sharply defined.

Percent legume

Statistical analyses. The analyses of variance of the percentage of legume dry matter in the swards (using arcsine transformed data) were significant ($P < 0\cdot05$) for all years at all sites (Table 3.5.5). Excluding the first-year results, which were more variable, *S. hamata* 8 was ranked first for percentage legume in 5 out of the 6 (3 sites $\times$ 2 years) remaining analyses. The one exception was at F in the third year when *S. hamata* 8 was ranked fourth behind *S. hamata* 9, *S. humilis* 15 and *S. hamata* 10; however, the 4 accessions did not differ significantly.

Pattern analysis. The percent legume in the total yield for 15 accessions $\times$ 10 site years (excluding year 2 at site S) was analysed by the principal coordinate method of Gower (1966, 1967). Most of the variation could be accounted for by the first two vectors. The plot of vector 1 against vector 2 is shown in Fig. 3.5.1. Vector 1 separated the accessions mainly on the percent legume in the sward in the first year at I, M and F and in the third

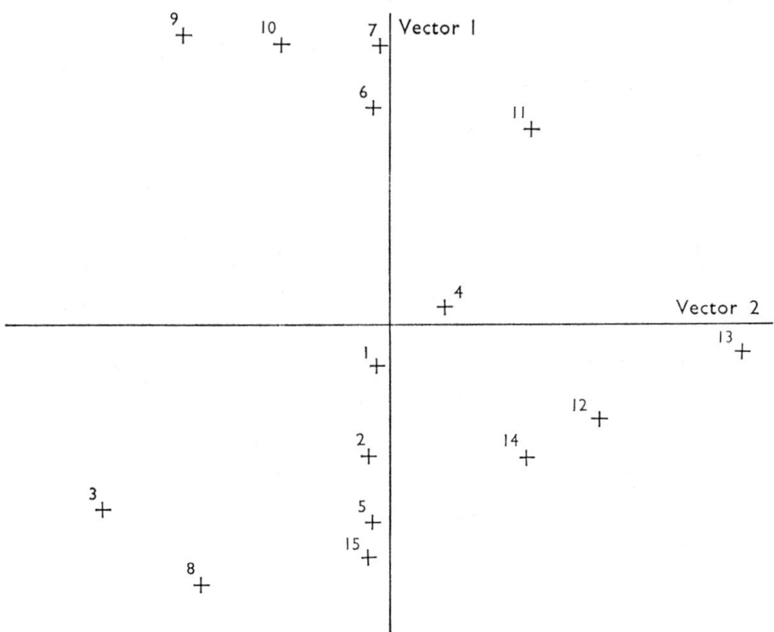

Fig. 3.5.1. Configuration in a Euclidean model of the 15 accessions for percentage legume in the swards $\times$ 10 site years: vectors 1 and 2.

year at M; accessions 3, 5, 8 and 15 gave a high percentage of legume whereas accessions 6, 7, 9, 10 and 11 were low. Vector 2 separated the accessions mainly on the percent legume in the swards in the second and third years at F and in the first year at S; accessions 3, 8, 9, 10 and 15 gave a high percentage of legume whereas accessions 12 and 13 were lower.

Yield of pods

Statistical analyses. There were significant differences ($P < 0.01$) between the accessions in the yield of pods recovered off the soil surface at each site (Table 3.5.6). The two *S. humilis* accessions and some *S. subsericea* accessions were generally higher-yielding than accessions of *S. hamata*, the two hybrids and *S. guyanensis*. At I, the 1971 yield was unreplicated and not analysed.

Pattern analysis. The 15 accessions were classified for pod yields at the 3 sites linked over 3 years. The resultant dendrogram was truncated at the 5-group level and the group means determined by GROUPER.

The pattern analysis showed:

Accession No.	Comment
11, 13	Produced few or no pods at all sites
12	As above but it did produce 32 kg ha^{-1} of pods at I in 1971 and 16 kg ha^{-1} at M in 1970
6, 7, 9, 10	Did not produce pods at M, gave low yields at F in both years
5, 8	Generally gave comparable yields of pods at all sites to the previous group of accessions except in 1970 at M and I where the yields were higher
1, 2, 3, 4, 14, 15	Produced the highest pod yields at all sites in all years except 1971 at F. The group included all accessions of *S. humilis* and *S. subsericea* with the exception of 5

Discussion

The agronomy of the *Stylosanthes* species and accessions included in this experiment has been discussed in detail (Edye, *et al.* 1975*a*). In this chapter, it is only necessary to consider the use of pattern analysis in plant evaluation.

In the preliminary evaluation of introduced plant material at a wide range of sites, large differences are generally immediately obvious and occur frequently. The agronomist is mainly interested in summarizing these differences by comparing the relative performance and range of adaptability of groups of new species or accessions with the objective of selecting promising accessions for further evaluation. Comparative work of this type often involves a large number of accessions and sites and the agronomist

Table 3.5.6. Yield (kg/ha) of *Stylosanthes* pods at three sites

Rank[A]	Musgrave 1970		Lansdown irrigated 1970		1972		Fanning River 1971		1972	
	Yield	Acc. Group[B]	Yield	Acc. Group	Yield	Acc. Group	Yield	Acc. Group	Yield	Acc. Group
1	875	15	651	15	207	4	559	3	445	3
2	610	14	598	4	165	14	558	15	278	15
3	446	4	481	14	163	2	493	1	232	1
4	372	3	402	3	151	15	357	5	216	14
5	284	2	389	1	134	1	180	7	187	7
6	272	1	268	2	130	3	171	8	160	2
7	192	8	239	5	117	9	163	9	142	5
8	173	5	117	8	101	10	147	14	138	4
9	16	12	24	9	96	8	132	10	131	9
10	15	13	15	6	75	7	128	6	112	8
11	4	11	7	7	73	5	120	4	106	10
12	0	6	9	10	52	6	96	2	30	6
13	0	7	4	11	3	12	0	11	0	11
14	0	9	3	12	0	11	0	12	0	12
15	0	10	3	13	0	13	0	13	0	13
Significant at[A]	***		***		***		**		***	

A Based on square root or log $(x+1)$ transformations.

B Linked (]) accessions do not differ significantly $(P = 0.05)$.

** $P < 0.01$;   *** $P < 0.001$.

will have only limited prior knowledge of which comparisons between accessions will prove to be worth while. For example, it can occur that a completely new species recently introduced and never previously tested will perform better than existing cultivars of other species at some sites but not at others. Such inconsistencies in relative performance make it difficult to assess the overall performance of new introductions.

Similar complex systems of preliminary evaluation were studied by Burt *et al.* (1974) who evaluated 27 accessions over 20 site-years by 4 attributes and Edye *et al.* (1975*b*) with 33 accessions over 24 site-years by 6 attributes. Classical statistical procedures were not chosen for use in these particular cases because of the large number and the uncertainty of the statistical comparisons that were possible: for any one attribute over 10^5 comparisons were possible and even within accessions there were more than 4500 comparisons. With such a large number of comparisons and with so many likely to show significance, it would prove difficult to sort out major and minor differences. Burt *et al.* (1974) and Edye *et al.* (1975*b*) only wished to elucidate the pattern of major differences so as to elicit the comparisons that would justify later statistical testing.

Simple data sets are more likely to occur in the more advanced stages of evaluating promising accessions. Such was the case with the experiment described in this paper.

The simple data set reported in this paper was adequately analysed by classical statistical procedures and by pattern analysis. The results showed that the two approaches do not conflict with one another and that they agreed substantially. Both methods were looking for large differences in a simple data set incorporating a lot of irrelevant environmental disturbance and although based on different mathematical models they produced similar results. The statistical comparisons that were possible were sufficiently limited to enable ready comprehension and interpretation. Yield trends with time were followed in this experiment using principal component analysis. One important difference between the two methods was that pattern analysis produced discrete groups of accessions whereas the statistical procedures showed that group separation was rarely sharply defined.

From our experience with both methods, we suggest that pattern analysis is more simple to carry out, interpret and present than statistical analyses when one is dealing with very large and complex data sets. However, this may not be immediately obvious from the simple data set analysed in this experiment. One of our main objectives in this paper was to demonstrate the degree of agreement between the methods so that pattern analyses of complex data sets can be more confidently accepted by agronomists.

L. A. Edye

References

Burt, R. L., Edye, L. A., Williams, W. T., Grof, B., and Nicholson, C. H. L. (1971). Numerical analysis of variation patterns in the genus *Stylosanthes* as an aid to plant introduction and assessment. *Aust. J. Agric. Res.* **22**, 737–57.

Burt, R. L., Edye, L. A., Williams, W. T., Gillard, P., Grof, B., Page, M., Shaw, N. H., Williams, R. J., and Wilson, G. P. M. (1974). Small-sward testing of *Stylosanthes* in northern Australia: preliminary considerations. *Aust. J. Agric. Res.* **25**, 559–75.

Edye, L. A., Burt, R. L., Williams, W. T., Williams, R. J., and Grof, B. (1973). A preliminary agronomic evaluation of *Stylosanthes* species. *Aust. J. Agric. Res.* **24**, 511–25.

Edye, L. A., Burt, R. L., Norris, D. O., and Williams, W. T. (1974*a*). The symbiotic effectiveness and geographic origin of morphological–agronomic groups of *Stylosanthes* accessions. *Aust. J. Exp. Agric. Anim. Husb.* **14**, 349–57.

Edye, L. A., Burt, R. L., Nicholson, C. H. L., Williams, R. J., and Williams, W. T. (1974*b*). Classification of the *Stylosanthes* collection, 1928–69. CSIRO Aust. Div. Trop. Agron. Tech. Pap. No. 15.

Edye, L. A., Field, J. B., and Cameron, D. F. (1975*a*). Comparison of some *Stylosanthes* species in the dry tropics of Queensland. *Aust. J. Exp. Agric. Anim. Husb.* **15**, 655–62.

Edye, L. A., Williams, W. T., Anning, P., Holm, A.McR., Miller, C. P., Page, M. C., and Winter, W. H. (1975*b*). Sward tests of some morphological–agronomic groups of *Stylosanthes* accessions in dry tropical environments. *Aust. J. Agric. Res.* **26**, 481–96.

Gower, J. C. (1966). Some distance properties of latent root and vector methods used in multivariate analysis. *Biometrika* **53**, 325–38.

Gower, J. C. (1967). Multivariate analysis and multidimensional geometry. *Statistician* **17**, 13–28.

Lance, G. N., and Williams, W. T. (1967). Note on the classification of multi-level data. *Comput. J.* **9**, 381–2.

Williams, W. T., Edye, L. A., Burt, R. L., and Grof, B. (1973). The use of ordination techniques in the preliminary evaluation of *Stylosanthes* accessions. *Aust. J. Agric. Res.* **24**, 715–31.

Case 3.6. The Use of Classification to Elucidate Nutrient Responses in *Stylosanthes*

The problem

The data in this case history were from a glasshouse experiment on the responsiveness to phosphorus of 30 accessions representing 7 species of the genus *Stylosanthes* (Jones 1974). The experiment was of randomized block design with 6 phosphorus treatments (5 rates of monocalcium phosphate and 1 of rock phosphate), 30 accessions and 2 replications. The primary data consisted of the dry weight of the tops and their nitrogen and phosphorus concentrations at each of the 6 treatments. The plant material had to be bulked across replicates for the chemical analyses, so conventional analyses of variance could be carried out only on the dry-weight data.

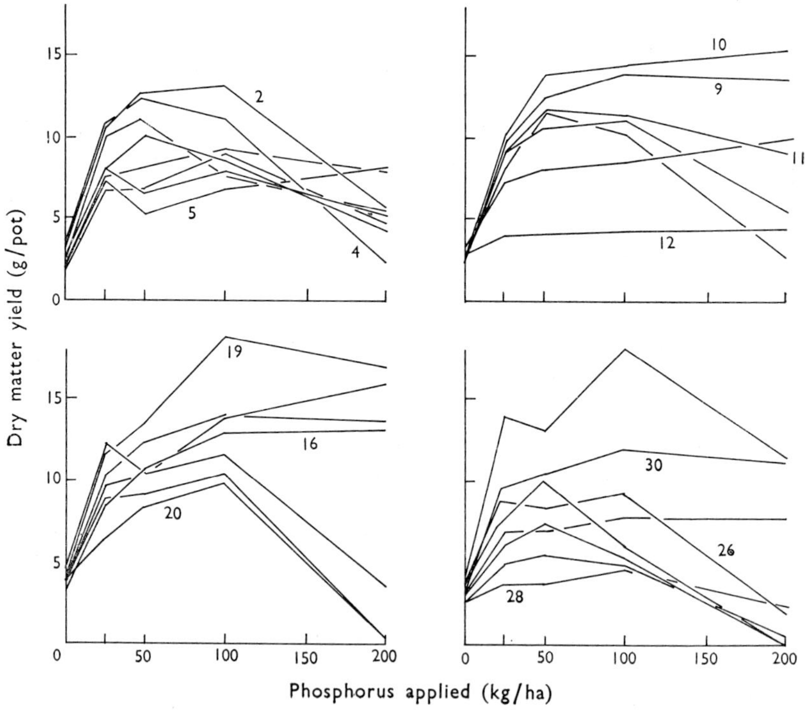

Fig. 3.6.1. Yield of dry matter in the plant tops plotted against phosphorus applied for each of the 30 accessions. The accessions numbered are referred to in the text.

Fig. 3.6.1 shows the dry-matter-yield curves for all accessions and Fig.3.6.2 the nitrogen and phosphorus concentrations of four of the accessions labelled in Fig. 3.6.1. In describing and interpreting data such as these, it is very convenient to be able to group accessions with similar patterns of response. Considering a single attribute at a time, one can do some grouping by eye. For example, in Fig. 3.6.1, one might group accessions with 'normal' response curves (numbers 9, 10, 16, 30), those with depressed yields at high phosphorus applications (4, 26) and so on. Even with this fairly small set of primary data, however, it becomes quite difficult to group accessions with similar responses when one must keep three attributes in mind.

Ordination

The primary data were, therefore, first subjected to an ordination procedure to see what degree of likeness there was between accessions.

The squared Euclidean distance between each pair of accessions over all measurements was calculated (Burr 1968) and these were mapped into Euclidean space by the principal coordinate analysis of Gower (1966, 1967).

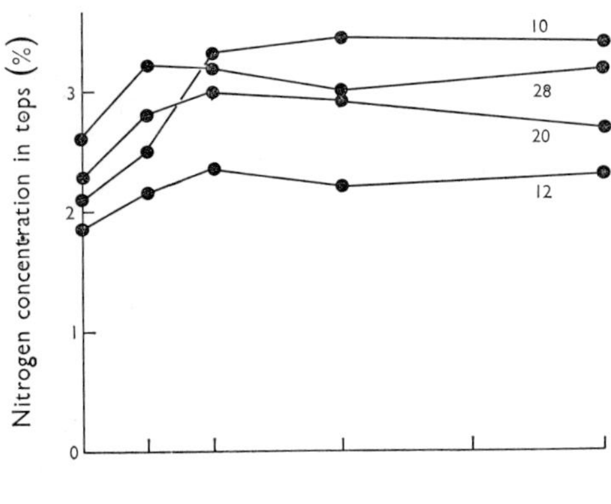

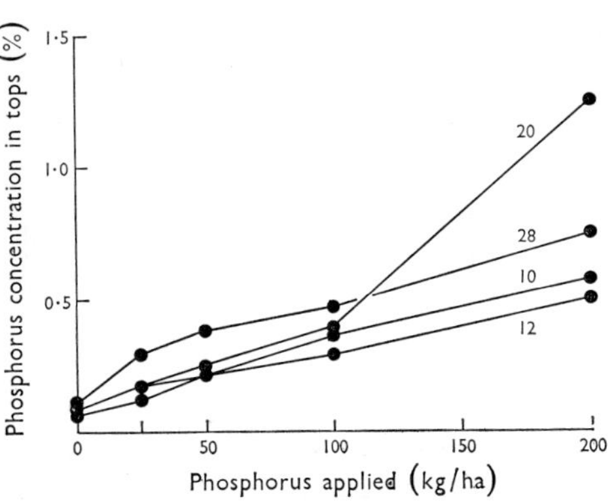

Fig. 3.6.2. The effect of level of applied phosphorus on the concentration of nitrogen and phosphorus in the plant tops of four accessions.

The first five latent roots (2·81, 1·62, 0·95, 0·66 and 0·27) showed a fairly slow drift downwards, indicating that there was a considerable amount of uncorrelated variation in the data. The first two roots are shown in Fig. 3.6.3. Although they account for 69% of the total variation, it is clear that these two vectors alone do not give very distinct groups (apart perhaps from one consisting of Nos 5, 13 and 26).

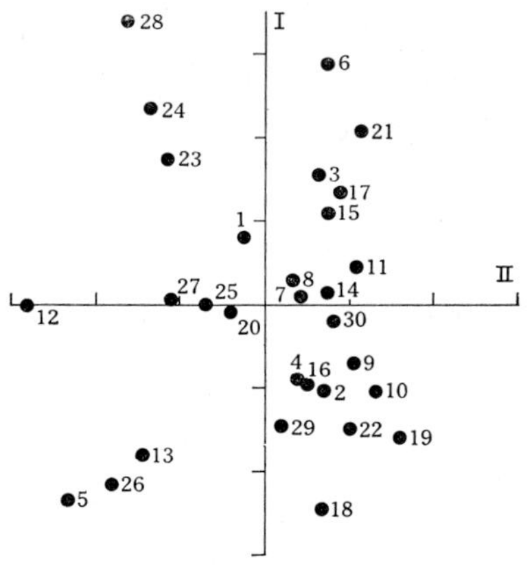

Fig. 3.6.3. Spatial arrangement of accessions using the first two vectors from ordination procedure.

Classification

In view of the results of the ordination, an intensely clustering strategy—whereby the squared Euclidean distances between accessions were grouped by the incremental sum of squares strategy of Burr (1970)—was then used. The resulting classification is shown in Fig. 3.6.4. Attributes that made the greatest contribution to the distances between these six groups were determined using the diagnostic program GROUPER (Lance *et al.* 1968).

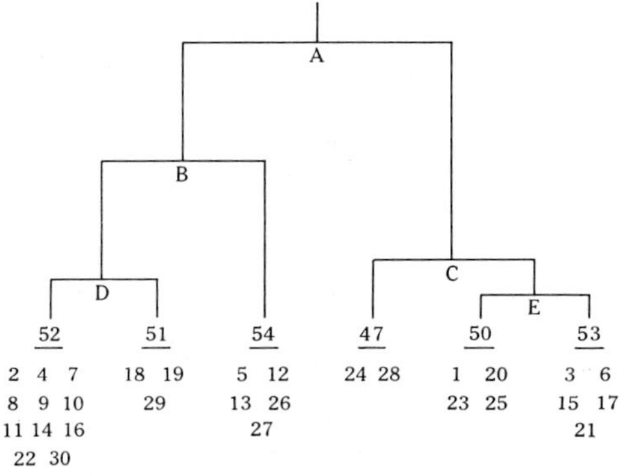

Fig. 3.6.4. Hierarchy for the classification of the data for dry weight and N and P concentrations. The numbers of the groups so formed are underlined.

Although the procedure used is agglomerative. it is convenient to adopt the convention of reading the hierarchy downwards and referring to the fusions as divisions. The division at A separated out to the right accessions with higher phosphorus concentrations in all treatments but lower yields at the highest P treatment (P_{192}) than those to the left. The former were further divided at C and group 47 formed. Compared with the remaining accessions, the group had higher phosphorus concentrations and lower yields at low to medium levels of applied P. The division at E was based on nitrogen concentrations in all treatments, accessions with lower values throughout going into group 50.

Returning to the left side, the division at B placed accessions with low yields and nitrogen concentrations in group 54. The final division at D separated out three accessions (group 51) which had higher dry-matter yields than the remainder at all P treatments.

The usefulness of the groupings is illustrated using both primary data (Fig. 3.6.5) and derived data (Fig. 3.6.6). Although there is considerable overlap of groups in the derived data, the classification does, nevertheless, provide a convenient framework for discussion.

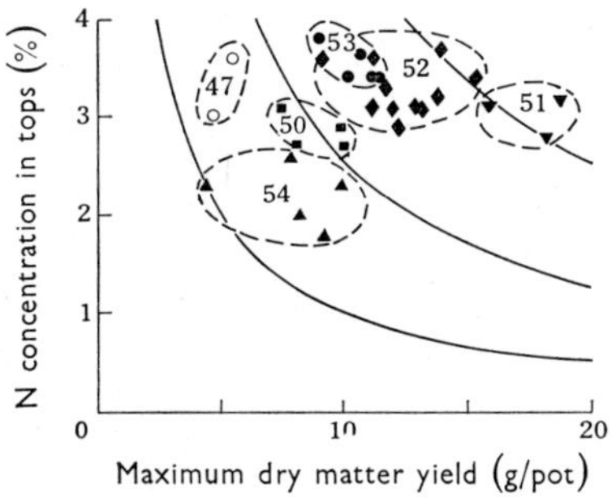

Fig. 3.6.5. The relation between nitrogen concentration in the tops at maximum yield of dry matter and that maximum yield. The various groups of accessions from the classification (Fig. 3.6.4) are outlined.

Fig. 3.6.6. Diagrammatic representation of the data on phosphorus uptake under conditions of severe to marginal phosphorus deficiency. Linear regressions were fitted to the uptake data for each accession on the mean uptake for all accessions for four of the treatments (P_0, P_{RP}, P_{24} and P_{48}). The figure shows the slopes of the regression line and the fitted values for phosphorus uptake at P_0, for all accessions in each of the six groups.

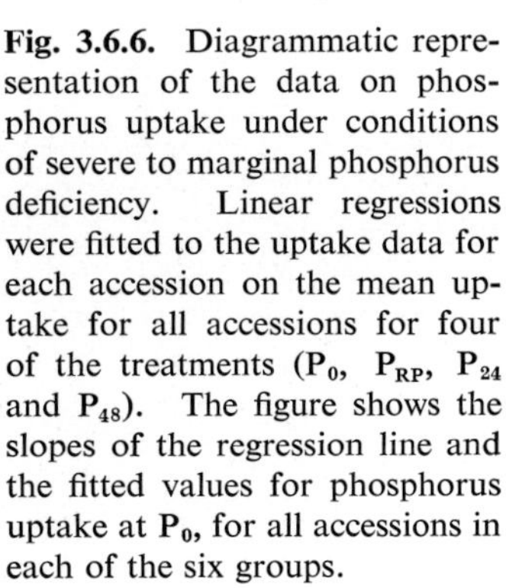
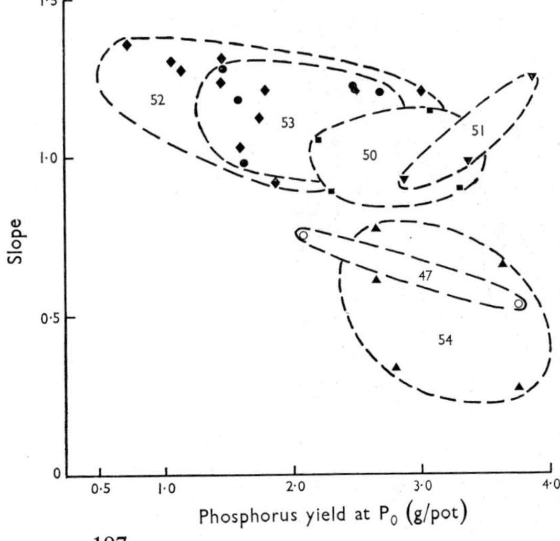

A second classification was also carried out using many of the primary data and the following derived data:

Yield of P in tops for all treatments

Yield of P in tops and roots for treatment P_0

P distribution between tops and roots for P_0

'Critical value' (P concentration in tops at 90% of maximum yield)

Treatment at which maximum dry matter yield was achieved.

Table 3.6.1. Comparison of groupings from the two classifications

		Classification 2			
Classification 1	1	2	3	4	5
52				2, 4, 7, 8, 11, 14, 30	9, 10, 16, 22
51					18, 19, 29
54	5, 12, 13, 27		26		
47		24, 28			
50		23	20, 25	1	
53			17, 21	3, 6, 15	

The resultant groups are compared with those from classification 1 in Table 3.6.1. Agreement between the two classifications was quite good, although both have their weaknesses. No. 26, for example, seems to be misplaced in classification 2, as it had quite low nitrogen contents in all treatments and was more similar to 5, 12, 13 and 27 in this respect. Classification 2 had a major division between groups 3 + 4 and group 5, the latter containing accessions with higher maximum yields and, at P_{192}, higher relative yields of both dry matter and phosphorus. Classification 1 did not make a division at this stage, but it did place the three highest-yielding accessions (18, 19, 29) in a small group on their own.

Conclusions

Classification procedures were very useful tools in this study since they helped simplify the description and interpretation of the data. Classification 1 was selected for use in the publication arising from this work (Jones 1974) because it used only primary data and was therefore easier to describe and because it gave similar groupings to those produced by the more complicated classification 2.

R. K. JONES

References

Burr, E. J. (1968). Cluster sorting with mixed character types. I. Standardization of character values. *Aust. Comput. J.* **1**, 97–9.

Burr, E. J. (1970). Cluster sorting with mixed character types. II. Fusion strategies. *Aust. Comput. J.* **2**, 98–103.

Gower, J. C. (1966). Some distance properties of latent root and vector methods used in multivariate analysis. *Biometrika* **53**, 325–38.

Gower, J. C. (1967). Multivariate analysis and multi-dimensional geometry. *Statistician* **17**, 13–28.

Jones, R. K. (1974). A study of the phosphorus responses of a wide range of accessions from the genus *Stylosanthes*. *Aust. J. Agric. Res.* **25**, 847–62.

Lance, G. N., Milne, P. W., and Williams, W. T. (1968). Mixed-data classificatory programs. III. Diagnostic systems. *Aust. Comput. J.* **1**, 178–81.

Case 3.7. Analysis of Electrophoretic Data using Pattern-analysis Techniques

Several species of the genus *Stylosanthes* have been shown to be successful pasture plants for the tropical and subtropical regions of Australia. Consequently, considerable effort has been put into the introduction and agronomic testing of additional accessions in the search for legumes superior to those at present in use, and some 600 *Stylosanthes* accessions have been introduced in the last 25 years. Many introductions have been found to be similar and some may well be identical, having been introduced by different collectors at different times. Because of the time and resources involved in the testing of potentially similar plants, both during and after the quarantine stage, a rapid means of characterizing new introductions was sought. Seed protein patterns, which are not sensitive to environmental changes, were considered to be potentially suitable and had the additional advantage that the characterization could be carried out on as little as one seed.

Experimental procedures

A total of 182 *Stylosanthes* accessions covering a range of types within 16 species were examined using polyacrylamide gel electrophoresis, and 40 different proteins were observed. Pattern-analysis techniques were then used in an attempt to group together accessions having similar protein patterns and to determine whether interspecific relationships obtained using biochemical data were similar to classifications using morphological and/or agronomic data.

Densitometer traces of the protein patterns of each accession were obtained and, after standardization between different gels, these traces were represented by 100 optical density readings along the migration axis of each sample (for example see Fig. 3.7.1). Thus the complete data set consisted

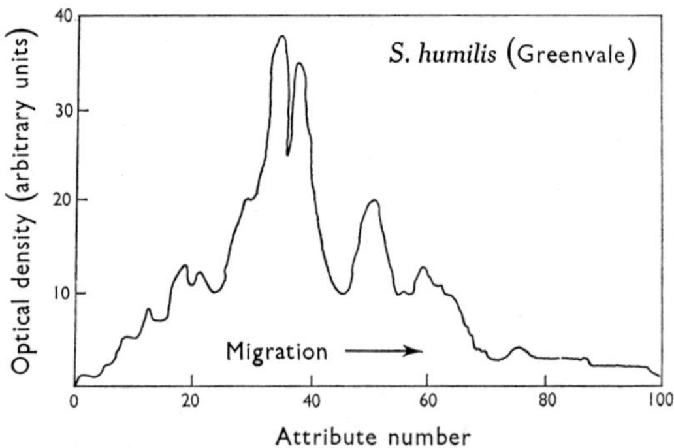

Fig. 3.7.1. Densitometer trace for seed-protein pattern of *Stylosanthes humilis* cv. Greenvale.

of 182 individuals (*Stylosanthes* accessions) defined by 100 numeric attributes (optical density readings). The attributes were continuous variables of unknown statistical distribution, and there was no previous experience to suggest which of the many possible similarity measures and fusion strategies would be appropriate for classification. It was therefore decided to treat the study as a methodological exercise and to investigate the results obtained by a number of different numerical models. Three measures were chosen: Euclidean distance, the Canberra metric and the correlation coefficient. Within Euclidean distance three further alternatives were examined:

(*a*) Unstandardized distance where the coordinates were used for calculation exactly as they stood in the raw data. In this case all minor variation would be swamped, only the major peaks making any appreciable contribution to the classification.

(*b*) Standardized distance, giving each of the 100 attributes the opportunity of making an equal contribution to the classification.

(*c*) Truncation of data. Method (*a*) may give too much emphasis to major peaks and method (*b*) too much emphasis to minor fluctuations. The data were truncated so that all attribute values less than a predetermined value (6 in our case) were put equal to zero, before standardization as in (*b*).

Table 3.7.1. Summary of classificatory procedures used and of data subsets

a, 35 accessions representing 12 species; b, 75 accessions of *S. guyanensis;* c, 110 accessions representing 15 species, excluding *S. guyanensis;* d, complete data set: 182 accessions representing 16 species

| | Similarity measure | | | | |
| | Euclidean distance | | | Canberra | Correlation |
Fusion strategy	Unstandardized	Standardized	Truncated	metric	coefficient
Group average	—	a	—	a	a
Flexible	—	d	—	a	a
Incremental sum of squares	a, b	a, b, c	b	—	—

Three fusion strategies were also chosen. These were (i) the intensely clustering 'incremental sum of squares', applicable only to Euclidean distance; (ii) 'flexible' sorting, with the intermediate level produced by setting the cluster-intensity coefficient, β, to the value of $\beta = -0.25$; and (iii) the almost space-conserving 'group average' strategy.

All combinations of measures and strategies involved nine analyses; it was, however, considered that to carry out all nine on the complete data set would involve a computational expense that would be difficult to justify. Preliminary classifications were therefore carried out on subsets of the data; the constitution of these subsets and the analyses for which they were used are summarized in Table 3.7.1.

Results

Presentation of dendrograms is precluded by lack of space, and only a brief summary of the species relationships suggested by the different methods is given below.

All accessions of the species *S. guyanensis* (with the exception of the fine-stem stylo type) separated from the other species when the Canberra metric and Euclidean coefficients were used with the small data set (a in Table 3.7.1.), regardless of fusion strategy. *S. capitata* and *S. montevidensis* accessions also separated at a high level in the hierarchy in most analyses. The relationships between species varied according to the method used, and these differences appeared to be more dependent on the similarity coefficient than on the selected fusion strategy. In all analyses using Euclidean and Canberra metric coefficients, close relationships were shown between the species *S. scabra* and *S. fruticosa*, between *S. sundaica* and *S. subsericea* and between *S. viscosa* and *S. hamata*. Analyses using the correlation coefficient separated the *S. guyanensis* accessions into two groups at a high level, and were not further considered.

Comparison of the analyses of *S. guyanensis* accessions (b in Table 3.7.1) revealed only minor differences, indicating that standardization did not magnify the importance of the background near-zero values which belonged to distance positions (attribute numbers) at which no protein bands were present (e.g. attributes 88–100 of Fig. 3.7.1).

The standardized Euclidean coefficient was superior to the other coefficients in forming monospecific groups, and was selected for the analysis of the full data set, using the flexible fusion strategy (d in Table 3.7.1). Agglomerative classifications are prone to a certain amount of misclassification, and at the time the work was undertaken no objective reallocation strategies had been developed. Cases where an accession appeared to have been markedly misclassified (10 cases out of 182) were redressed intuitively by comparison with the subset classifications.

It was first noted that the species relationships suggested by this analysis were not in agreement with the accepted taxonomic division of the genus into the two sections *Styposanthes* and *Stylosanthes*. *S. scabra* was the only species where the groups obtained agreed with the M–A classification of Burt (Case 3.3). Accessions of *S. humilis* and *S. fruticosa* showed little variation in protein pattern and M–A groups of these species could not be distinguished. The 32 *S. humilis* accessions, however, were divided into introductions from Brazil and Central America, with a few exceptions.

In the densitometer traces of the data, proteins of low concentration were represented by fewer attributes than proteins of high concentration, and hence had less influence on the groups obtained by pattern analysis.

Further characterization of some accessions was possible by close visual inspection of the densitometer traces. For example, in species *S. subsericea* and *S. hamata*, M–A groups could be distinguished by several minor bands, which had little or no effect on the groupings obtained by pattern analysis.

To summarize, pattern-analysis techniques were useful in the initial separation of the large number of accessions examined, where visual matching of densitometer traces would have been impossible. The groups obtained did not correspond well with M–A variants within species, nor did the species relationships correspond with the accepted taxonomic relationships. Minor variation in densitometer traces, detected by visual means but not large enough to influence the pattern analyses carried out, could be used to detect M–A variation in some species. Further work is required to determine whether this latter problem can be overcome by different pattern-analysis methods.

P. J. ROBINSON

Case 3.8. Ordination of Soil Data

Ordination techniques may be used with one or more of the following objectives in mind: (1) summarization and/or display of information in a convenient form by reducing the dimensionality of a set of intercorrelated variables; (2) hypothesis generation, i.e. attempting to identify a pattern in, or factors underlying, a set of variables; (3) achievement of maximum discrimination among entities (rather than groups as is the case with discriminant function analysis) in a small number of dimensions; and (4) data transformation prior to analysis, e.g. the generation of uncorrelated variables for use in regression to avoid problems of interpretation of regression coefficients resulting from covariance. Almost invariably more than one objective is involved in considering any particular set of data.

Problems of ordination (and classification also) which are specific to a particular area of research are those associated with choosing a suitable model for the entities being considered and deciding which attributes should be used to describe the entities. This is discussed in Case 3.10 with regard to soils; by and large, most of the discussion below could apply to other research areas also.

Most published accounts of ordination of soil data involve principal component analysis, although one of the earliest papers in this field (Hole and Hironaka 1960) used a technique based on the Bray–Curtis dissimilarity coefficient plus a 'geometric' ordination, i.e. an analogue of principal coordinate analysis (Gower 1966). The use of principal component analysis has been criticized by a number of authors (e.g. Gauch and Whittaker 1972; Beals 1973) where vegetation or animal communities are the domains being examined, primarily on the grounds that the model of species-dimensional space and principal component analysis based on it are ecologically unsound. This may well be so, as in general the response of a species (an attribute of a community, the latter constituting an entity) to an environmental gradient is far from linear or even uni-directional. In the case of soils, however, attributes often do show uni-directional, if not strictly linear, responses to environmental gradients; hence principal component analysis is probably a reasonable ordination technique in many situations involving soil data.

The results of subjecting several sets of soil data to principal component analysis (Cooley and Lohnes 1971) or factor analysis (Harman 1960) are presented and discussed below in an attempt to illustrate how some of the objectives mentioned above may be achieved and the difficulties that may arise.

CASE 3.8

Example 1: Trace elements in some Queensland profiles

The first example deals with a very simple raw data matrix of 28 entities (soils) × 10 attributes (trace elements). The soils were representative profiles of nine great soil groups occurring in Queensland and the attributes were weighted-mean trace-element concentrations (Oertel and Giles 1963). The data were transformed as follows: none (Mo), square root (Ga, V, Zr) and logarithm (Co, Cu, Mn, Ni, Zn, Fe). The similarity matrix of Euclidean distances has been published elsewhere (Moore and Russell 1967, Table III).

Table 3.8.1. Principal component pattern matrix[A] for trace-element suite of 28 Queensland soil profiles[B]

Attribute	Loading for PC1	Loading for PC2	Loading for PC3	Loading for PC4	h^2 [C] PC1–3	PC1–4
Cobalt	0·81	0·46	0·08	0·14	0·875	0·895
Manganese	0·82	0·41	0·19	0·15	0·868	0·890
Copper	0·90	0·31	0·18	−0·14	0·942	0·961
Zinc	0·85	0·27	−0·00	0·12	0·789	0·802
Nickel	0·91	0·16	−0·03	0·02	0·848	0·848
Iron	0·86	−0·39	−0·17	−0·20	0·928	0·970
Gallium	0·81	−0·45	−0·18	0·02	0·897	0·897
Vanadium	0·82	−0·36	−0·09	−0·40	0·807	0·967
Zirconium	0·46	−0·68	−0·08	0·52	0·686	0·960
Molybdenum	0·09	−0·46	0·88	−0·05	0·993	0·996
Vp	5·975	1·729	0·929	0·533	8·633	9·186
Percentage of total variance	60	17	9	6	86	92

[A] Principal components 5–10 not significantly different, $P = 0·01$.
[B] Original data from Oertel and Giles (1963).
[C] Communality for each attribute, representing the proportion of its variance retained in the reduced number of principal components.

Principal component analysis of the correlation matrix or principal coordinate analysis of the Euclidean distance similarity matrix result in the same ordination. The principal component pattern matrix is shown in Fig. 3.8.1. The value Vp is the eigenvalue or variance accounted for by a principal component. The value h^2 (sum of squares of coefficients) represents the contribution of a particular variable to the set of principal components under consideration. The sum of Vp over principal components or h^2 over variables is a measure of the proportion of the total variance accounted for by the principal components under consideration. In this example total

205

variance is 10 and thus the first three and first four components account for 86% and 92% respectively of the total variance.

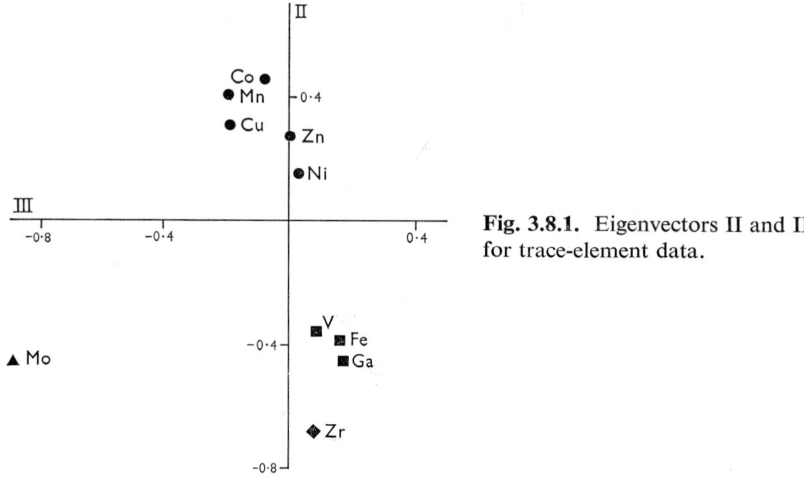

Fig. 3.8.1. Eigenvectors II and III for trace-element data.

The Bartlett–Lawley test (Seal 1964, p. 115) indicates that the last six principal components are not significantly different from each other. This significance test is, however, a function of the size of the sample of entities (cf. significance test for correlation coefficients) and is of limited use. In particular, in going a step further and applying factor analysis it is not a satisfactory criterion to use in the choice of number of factors to rotate (see below).

Principal component analysis may be considered a unique instance of factor analysis in which the (common) factors are orthogonal, equal in number to the variables and account for all the variance in the set of data (i.e. in factor analytic terms each of the communalities equals unity). Thus it is mathematically clear-cut, but interpretation of the components in real-world terms is frequently difficult. The following quotation from Holland (1969) is pertinent here: 'While the results may define intelligible biological concepts they will not always do so, and the method cannot be relied on to derive biological laws *ab initio*. If these restrictions are not recognized and attempts are made to force a biological interpretation on to each and every principal component, or insufficient account is taken of the nature of the original data in relation to the mathematical properties of the derived variates and the ultimate purpose of the investigation, the results can be most misleading.'

In this first example, however, reasonable interpretations of at least the first three components are possible. As is usual with a fairly homogeneous correlation matrix, the first component can be considered a general magnitude or size measure, i.e. a measure of overall trace-element status. Molybdenum makes virtually no contribution and zirconium little, but the remaining eight elements contribute about equally. In Fig. 3.8.1 the relationships between the 10 elements in a coordinate system of the second and third principal components are shown. The second component is a contrast between Co, Mn, Cu, Zn and Ni (with contributions in that order) and Zr, Ga, Fe, V and Mo (with Zr making the largest contribution). In general, the first five elements are found in minerals that weather more easily than those containing the remaining elements. Zirconium, which is generally present in very resistant minerals, makes the largest contribution. Thus the second component may be regarded as being related to past weathering history. The third principal component is dominated by molybdenum, the remaining nine trace elements making a contribution of only 17% to that particular component. The fourth (and remaining statistically significant) component is essentially a contrast between vanadium and zirconium. We have no suggestion for concrete interpretation of this. Fortunately, however, it accounts for only 6% of the total variance and for most purposes could probably be ignored.

In Fig. 3.8.2 are shown the relationships between the 28 soils in a coordinate system consisting of the first two principal components, which between them account for 77% of the variation in the data. It can be seen,

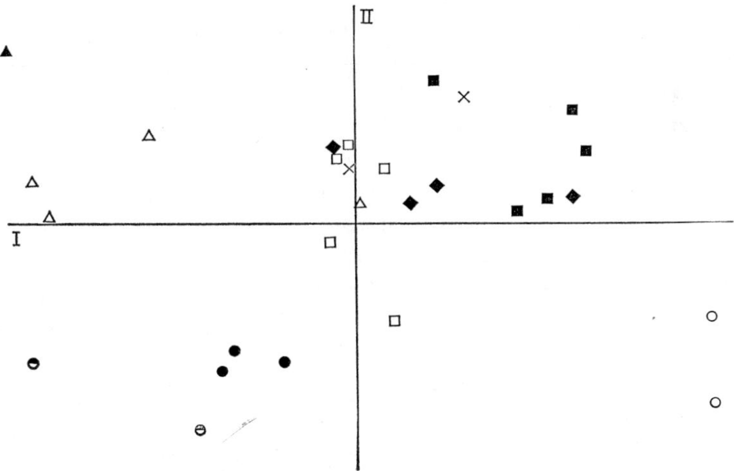

Fig. 3.8.2. Soils plotted on plane of principal components I and II. ▲ Red podzolic; △ solodized solonetz; ● red earth; ◓ lateritic podzolic; ○ krasnozem; ◼ black earth; ◆ red-brown earth; □ grey or brown soil of heavy texture; × rendzina.

for example, that the krasnozems are at the 'high' end of the first (or size) component (X-axis), the podzolic soils at the 'low' end. Red earths are towards the 'more weathered' end of the second component (Y-axis), black earths towards the 'less weathered' end. It is of interest to note that one of the soils in the red earth cluster (Fig. 3.8.2) was originally identified as a krasnozem in the field as its morphological characteristics made it a borderline case. Its trace-element status, as expressed by its position in the plane of the first two principal components, places it unequivocally with the two other red earths.

We can see in this example that principal component analysis has succeeded in satisfying at least two of the objectives listed earlier in the paper. First, dimensionality has been reduced from 10 attributes to 2 or 3, provided we are prepared to accept the loss of 23% or 14% respectively of the information present in the original set of data. Note, however, that the discarded information is not error or 'noise', and any one of the remaining components may be of interest; nevertheless, the interest would be misplaced if the component were not statistically significantly different from the succeeding components. Secondly, interpretation of the first three components in terms of the concrete world seems reasonable and we can hypothesize about the overall trace-element status, weathering history and molybdenum status of the soils.

This particular illustrative example is rather trivial in that well-defined ideas about pattern in the data existed before the analysis was done, e.g. the soils had already been identified in terms of the great soil group classification. Nevertheless, in soil science the researcher is often confronted with a mass of data in which the structure is not obvious and about which he has few preconceived notions.

Factor analysis is a technique that has a history rooted in the social sciences, particularly psychology. Originally it was used in the search for a single underlying factor which could explain the results of all intelligence tests. With hindsight, it is not surprising that it has not succeeded in accomplishing this. However, factor analysis is still directed primarily towards trying to detect underlying factors that achieve expression through a suite of measured attributes. Briefly, the variation associated with each attribute is partitioned into that due to one or more *common* factors, shared with other attributes, and that due to a *specific* factor associated with that particular attribute alone. In practice the specific factors are usually ignored and only the common factors considered. Operationally, the axes derived from an analysis such as principal component analysis are rotated according to rules which in some way maximize the loadings (coefficients) of some of the attributes on each of the axes. Problems that arise are choice of number of common factors to extract (assuming that there are fewer than the number of original attributes) and how to estimate communality, i.e. the extent to which the attributes are interrelated.

Choice of the number of common factors to extract is usually based on the number of eigenvalues greater than (approximately) 1 appearing in the principal component analysis. The contribution of each principal component to overall variance and significance or otherwise of eigenvalues also provides some assistance in making a choice. Although arbitrary, the 'greater than one' rule of thumb appears to have worked satisfactorily in many factor analyses (Harman 1960, p. 363). As mentioned above, the use of statistical significance (between eigenvalues) alone as a criterion is not satisfactory.

Commonly communality is represented by squared multiple correlation coefficients in the diagonal of the correlation matrix. As mentioned previously, if the number of factors is made equal to the number of attributes and the communality is represented by unities in the diagonal the factor analysis reduces to a principal component analysis.

In the present instance three and four factors have been extracted in two separate analyses. Table 3.8.2 is a simplified way of presenting the resulting matrices. Although rotational procedures may be extremely sensitive to the number of factors retained (Kaiser and Caffrey 1965), it is evident that in this case the three factors of the first analysis and the first three (of four) factors of the second are similar.

Table 3.8.2. **Simplified orthogonal factor structure matrices (principal factor–varimax solution, with squared multiple correlation coefficients in the diagonal) for trace-element suite of 28 Queensland soil profiles**

Variable	Three factors extracted[A]			Four factors extracted[A]			
	1	2	3	1′	2′	3′	4′
Manganese	++++			++++			
Cobalt	++++			++++			
Copper	+++			+++			
Zinc	+++			+++			
Nickel	+++			++			
Vanadium		+++			+++		
Iron		+++			+++	+	
Gallium		+	++		+	++	
Zirconium			++			++	
Molybdenum							+
Percentage of total variance	42	21	14	40	20	15	4

[A] ++++, >0·90; +++, 0·76–0·90; ++, 0·61–0·75; +, 0·46–0·60; blank, <0·46.

It is obvious that the pattern which has emerged is different from that of the corresponding principal component analysis, although this may not necessarily be the case where data are very strongly patterned. In the present case, however, different interpretations are necessary and each pattern may serve a different purpose or perhaps no purpose at all. 'Size' and molybdenum have become insignificant in the factor analysis pattern—the first factor cannot be interpreted as a reflection of overall trace-element status but rather of status of elements derived from easily weatherable minerals, while molybdenum is involved only if we include a fourth factor by not following the rule of thumb and even then it is only marginally involved (factor 4 only accounts for 4% of the total variance). Factors 2 and 3 could be interpreted in terms of trace elements from moderately weatherable minerals and resistant minerals respectively. In the factor analysis weatherability of minerals dominates the pattern, in the principal component analysis it does not.

Example 2: Chemical elements in some Queensland surface soils

Although pattern analysis in general is frequently considered to be non-probabilistic, probability is involved unless we are prepared to confine our interest exclusively to the set of data that we are analysing; in many cases, however, this would result in the exercise being rather trivial. In other words, although many of the notions of statistics are given little or no consideration when a pattern analysis is carried out, there is usually implicit in it the idea that the set of entities involved in the data matrix is a sample from a larger population, i.e. we are dealing with a probabilistic situation. This being the case, one of the measures of 'goodness' of an elucidated pattern is its stability, i.e. the extent to which the same analysis in a second sample of the same population produces a similar pattern (Macnaughton-Smith 1965). Duplicate samplings were not available in the present instance but we did have a second data matrix of surface soil data whose 15 attributes include the 10 trace elements of the first data matrix. It is of some interest to compare briefly the pattern derived from a principal component analysis of these data with that discussed previously.

Table 3.8.3 shows a moderately similar pattern, with the first component interpretable as a 'size' factor, the second as a 'weathering' factor and the third a 'molybdenum' factor. There are differences, however; for the first principal component Mo and Zr make no contribution for data sets 1 and 2 respectively. For the bipolar second principal component the introduction of elements lost easily by weathering (Ca, Mg, Na) has shifted the contrast with Zr from Co, Ni, etc. to these elements. Common factors in soil data appear to be as elusive as they were in psychology data.

Table 3.8.3. Principal component pattern matrix[A] for chemical analysis of 124 Queensland surface soils[B]

Attribute	Loading for PC1	Loading for PC2	Loading for PC3	h^2
Zirconium	−0·02	0·83	0·06	0·69
Titanium	0·72	0·54	−0·21	0·84
Iron	0·86	0·25	0·06	0·81
Vanadium	0·89	0·24	−0·03	0·84
Gallium	0·91	0·20	0·05	0·86
Nickel	0·87	0·10	−0·22	0·82
Cobalt	0·88	0·01	−0·23	0·83
Copper	0·92	0·06	−0·02	0·85
Zinc	0·84	−0·08	−0·06	0·72
Manganese	0·82	0·08	0·05	0·68
Aluminium	0·89	0·04	0·18	0·83
Magnesium	0·80	−0·46	−0·12	0·86
Calcium	0·64	−0·66	−0·01	0·84
Sodium	0·46	−0·67	0·20	0·70
Molybdenum	0·42	0·16	0·83	0·89
Vp	8·87	2·29	0·93	12·09
Percentage of total variance	59	15	7	81

[A] Principal components 11–15 not significantly different, $P = 0·01$.
[B] Original data from J. B. Giles (unpublished).

Example 3: Chemical elements in two solodized solonetz profiles

The use of principal component analysis to reduce the dimensionality of data is well illustrated by an analysis of the solodized solonetz data of Brodie (1967). This consisted of chemical data for 10-cm layers of two profiles to depths of 190 cm. Mn and P data were rejected as their variation fell within the limits of measurements. Using the remaining 11 elements (Si, Fe, Ti, Al, Ga, K, Na, Sr, Ca, Mg, Zr), the first component accounted for 78% of the variance and the second 7%.

The first component was bipolar with Si contrasted with the remaining 10 elements. This could be interpreted as a 'texture' factor; note, however, that omission of Si from the analysis hardly changed the pattern for the remaining elements, so that once again the interpretation of the first component reverts to one of 'size'. As might be expected, layers in the surface 0–40 cm had the highest scores on this component, layers between 40 and

160 cm had the lowest, and layers between 160 and 190 cm were intermediate, reflecting the morphology of the solodized solonetz profile. The second component, also bipolar, was a contrast between the alkali earths (Ca, Mg, Sr) and Zr—again a 'weathering' factor of some kind.

For some purposes the first principal component, containing as it does over three-quarters of the information, could be an adequate way of describing the entities (layers) in this data set.

Example 4: Brigalow soil data

When a fourth set of data, consisting of 21 chemical and physical attributes of brigalow soil from Meandarra, Queensland (Horton *et al.* 1968) was subjected to principal component analysis the first five eigenvalues were 11·72, 3·61, 1·68, 1·02 and 0·76, corresponding to 56, 17, 8, 5 and 3%

Table 3.8.4. Simplified oblique factor structure matrix for brigalow soil data[A]

Attribute	Factor 1[B]	Factor 2	Factor 3	Factor 4
1. Organic matter	+++			
2. Nitrogen	+++			
3. Total phosphorus	++++			
4. Available phosphorus	++++			
5. Exchangeable calcium	++			
6. Soluble magnesium	×			
7. Soluble sodium	×			
8. Conductivity	×			
9. Total soluble salts	×			
10. Exchangeable sodium	××			
11. Bulk density	×××			
12. Soluble calcium	×	+		
13. Exchangeable magnesium	×	++		
14. Cation exchange capacity		++++		
15. Sum of metallic cations		++++		
16. Pw (air-dry)		++++		
17. Pw (saturation)		++++		
18. pH			++	
19. Exchangeable hydrogen			××××	
20. Exchangeable potassium				+++
21. Soluble potassium				++

[A] Original data from Horton *et al.* (1968); number of entities = 48.

[B] + represents positive, × represents negative loadings, with absolute values as follows: >0·69, 4 symbols; 0·60–0·69, 3 symbols; 0·50–0·59, 2 symbols; 0·35–0·49, 1 symbol; <0·35, blank.

of the total variance (the trace of the correlation matrix, i.e. 21·00). Applying the rule of thumb to consider only those eigenvectors with associated eigenvalues greater than 1, we next rotated orthogonally (varimax rotation) the first four principal factors, using squared multiple correlation coefficients as estimates of communality. This resulted in factors accounting for 46, 16, 12 and 10% of the variance, but with essentially the same pattern as that for principal component analysis emerging (unlike the first example above). To try to 'clean up' the factor structure matrix, i.e. reduce the overall number of high coefficients, an oblique rotation was carried out (Table 3.8.4). In this procedure the restraint of orthogonality between factors is relaxed.

Interpretation of the factors could be as follows. The first, bipolar factor is a depth factor related to decreasing amounts of nitrogen, phosphorus and exchangeable calcium and to increasing salts and bulk density down the soil profile. The second is a mineral colloid factor, the third soil acidity and the fourth a potassium status factor.

The intercorrelations among these non-orthogonal factors may be calculated (Table 3.8.5). Factor 2 does not appear to be correlated with factors 1, 3 and 4. Correlation among these last three factors ranged from 0·49 to 0·70, suggesting that all three are in fact related to some extent to profile anisotropy.

Table 3.8.5. Matrix of correlations among four oblique factors for brigalow soil data

	F1	F2	F3
F2	−0·23		
F3	0·49	−0·14	
F4	0·70	−0·06	0·61

The factor scores for these 48 soil samples have not been calculated. In fact, this is not possible because the calculation would involve an attempt to invert a singular matrix. A glance at the attribute list (Table 3.8.4) shows that attribute 15 is the sum of attributes 5, 10, 13 and 20. Thus the data matrix and the corresponding correlation matrix are singular. The analysis needs to be re-run with attribute 15 omitted. This type of situation is often encountered in soil data; further examples are the summation of measurements of particle-size groups and oxides to 100%, the calculation of ratios (e.g. C : N ratio) etc.

The examples that have been discussed are largely illustrative but the application of techniques of ordination to soil data has proved useful in a number of instances that we are acquainted with. Examples are the elucidation of patterns in the following situations: (1) conventional pedological data for northern Queensland profiles identified as belonging to one or other of three closely related great soil groups; (2) nitrogen mineralization data for soils from crop rotation plots on the Darling Downs (see Case 3.13); (3) soil and fertilizer response data for a wide range of Queensland soils; and (4) standard cell data for four Queensland profiles used in a weathering study.

A. W. MOORE
J. S. RUSSELL

References

Beals, E. W. (1973). Ordination: mathematical elegance and ecological naiveté. *J. Ecol.* **61**, 23–35.

Brodie, J. G. (1967). Analyses of duplicate solodized solonetz profiles M128, M129 for Fe, Mn, Ti, Ca, K, P, Si, Al, Mg and Na. CSIRO Aust. Div. Soils Tech. Memo. No. 23/67.

Cooley, W. W., and Lohnes, P. R. (1971). 'Multivariate Data Analysis'. (John Wiley: New York.)

Gauch, H. G., and Whittaker, R. H. (1972). Comparison of ordination techniques. *Ecology* **53**, 868–75.

Gower, J. C. (1966). Some distance properties of latent root and vector methods used in multivariate analysis. *Biometrika* **53**, 325–38.

Harman, H. H. (1960). 'Modern Factor Analysis'. (Univ. Chicago Press: Chicago.)

Hole, F. D., and Hironaka, M. (1960). An experiment in ordination of some soil profiles. *Soil Sci. Soc. Am. Proc.* **24**, 309–12.

Holland, D. A. (1969). Component analysis: an aid to the interpretation of data. *Expl Agric.* **5**, 151–64.

Horton, I. F., Russell, J. S., and Moore, A. W. (1968). Multivariate-covariance and canonical analysis: a method for selecting the most effective discriminators in a multivariate situation. *Biometrics* **24**, 845–58.

Kaiser, H. F., and Caffrey, J. (1965). Alpha factor analysis. *Psychometrika* **30**, 1–14.

Macnaughton-Smith, P. (1965). Some statistical and other numerical techniques for classifying individuals. Home Office Studies in the Causes of Delinquency and the Treatment of Offenders. (H.M.S.O.)

Moore, A. W., and Russell, J. S. (1967). Comparison of coefficients and grouping procedures in the numerical analysis of soil trace element data. *Geoderma* **1**, 139–58.

Oertel, A. C., and Giles, J. B. (1963). Trace element contents of some Queensland soils. *Aust. J. Soil Res.* **1**, 215–22.

Seal, H. L. (1964). 'Multivariate Statistical Analysis for Biologists'. (Methuen: London.)

Case 3.9. Problems in Numerical Classification of Soil Data

The problems involved in pattern analysis of soil data are, in part, those which have to be faced by the classifier of soil data no matter what approach he takes to classification and, in part, those associated with the choices that have to be made in the course of a numerical analysis no matter what data are being used. In the first category we are concerned with such questions as what constitutes an entity (or individual) and what attributes should be used, and so on; in the second, we are concerned with what strategy should be used. This course is directed primarily towards the latter but there are interactions between the two types of problems that make it desirable to consider both.

The entity

Soil is a three-dimensional continuous mantle covering much of the Earth's land surface and before any type of classification or pattern analysis can be attempted it is necessary that the individual be clearly defined. Recognition of the three-dimensional nature of soil is explicitly expressed in such concepts as the pedon or mapping units or mapping faces. The soil profile purports to be a two-dimensional picture of soil; in fact, it is often little more than a one-dimensional representation, particularly when observed or sampled by means of coring or augering. Finally, at the opposite end of this range of 'conceptual soils' is soil material, without reference to its location in space (e.g. the contents of the chemist's sample bottle), where the entity is isotropic.

While the problems of the nature of the entity with which we are dealing are always with us, they are particularly acute when numbers have to be put on it for purposes of pattern analysis. We shall not pursue this further here: suffice it to say that as the dimensionality of the entity increases so does the difficulty of describing it mathematically. As far as the profile is concerned, we have suggested several approaches to this elsewhere (see Case 3.10). The interaction of this problem of the nature of the entity with numerical analysis is simply that present mathematical models inadequately describe very complicated entities and available computer facilities are inadequate to handle them.

Nature of soil data

Most 'traditional' soil data are quantitative and those which are not can in many cases be considered as ordered multistate data. A few attributes, e.g. shape of peds, of inclusions or of slope, are disordered multistate; undoubtedly these could be quantified but it is unlikely that the effort involved is justified in most instances.

The nature of the data dictates to some extent the strategy used in a numerical analysis. Generally, if we wish to include disordered multistate data it is necessary to use an information statistic type of similarity coefficient. In our work we have chosen to use various metrics, mostly Euclidean distance, and have got around the problems of multistate data by treating ranked data as quantitative and excluding the relatively small amount of disordered multistate data. The reasons for this are threefold: (*a*) most of our data have been quantitative or quasi-quantitative and we felt that an unnecessary loss of information could occur if we used an information theory measure, (*b*) computing costs are higher for analyses involving an information theory measure, and (*c*) there are fewer pitfalls, e.g. problems due to space dilation are probably avoided to a large extent.

Another aspect of data which must be considered is their degree of permanence. Provided we are not looking for temporal patterns, the attributes used as a basis for grouping soil entities need to be fairly constant within time spans of years or decades. Thus for most purposes it is pointless to use nitrate concentration, temperature or water content measured at particular points in time. The type of attribute used in these cases needs to be a parameter such as mean (e.g. mean of weekly nitrate concentrations over one or more years), parameters of time functions (e.g. those of a Fourier series fitted to daily temperature measurements) or 'derived' parameters (e.g. 'available' soil water).

Weighting of attributes

The so-called Adansonian concept that all attributes should carry equal weight in calculating similarity coefficients was virtually an integral part of early numerical taxonomic work. It is now generally conceded that there is no particular virtue in this approach. Nevertheless, there still appears to be considerable reluctance to apply 'arbitrary' weights to attributes, with two exceptions: the application of a weight of 1 or 0 to an attribute and standardization.

The use or non-use of an attribute in any particular analysis is equivalent to weighting it by 1 or 0. Zero weight may be allocated because (1) the attribute is unknown, measurements of it are not available or variation falls within the limits of measurement, or (2) we deliberately exclude that particular attribute for a given purpose, e.g. the rejection of an attribute that is considered to have no bearing on plant growth when we are interested in soil as a medium for plant growth.

An examination of one procedure for deliberate exclusion was made by Grigal and Arneman (1969) who grouped 40 forested Minnesota soils by numerical analysis using (1) all 22 attributes measured, (2) field attributes only, (3) moisture-related attributes only, (4) nutrient-related attributes only and (5) horizon texture and thickness. Although none of the 'numerical classifications' was closely related to classifications based on the Seventh Approximation, all five were closely related to each other in terms of both ultimate groups and the structure of the dendrograms produced. On the other hand, Sarkar (1965) found that clustering patterns were more satisfactory and consistent with all 61 attributes measured than with the more limited morphological characters alone.

In biological taxonomy the degreee of similarity between two entities is generally related to their phylogeny. Since their measured attributes reflect a finite genetic complement it is reasonable to expect that as the number of attributes used increases, so the amount of information approaches an asymptote. Sokal and Sneath (1963), for example, felt that little further information was gained by adding further attributes beyond about 60. With soils this cannot be argued *a priori* because there is nothing comparable to biological evolution which can be evoked to explain affinities between entities.

There is a dilemma here in that, while it seems logical that attributes should be as independent as possible to give the best overall measure of similarity, traditionally classification tends to be based on criteria that are 'cardinal' attributes, i.e. those which are correlated with a number of others. Again, weighting is involved since the inclusion of two highly correlated attributes is equivalent to giving one or other on its own a weight approaching twice that of associated attributes. Using 26 profiles representative of nine Orders of the Seventh Approximation, Sarkar *et al.* (1966) examined the effect on numerical taxonomic groupings of eliminating attributes on the basis of their intercorrelations. In reducing the number of attributes from 61 to 22 they found relatively small changes in the soil groupings produced, suggesting that the original attributes were clustered in a number of groups of approximately equal size and level of intercorrelation. Under these circumstances the weights attached to these groups of attributes would show little change as the number of attributes was reduced. A similar approach was that of Arkley (1971) who used seven factors derived from a factor analysis of 53 soil properties as the attributes in a numerical analysis of 86 soils from around the world.

The second type of weighting referred to above is associated with standardization. With quantitative data it is desirable that units be eliminated as otherwise arbitrary weighting of attributes occurs. This has been accomplished in most numerical studies by converting data to standard variates (zero mean and unit variance) and this convention has been followed by us in most of our work. An alternative procedure is division by range, which is similar to standard deviation as a measure of dispersion. Note that this division by a standard deviation or range could be considered to be 'arbitrary' weighting also. Also by definition, invariant data cannot be included; it is necessary to realize that we are not dealing with absolute values. 'High' and 'low' are meaningful terms only within the context of the data being handled, or at most within data relating to the universe of soils as we know them. It is particularly important that an attribute be excluded if its range of values approximates that expected to arise from error; inclusion in these cases would give error in one attribute a weight equal to that given to real differences in another.

Transformation of data

The removal of dimensions by standardization in effect enables all attributes to be expressed on the same scale (this is not necessary in the case of Canberra metric where dimensions are removed by including attribute values in both denominator and numerator). The distribution of the set of values for a particular attribute requires further consideration, however. From examination of dendrograms it appears that transformation of data to approximately normal distribution prior to analysis helps to stabilize the hierarchies produced by different procedures. The Euclidean distance coefficient, in particular, is very sensitive to the nature of the distribution within an attribute (Fig. 3.9.1 (A, B)).

The decision whether transformation should be carried out for non-probabilistic numerical analysis is not based on orthodox statistical grounds. Transformation should result in each attribute varying arithmetically. Logarithmic transformations are already widely used in soil science, e.g. pH, pF, pCa etc. Other soil properties, although not varying arithmetically, are commonly expressed on arithmetic scales, although it is doubtful whether these are the most satisfactory for many purposes, since for many quantitative soil data the possible range of values is closed at the lower end but virtually open at the upper end (Heath 1967).

Choice of strategy

Choice of standardization procedure, of transformations (if any) and of similarity coefficient defines the model used in numerical analysis. These are all subjective choices but once they have been made objectivity is attained in the sense that the same matrix will be produced consistently from a given set of data. Delineation of groups is a further subjective choice which may be assisted by various hierarchical sorting strategies or visual inspection of the matrix or an ordination with a small number of axes.

It is not possible to examine the differences between the large number of overall strategic options available. Moore and Russell (1967) have examined some options, using trace-element contents of 23 Queensland soils (Oertel and Giles 1963) as a set of 'test' data. One aspect dealt with in the former paper, viz. the effect of choice of similarity coefficient, is presented there because it is fairly simple to see the difference visually. The distribution of coefficients within a matrix is indicated by a visual pattern, darker shading indicating a greater degree of similarity (Fig. 3.9.1). Six matrices are presented in this manner to illustrate similarities and differences between coefficients and also the effect of transformation of data on three of the coefficients.

It is evident that matching coefficient, Canberra metric, mean character difference and Euclidean distance generate similar matrix patterns in general. However, there is a marked variation from matrix to matrix in the relationship between entity 1 and the remaining entities because of the presence of outlier value in the former. Mo concentration for entity 1 is 12 ppm while the range for the remaining entities is 2·0 to 6·3 ppm. For simple matching coefficient the weight of this extreme value is largely lost (Fig. 3.9.1(C)) since for this particular attribute the entity is merely allocated to the highest class of the attribute. Similarly the 'internal reference' property of the Canberra metric (attribute value occurs in both denominator and numerator) attenuates the weight of an extreme attribute value (Fig. 3.9.1(E,F)). In the case of Euclidean distance, however, where differences between attribute values for each pair of entities are squared during computation, the extreme attribute value has a pronounced influence clearly separating entity 1 from all other entities (Fig. 3.9.1(A,B)). As might be expected, mean character difference gives an intermediate situation (Fig. 3.9.1(D)) since the attribute differences are not squared.

In a similar fashion exceptionally high Co and Mn values for entity 11 and low Co, Mn, Cu, Zn and Ni values for entity 24 have resulted in these two being separated from other entities on the basis of Euclidean distances.

Some of the pattern differences between various 'magnitude' coefficient matrices result from skewness of distribution of attribute values. With the exception of Mo all the trace elements considered in this study showed distributions skewed to the left; in some cases this was extreme. Again, the 'internal reference' property of the Canberra metric meant that transformation had little effect on matrix pattern (Fig. 3.9.1 (E,F)) also. On the other hand, for Euclidean distance transformation resulted in a pronounced shift of smaller coefficient values (greater similarity) from the lower right-hand corner of the matrix to the upper left-hand corner (Fig. 3.9.1(A,B)), because arithmetic differences between attribute values for pairs being compared in the former area are smaller than for those pairs in the latter (Fig. 3.9.1(A,B)). Thus, although the same groups tended to appear, their compactness and relationships to each other were changed by data transformation.

Of other coefficients of similarity experimented with, correlation coefficient and Goodall's probabilistic coefficient (Goodall 1964) gave completely different patterns which bore little relationship to great soil groups,

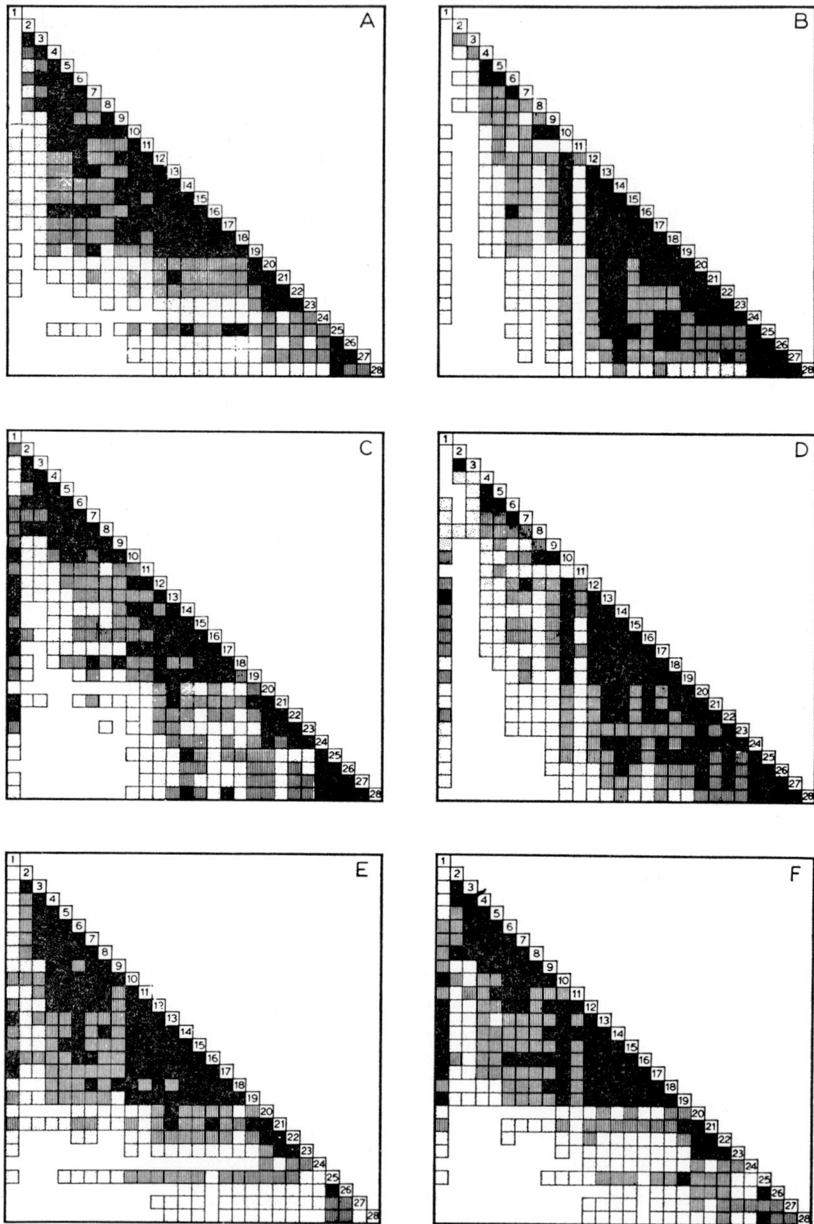

Fig. 3.9.1. Matrix patterns for four 'magnitude' coefficients of similarity. Four types of shading indicate the portions separated by quartiles for each matrix (darker shading indicates a higher degree of similarity). A, Euclidean distance (transformed data); B, Euclidean distance; C, simple matching coefficient (chopped data); D, mean character difference; E, Canberra metric (transformed data); F, Canberra metric. Arrangement of all matrices based on flexible sort of A.

unlike the metrics discussed above. The information statistic, however, gave a similar pattern to these metrics, particularly those other than Euclidean distance. This is because of the 'chopping' or coding which is done before the information statistic is calculated.

Geographical considerations

At various stages of pattern analysis choices, often multiple in nature, are presented to the user and the number of possible strategic programs as a consequence is virtually infinite. In general, the strategy favoured is that which gives the user the answer that 'makes most sense' to him, i.e. options are chosen so that the strategy simulates the pattern-detection procedures in the user's mind as nearly as possible. For most data the pattern sought is not related to where the entities came from; in some cases, however, geographic location of the entities has some relevance in the derivation of the final pattern.

Thus in mapping soils the surveyor presumably takes into account intuitively the location of each (undefined) parcel of soil in relation to every other parcel within his current sphere of operation when he delineates an area on a map. In general, he is trying to delineate areas that are reasonably homogeneous, geographically compact and continuous. In pattern analysis the problem is to simulate this process in the computer.

This problem has been recognized for some time by geographers (Spence and Taylor 1970), who have proposed a number of ways of attacking it. There are broadly two approaches: (*a*) application of a contiguity constraint during grouping, after calculation of a similarity matrix, and (*b*) incorporation of spatial coordinates as an additional pair of attributes. In the first instance, when entities are compared, fusion is permitted only if they are geographically contiguous. Long narrow parcels may result; however, this is not necessarily undesirable, e.g. it could be appropriate in the delineation of an area along a valley floor. Webster and Burrough (1972), working with soils in south central England, proposed a compromise, viz. the use of a function that weighted location heavily at the local level but still recognized long-distance similarities. In essence, the weight of the contribution of geographical distance to similarity was varied according to an inverse square or exponential function. They concluded that this procedure produced smoother maps that were as good as those produced without taking location into account, in that the soil was no more variable within the groups so mapped.

In a similar exercise we have applied a simple inverse function in a pattern analysis of the crop rotation plots (experiment A260) at Narayen Research Station. In this analysis a triangular matrix of distances between all plots was determined using plot coordinates. The maximum value in this matrix was determined. Then the values of the similarity coefficients were weighted for distance by means of the linear relationship,

$$S_w = S\left[k + (1 - k)(d/t)\right],$$

where S is a similarity coefficient, d is the corresponding element from the matrix of distances between pairs of plots, t is the maximum distance between any pair of plots, k is a coefficient which can be varied to give a different weighting for distance and S_w is the weighted similarity coefficient.

The experimental area consists of two sections; the northern section only is used in Fig. 3.9.2 to show the effect of contiguity weighting. Using the 10-group level for both weightings illustrated, 7 groups occurred in the northern section at $k = 0 \cdot 8$ (i.e. with slight weighting) compared with 5 groups at $k = 0 \cdot 0$ (i.e. with maximum weighting). It can be seen that the introduction of contiguity weighting has resulted in a reduction of fragmentation of groups and in more geographically compact soil groups.

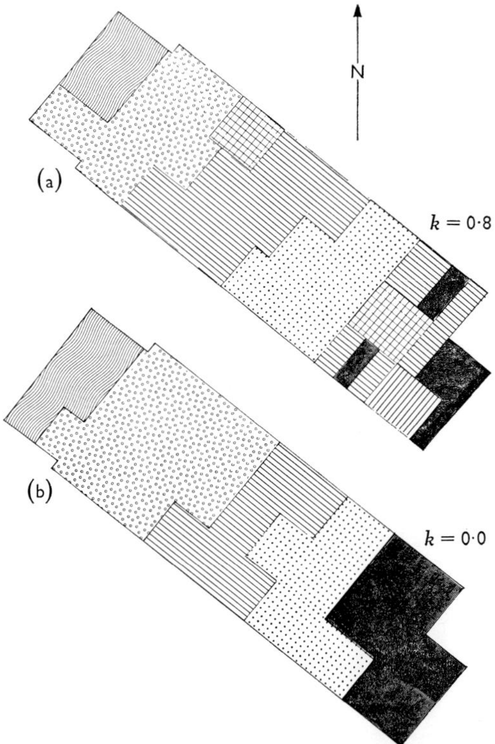

Fig. 3.9.2. Grouping of crop rotation plots at CSIRO Narayen Experimental Station, central Queensland. (a) With little geographical location weighting. (b) With maximum geographical location weighting. The coefficient k is explained in the text.

Conclusion

The application of numerical classification to soil data has probably been explored sufficiently over the last decade for some evaluation of its usefulness to be made. In situations where there are no preconceived ideas about pattern in the data numerical analysis may prove to be a useful technique in setting up hypotheses.

Pattern analysis of soil data has associated with it the usual problems arising from the virtually infinite number of strategic options available, some of which have been discussed above. In addition, it has its own peculiar problems associated with defining the nature of the entities involved and the spatial relationships between them in the field.

A. W. MOORE

J. S. RUSSELL

References

Arkley, R. J. (1971). Factor analysis and numerical taxonomy of soils. *Soil Sci. Soc. Am. Proc.* **35**, 312–15.

Goodall, D. W. (1964). A probabilistic similarity index. *Nature (Lond.)* **203**, 1098.

Grigal, D. F., and Arneman, H. F. (1969). Numerical classification of some forested Minnesota soils. *Soil Sci. Soc. Am. Proc.* **33**, 433–8.

Heath, D. F. (1967). Normal or log-normal: appropriate distribution. *Nature (Lond.)* **213**, 1159–60.

Moore, A. W., and Russell, J. S. (1967). Comparison of coefficients and grouping procedures in numerical analysis of soil trace element data. *Geoderma* **1**, 139–58.

Oertel, A. C., and Giles, J. B. (1963). Trace element contents of some Queensland soils. *Aust. J. Soil Res.* **1**, 215–22.

Sarkar, P. K. (1965). Numerical classification of soils. *Diss. Abstr.* **26**, 2946.

Sarkar, P. K., Bidwell, O. W., and Marcus, L. F. (1966). Selection of characteristics for numerical classification of soils. *Soil Sci. Soc. Am. Proc.* **30**, 269–72.

Sokal, R. R., and Sneath, P. H. A. (1963). 'Principles of Numerical Taxonomy'. (W. H. Freeman: San Francisco.)

Spence, N. A., and Taylor, P. J. (1970). Quantitative methods in regional taxonomy. *Prog. Geogr.* **2**, 1–64.

Webster, R., and Burrough, P. A. (1972). Computer-based soil mapping of small areas from sample data. II. Classification smoothing. *J. Soil Sci.* **23**, 222–34.

Case 3.10. Numerical Models for the Analysis of Soil Profile Data

The application of pattern analysis to certain aspects of soil and soil–plant relations appears promising in theory. But in practice difficulties arise due to the layered or structured nature of soils and to the fact that plant growth occurs at the soil–atmosphere interface. These difficulties in the numerical expression of structure are not unique to soils. They occur in other fields such as oceanography and limnology but in relation to plant growth, in particular, they appear formidable.

In attempting to cope with the structured nature of soils a number of models have been proposed to describe them prior to analysis of soil data. Four such models are discussed below: the soil profile as (1) an isotropic body; (2) an array of layer attributes; (3) a set of depth functions; and (4) a sequence of layers. All of these models have been used in pattern analysis of soils data. There are advantages and disadvantages associated with each approach.

The soil profile as an isotropic body

The simplest approach to layering of soils is to ignore these effects and assume that soils or the soil materials being considered are isotropic. There are a number of situations where such an assumption is justified.

A similar approach is to use the data for the surface horizon and to assume that this provides adequate information about the soils. This approach was used in the analysis of deep sandy soils in South Australia (Russell and Moore 1967) where layering effects were minimal. In this paper relationships between 43 sites with deep sandy soils sampled from southern Australia were determined using all variant field and laboratory data available for the surface layer.

Information on 28 variant characters was available for the 0–8-cm layer. The qualitative and semi-quantitative data were ranked and, with the quantitative data, were then standardized (unit variance). Four main groups were delineated and these showed a geographical pattern indicating that the character patterns on which the affinities were based were associated largely with geographical location.

This model can be of use in a number of situations, e.g. in the comparison of surface soils, but it has obvious limitations as soon as deeper soil profiles have to be considered or where soils with marked anisotropy are compared.

The soil profile as an array of layer attributes

This model and variants of it have been widely used. Thus various authors (Bidwell and Hole 1964; Rayner 1966; Lance and Williams 1967) have considered profiles as made up of three layers, equivalent in most cases to traditional A, B and C horizons. Regardless of subsequent computation, possible objections to such layer selection include the subjective choice of horizon designation, the lack of comparative horizons in certain soils and the fact that depth differences within and between horizons are not considered. Sarkar *et al.* (1966) regarded properties of the A, B and C horizons as independent but included other characters, such as the thickness of horizons, ratios between horizon properties and so on. However, there are theoretical objections to regarding properties of layers as independent (Lance and Williams 1967).

Rayner (1966) proposed the interesting approach of comparing a set of horizons of a soil profile with every other set and taking a mean value of the similarity coefficients of the horizons using the highest value for each comparison. In this way the average of all horizon similarities was taken as the similarity between the profiles.

One variant of the layer attribute approach is to consider the profile as being made up of numerous layers of equal thickness. Individual layers are assumed to be isotropic and a comparison between profiles involves the summation of a large number of comparisons between layers at equivalent depths.

Assessment of the relative significance of layers is more difficult. It has been pointed out (Moore and Russell 1966) that implicit in published soil literature is the belief that at some depth beneath the surface properties are no longer of any significance to any particular investigation. Even if all soil layers are given equal weight, in any profile summation it is still necessary to make a decision regarding the lower limit of the profile.

It is possible to reduce the difficulties involved in the choice of depth of soil profile by the use of weighting factors. Russell and Moore (1968) used a relationship of the form

$$y = ce^{-cx}$$

where y = weighting function, x = depth in cm and c = constant. Integration with respect to x gives the area beneath the curve. Between 0 (soil surface) and ∞ (infinite depth) the area is unity regardless of the value of c. This function was chosen because of its mathematical simplicity and its apparent similarity to the profile depth changes found for biological and related properties.

The effect of varying the value of c on the pattern obtained with a set of widely different soils was examined. The soil data were stored as a three-dimensional array of 20 profiles $\times$ 20 properties $\times$ 40 2·5-cm layers. The properties were standardized (unit variance) to remove dimensions. The profiles were then compared layer by layer using Euclidean distance ($\triangle$) as the similarity coefficient, i.e. for the 0–2·5-cm layer each soil was compared with every other soil using 20 properties and this was repeated with the 2·5–5-cm layer, etc. In this way 40 similarity matrices (one for each layer) were obtained.

From these layer matrices profile matrices were obtained which, in effect, were combinations of the layer comparisons. One profile matrix was obtained by giving all layers equal weight, i.e. by obtaining a mean value of Euclidean distance

$$\triangle k_i k_j = (\triangle k_i k_j l_1 + \triangle k_i k_j l_2 + \ldots + \triangle k_i k_j l_n)/n,$$

where $\triangle k_i k_j l_n$ represents the Euclidean distance between the nth layers of two soils K_i and K_j. In this approach the 40th layer is given weight equal to that of the first layer.

Three other profile matrices were obtained using weighting factors. These were calculated from the layer matrices as follows:

$$\triangle k_i k_j = W_{1,1}^c \, \triangle k_i k_j l_1 + W_{1,2}^c \triangle k_i k_j l_2 + \ldots + W_{1,n}^c \, \triangle k_i k_j l_n,$$

where $\triangle^c k_i k_j l_n$ has the same connotation as above and $W_{1,n}^c$ represents a weighting factor where this has the value for any layer, n, of

$$W_{1,n}^c = \int_{n-1}^{n} cq e^{-cqx} \, \mathrm{d}x$$

and where q is the conversion factor from inches to centimetres. Values of $c = 0\cdot01, 0\cdot02$ and $0\cdot04$ were used and, since the depth of profile comparison was only 100 cm, a term was introduced so that the summation of weighting factors for the profile was unit for each value of c. The numerical analyses that most closely approximated field classification used the weighting factors $c = 0\cdot02$ and $c = 0\cdot04$.

This multilayer expression of profile characters overcomes the problems associated with anisotropy and has the advantage that variously sampled profile data can be used. But the multilayer strategy involves the use of a mathematical model to link the layers. Lance and Williams (1967) assumed layers to be of equal significance and averaged similarities. However, this approach is likely to be inadequate when extended to many layers and considerable soil depths and it is noteworthy that in the above study groupings obtained by averaging showed least correspondence with field groupings. In using the present function, change in the constant, c, from $0\cdot02$ to $0\cdot04$ did not appear to be critical and similar groupings were obtained corresponding broadly to certain great soil groups.

The soil profile as a set of depth functions

It is possible to consider each attribute of a soil profile as a depth function and then use the parameters of the depth functions as attributes in comparing different soil profiles.

Moore *et al.* (1972) expressed depth functions as polynomia. Orthogonal polynomials were fitted to the series of values of attributes of the 5-cm layers of a number of soil profiles using tables which extend to the quintic degree (Fig. 3.10.1). Orthogonal polynomials are equations such that each coefficient is associated with a power of the independent variable (in this case depth) and all coefficients are independent of one another. This permits independent testing of the variance contribution of each term in the equation.

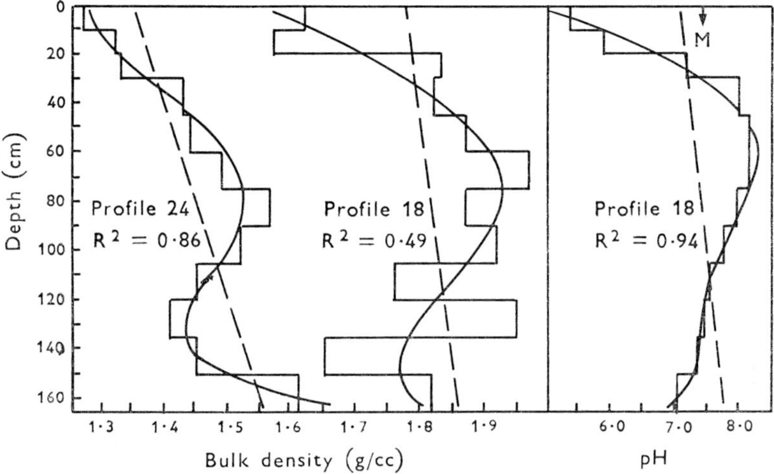

Fig. 3.10.1. Profiles for bulk density (soils 18 and 24) and pH (soil 18), represented in each case by a sequence of values for sample layers and an orthogonal polynomial of quintic degree. Mean is indicated by M and linear trend by a broken line. R^2 is the variance due to regression for the polynomial of quintic degree as a proportion of the total variance due to deviation of variates about their mean.

The mean value of the property for the profile was considered to be a measure of size, the coefficients measures of shape. For the purpose of pattern analysis the advantage of using orthogonal polynomials is that their orthogonality enables the set of polynomial coefficients to be varied by the addition or subtraction of one or more coefficients. Each set (all attributes) of depth functions was used to describe the soil profile as a whole and the 14 attribute means per profile and 70 coefficients were used as variables for analysis. Each variable was standardized to unit variance (across profiles) and a similarity matrix of Euclidean distance was calculated. The contributions of the size and shape components were examined by changing their relative weights and repeating the analysis several times. In fitting orthogonal polynomials to the attributes varying with depth a clear separation of overall size and shape is inevitably obtained. The profile mean contains no information about the relationships between layers, but the coefficients are wholly dependent upon these relationships.

The soil profile as a sequence of layers

The technique of sequential analysis described by Dale *et al.* (1970) can be used to group the profiles on the basis of the sequences in which layers follow each other in the profiles, a method that has some similarities with field classification by horizon sequence. The successive steps involved in sequential analysis (Norris and Dale 1971) are as follows.

(1) The profile layers (sample) are split into groups using pattern analysis.

(2) Each profile layer is assigned a number indicating to which group it belongs.

(3) Each profile is thus represented as a sequence of numbers.

(4) A transition matrix is derived from the sequence of numbers describing each profile.

(5) The transition matrices are grouped hierarchically. Considering each entry in a matrix to be a single state of a multistate attribute, hierarchical grouping based on the information statistic is the most suitable strategy (Dale *et al.* 1970). The information content, *I*, for an $m \times m$ transition matrix is given by

$$I = g\ln g - \sum_{i=1}^{m} \sum_{j=1}^{m} f_{ij}\ln f_{ij},$$

where f_{ij} is the *j*th entry in the *i*th row of the matrix and g is the sum of the f_{ij}s. The information gain on fusing matrices is calculated for all possible pairs of matrices and the pair giving the minimum information gain is fused to form a new composite matrix. This process is repeated until a complete hierarchy is built up.

The grouping of profiles on the basis of transition matrices is dependent entirely on the relationship between each layer and the one preceding it. The actual attribute values are used only to decide to which group (transition state) a particular profile layer belongs.

In one study (Moore *et al.* 1972) models 2, 3 and 4 were compared with each other and with field classification. Dendrograms derived from Euclidean distance similarity matrices were constructed for 49 profiles from the solodic soil area of the western Darling Downs. Very similar dendrograms were obtained using model 2 layer-by-layer comparisons (with layers weighted by an exponential depth function) and model 3 orthogonal polynomial coefficients (shape parameters). The dendrogram derived from sequential analysis (model 4) differed markedly from the others and seemed less meaningful.

An independent field study which related the soils to geomorphic units and to differences in parent material provided a further, separate classification. The groups derived by numerical analysis, using models 2 and 3, agreed well with the field classes even though the strategies used were different and independent.

Norris and Dale (1971) on the other hand saw certain advantages in a sequential analysis approach. These advantages included a convenient means of defining horizons and the inclusion in the classifications of information about the relative position of samples. A third advantage was that the number and thickness of depth increments need not be identical in each profile.

Conclusions

In the pattern analysis of soil data one of the important decisions is the model of the soil profile which is used. As this paper has shown, the mathematical problem is that the soil profile has structure and this has to be expressed in an explicit form.

On the one hand this structure can be ignored and the soil treated as an isotropic body. Although this may be suitable in some cases it is obviously not adequate for more than a few simple situations.

On the other hand there are various mathematical models which take into account the depth factor in a soil profile. Some of these models appear more relevant to the real world than others.

Present experience suggests that different models and a number of different clustering strategies should be applied to any given set of data in a search for stability of groups. The cumulative results of this approach may eventually enable a particular model for strategy to be selected for use on a routine basis.

J. S. RUSSELL
A. W. MOORE

References

Bidwell, O. W., and Hole, F. D. (1964). An experiment in the numerical classification of some Kansas soils. *Soil Sci. Soc. Am. Proc.* **28**, 263–8.

Dale, M. B., Macnaughton-Smith, P., Williams, W. T., and Lance, G. N. (1970). Numerical classifications of sequences. *Aust. Comput. J.* **2**, 9–13.

Lance, G. N., and Williams, W. T. (1967). Note on the classification of multi-level data. *Comput. J.* **9**, 381–2.

Moore, A. W., and Russell, J. S. (1966). Potential use of numerical analysis and Adansonian concepts in soil science. *Aust. J. Sci.* **29**, 141–3.

Moore, A. W., Russell, J. S., and Ward, W. T. (1972). Numerical analysis of soils: a comparison of three soil profile models with field classification. *J. Soil Sci.* **23**, 193–209.

Norris, J. M., and Dale, M. B. (1971). Transition matrix approach to numerical classification of soil profiles. *Soil Sci. Soc. Am. Proc.* **35**, 487–91.

Rayner, J. H. (1966). Classification of soils by numerical methods. *J. Soil Sci.* **17**, 79–92.

Russell, J. S., and Moore, A. W. (1967). Use of a numerical method in determining affinities between some deep sandy soils. *Geoderma* **1**, 47–68.

Russell, J. S., and Moore, A. W. (1968). Comparison of different depth weightings in the numerical analysis of anisotropic soil profile data. Trans. 9th Int. Congr. Soil Sci. Vol. 4, Pap. No. 22.

Sarkar, P. K., Bidwell, O. W., and Marcus, L. F. (1966). Selection of characteristics for numerical classification of soils. *Soil Sci. Soc. Am. Proc.* **30**, 269–72.

Case 3.11. Classification of Climatic Data for Northern Australia

Pattern analysis has not been widely used in the comparison of classification of climate. This is surprising since climate is a field in which many classifications have been proposed in the past and some controversy has surrounded their relative merits. Further, climate is a complex system varying in time and space, for which large amounts of data are available on both a historical and a geographical basis.

We have made a number of attempts to use pattern analysis to compare climate. The first of these (Russell and Moore 1970) was the use of pattern analysis to search for homoclimates of specified areas; such an approach is clearly of value in plant geography. The second approach was to use pattern analysis to classify climate. One example (Russell and Moore, unpublished) was to classify the climate of two areas (in this case Australia and Africa) by treating the data of the two regions as a single set. By analogy this could be extended to a global classification.

These attempts to use pattern analysis to compare and classify climate have pointed up a number of aspects which have to be considered in using this technique. These are choice of attributes, choice of numerical methods and the specific advantages of pattern analysis in climatic classification.

Choice of attributes

The choice of which attributes should be used to describe climate inevitably arises in any pattern analysis study. In the first instance, where a general description of climate is sought, it is probably desirable to use as large a number of attributes as possible.

The data available obviously influence the attributes used. Geographically the wider and more extensive the study, the less is the chance that a large number of equivalent measurements will be available for all stations. The actual number of attributes used will also be influenced by the extent to which derived data will be used.

In the two studies mentioned above, the same source data were used. These were from the world meteorological tables of the United Kingdom Air Ministry supplemented by some additional Queensland data. For most stations information was available on a monthly basis for 11 different attributes giving a total number of 132 values for each station.

In the homoclimate study the source data (Table 3.11.1) were used without change. Of the 11 attributes 6 were temperature measurements, 2 were relative humidity and 3 were rainfall. In the classification study several changes were made. Two of the temperature event measurements were modified. Absolute maximum temperature was omitted and absolute

Table 3.11.1. Recorded and derived climatic attributes used in two studies based on the same set of source data

Type	Attribute no.	Recorded measurement	Derived attribute	Homoclimate study	Classifica- tion study	Comments
Temperature	1	Daily max. temp.				
	2	Daily min. temp.				
	3	Highest each month mean				
	4	Lowest each month mean				
	5	Absolute maximum			X	Event measurement
	6	Absolute minimum			X	Event measurement
	7		Absolute degrees of frost			From 6
	8		Temperature range			From 1 and 2
No.				6	6	
Relative humidity	9	Rel. humidity at 0900 hr				
	10	1500 hr				
No.				2	2	
Rainfall	11	Monthly mean				
	12	Max. fall in 24 hr				
	13	Mean no. of rain days				
	14		Rainfall/wet day			From 11 and 13
No.				3	4	
Evaporation and water balance	15		Free water evaporation			From 1 and 7
	16		Rainfall/evaporation			From 11 and 15
	17		Soil water storage			From 11 and 15
No.					3	
Day-length	18		Day-length			From lat. of station
No.					1	
Monthly total				11	16	
Yearly total				132	192	

minimum temperature was modified to measure degrees of frost. Additional monthly attributes calculated included rainfall/wet day, free water evaporation, rainfall/evaporation ratio and soil water balance. In addition monthly day-length was also calculated for each station from its latitude. Thus in all 16 monthly attributes were used including 6 of temperature, 2 of relative humidity, 4 of rainfall, 3 of evaporation and water balance and 1 of day-length.

The modification of the event temperature attributes in the classification study was carried out because of the effect of length of record on these measurements and the fact that length of records varied widely between stations. In such a situation absolute maxima or minima can be markedly affected. To avoid this the data on absolute maxima were not used. Also the data on absolute minima were modified to express only temperatures below zero—in effect, to show degrees of frost.

The balance of attributes, particularly between moisture and temperature, could be important in the groupings obtained. The numbers of each type of attribute used in the classification study represent an attempt to achieve a satisfactory balance.

There is no information on the effect of different suites of attributes on the classification obtained. To examine this effect two separate classifications of northern Australia have been made using two quite different sets of data.

The first set of data used was that of 16 monthly attributes or 192 values (Table 3.11.1) similar to that of the classification study. These data were obtained for 59 stations in northern Australia. Some of these data were available from the World Meteorological Tables (Anon. 1958); the additional stations were obtained from raw data at the Bureau of Meteorology (1968). This set of data included a wide range of measurements—mainly means but with several event measurements. In effect this suite of attributes attempts to express climate by means of a broad spectrum of climatic measurements.

The second set of data used was that of rainfall measurements only—the 1st, 3rd, 5th, 7th and 9th deciles of monthly rainfall for 254 stations in northern Australia. In effect there were 5 monthly attributes used or 60 in all per station. This set of data provided information on the amount of rainfall, its distribution throughout the year and its variability from year to year.

The data were standardized in both cases and the pattern analysis carried out using the Canberra metric and flexible sorting. Comparisons have been made at two levels: 20 groups and 9 groups (Figs 3.11.1 and 3.11.2). The geographic groupings of the broad-spectrum classification are shown in Figs 3.11.3 and 3.11.4 and the geographic groupings of the rainfall classification in Figs 3.11.5 and 3.11.6.

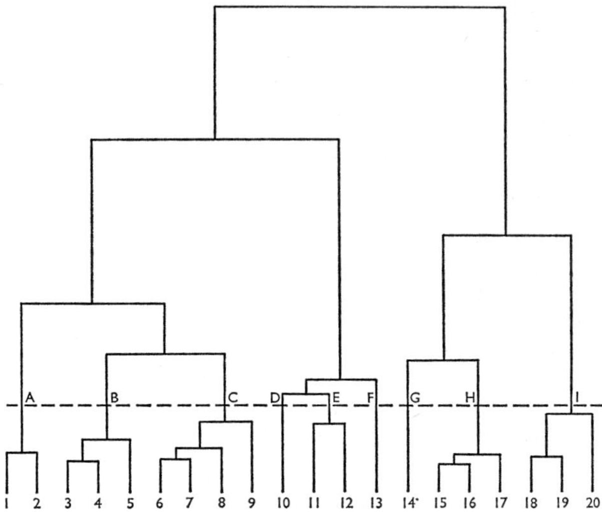

Fig. 3.11.1. Dendrogram of groups based on a broad spectrum of climatic measurements.

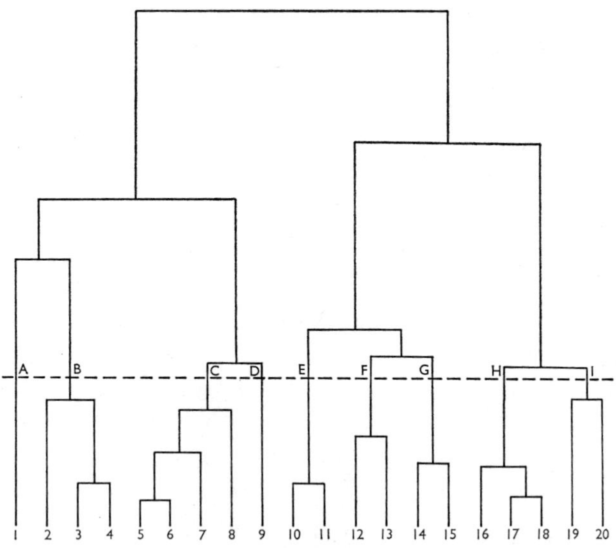

Fig. 3.11.2. Dendrogram of groups based on rainfall data only.

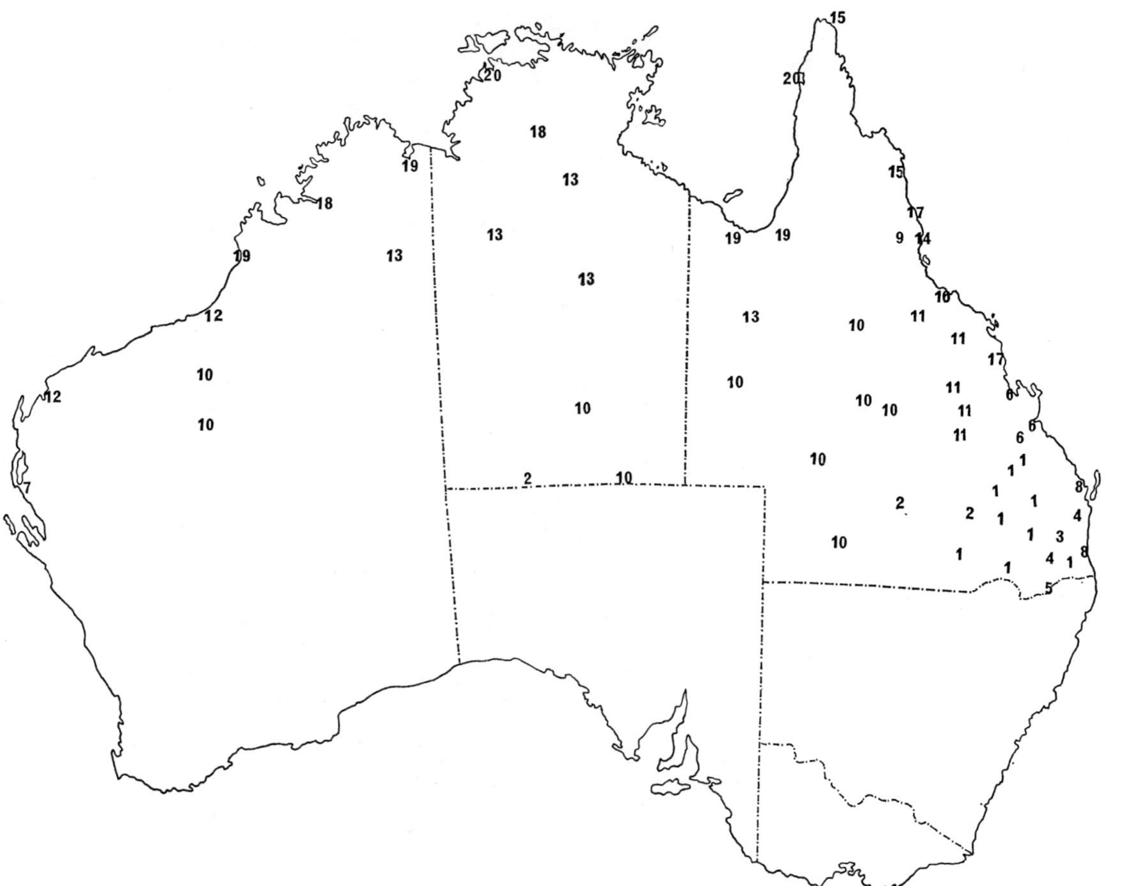

Fig. 3.11.3. Groups 1–20 based on the broad-spectrum classification.

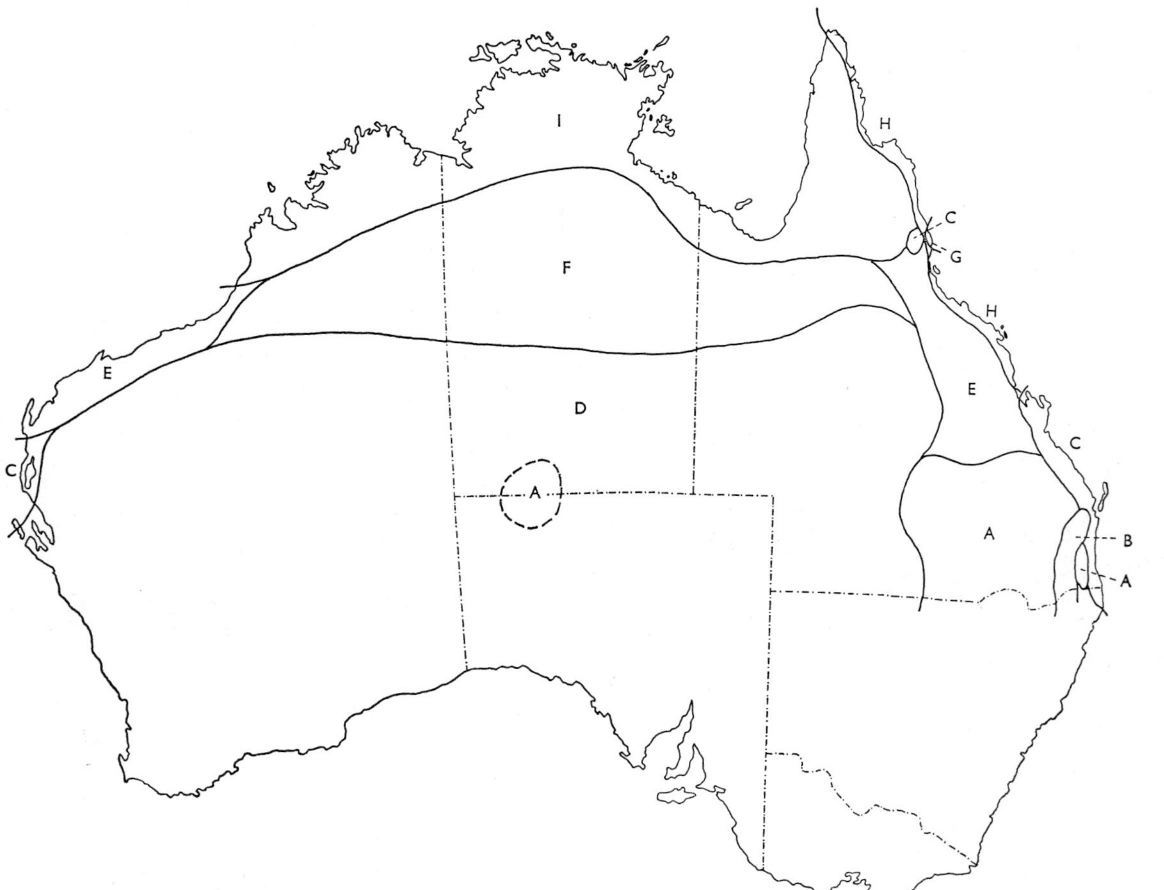

Fig. 3.11.4. Groups A–I based on the broad-spectrum classification.

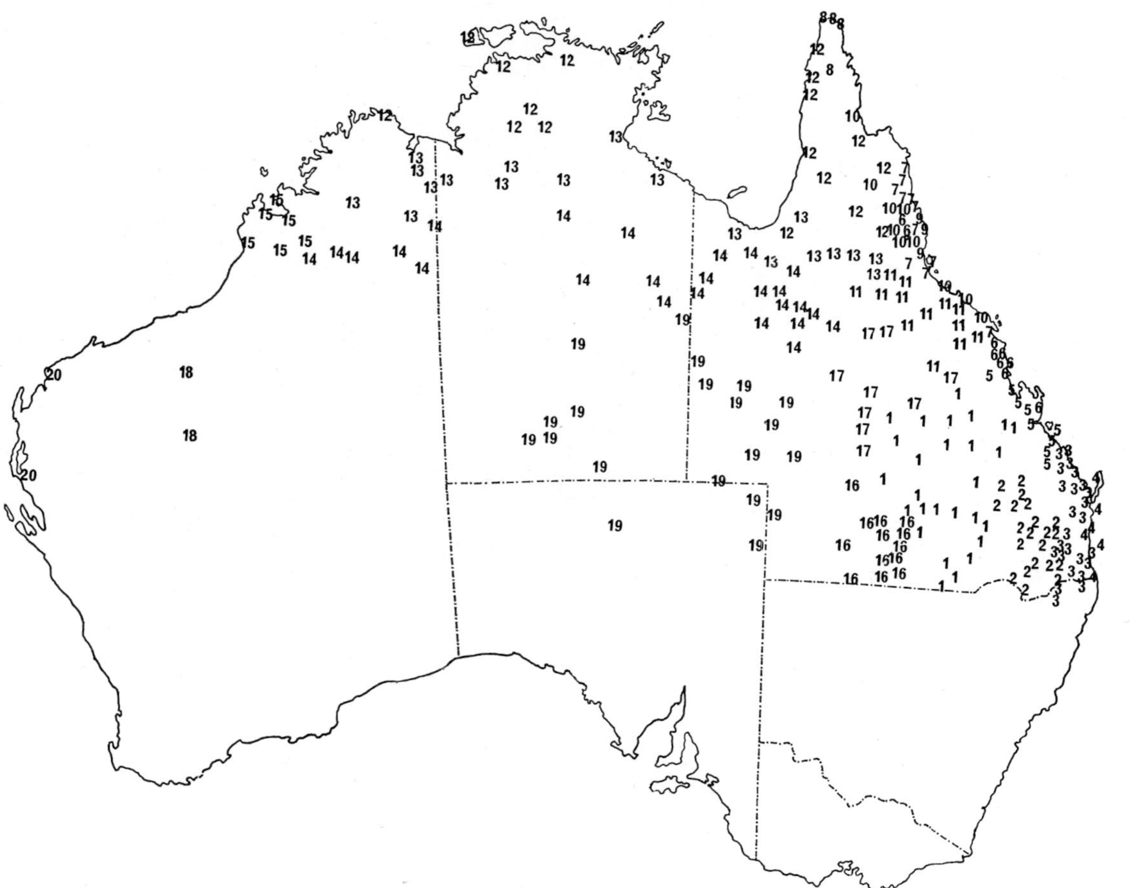

Fig. 3.11.5. Groups 1–20 based on rainfall classification.

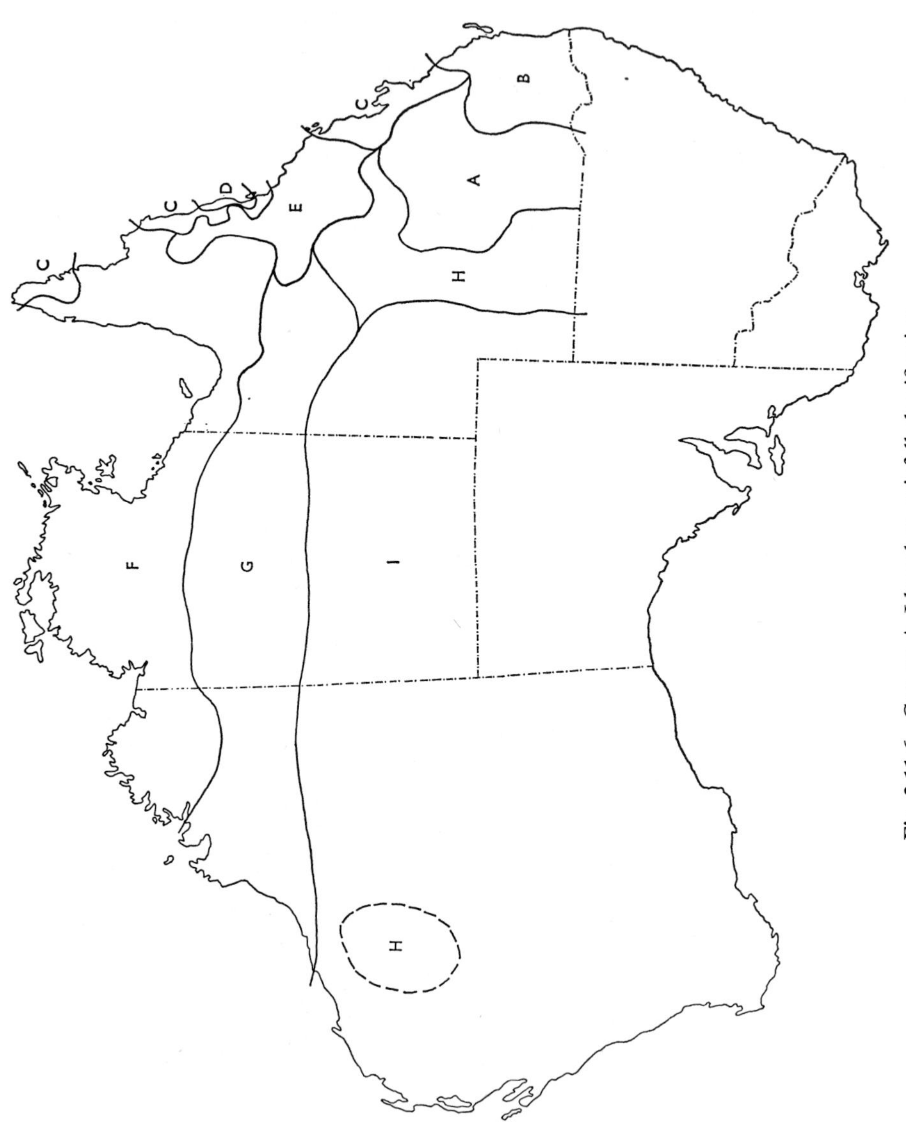

Fig. 3.11.6. Groups A–I based on rainfall classification.

At the 20-group level there are certain similarities between the two analyses. Both show a general east–west grouping across the central and western parts of the continent but a much more complex pattern on the east coast. Both also show the effect of coastal influences and a distinctive inland climate for northern, central and southern Queensland. The rainfall analysis, with its much greater number of stations, shows a more complex pattern of interspersed groups particularly on the coast and near mountain areas such as inland from Brisbane and Cairns.

At the 9-group level there were broad similarities in the groupings. Both analyses showed three continental east–west groups. The most northerly of these in both cases included much of Cape York Peninsula. The boundaries of the three groups differed slightly. In both cases there were coastal groups which showed similarities in relation to the tropical coast adjacent to Innisfail but differences particularly in relation to the area inland from Townsville and Bowen.

In both analyses there were subcoastal Queensland groups. In the broad-spectrum analysis it is of interest that two of the groups approximate the distribution of *Acacia harpophylla* (Isbell 1965).

Differences in the two analyses are evident in relation to south-east Queensland. In the rainfall analysis this area is included in one group but in the broad-spectrum analysis other groups are recognized including a mountain group containing Toowoomba.

From these groupings it is, of course, possible to obtain means and standard deviations of various climatic attributes. This has not been done in this paper. In considering the choice of attributes it is likely that there will be a steady evolutionary change in the suite of attributes used. It can be argued that for a general classification all relevant data should be included. Thus in the present instance perhaps an analysis combining the broad-spectrum and rainfall amounts, seasonal distribution and variability should be carried out.

For a specific classification, such as that concerned with the growth of a certain plant, a more restricted set of attributes may be required. The questions arise of which attributes should be used and whether any weighting should be carried out. Empirical studies may provide answers.

Choice of numerical methods

Since almost all climatic data are numerical and continuous this affects the choice of coefficient. In the similarity analyses we have carried out two coefficients have been used, Canberra metric and Euclidean distance. Comparison of these coefficients was made in the homoclimate study.

One of the limitations of Euclidean distance is the large effect that single outlier values have. This was noted in the homoclimate study and a specific example given where a single event measurement (absolute maximum rainfall in 24 hours) for Emerald had a value for December of 125 mm. In Clermont it was 412 mm. Values for the Euclidean distance between the two stations were affected much more than those for the Canberra metric because of the influence of this single value. In some instances, of course, such outlier values may be highly significant in a climatic classification and in a coefficient such as Euclidean distance may be more suitable than the Canberra metric.

The Canberra metric may require much less computing time, e.g. in subroutines such as screening stations for homoclimates, due to the fact that analyses can be carried out without standardization. Restrictions on number of stations are also overcome in such situations.

Sequential analysis, principal coordinate analysis and principal component analysis provide information on the structure of the data. They may be most useful where portions of the year are to be considered or where rapid climatic changes throughout the year make the sequence of events important.

Specific advantages of pattern analysis in climatic classification

Pattern analysis of climate has been used in two ways, firstly in the search for homoclimates of certain stations and secondly for the classification of climate of a set of stations.

Advantages of this method in the search for homoclimates are that specific times of the year may be used, e.g. the growing period of a plant say from December to March, and that the attributes used may be those known to affect plant growth, such as continuity of moisture and occurrence of frost.

Pattern analysis shows promise on a macro-scale for the classification of climate. The groupings derived using conventional meteorological measurements and derivatives from these are sensible, and suggest that the method could be applied more widely. Computing time increases quadratically with station number, however, and beyond 300 stations computing costs tend to become prohibitive. This may limit current methods to comparing different areas rather than providing a global classification. The development of group allocation procedures and the gradual increase in computer capability may allow global classifications in the near future.

Although pattern analysis of climate is of value on a macro-scale, possibly its greatest potential lies in the classification of microclimates as in fields, forests, watersheds, cities, buildings or, using an agricultural research example, experiment stations. Traditional methods are of less value in these situations, yet provided data are available there is no reason why stable and meaningful climatic groupings cannot be produced.

The special advantages of pattern analysis include reproducibility, use of large numbers of attributes and the possibility of considering particular periods of the year only. Hierarchical relationships between groups are explicit.

J. S. RUSSELL
A. W. MOORE

References

Anon. (1958). Tables of temperature, relative humidity and precipitation for the World. Nos 1–6. (H.M.S.O.)

Bureau of Meteorology (1968). Review of Australia's water resources. Monthly rainfall and evaporation. (Melbourne.)

Isbell, R. F. (1965). Soils and vegetation of the brigalow lands, eastern Australia. CSIRO Aust. Soils and Land Use Ser. No. 43.

Russell, J. S., and Moore, A. W. (1970). Detection of homoclimates by numerical analysis with reference to the brigalow region, eastern Australia. *Agric. Meteorol.* **7**, 455–79.

Case 3.12. Delimitation of World Pasture Types by Numerical Methods

Orderly plant introduction aimed at providing productive and adapted pastures in the tropics and subtropics is a relatively new phenomenon. The need for realistic classification into areas of similar pastoral potential has long been recognized but until recently no global classification of sown tropical pastures by objective multivariate methods had been proposed.

A simple division of the tropics into wet and dry by agrostologists provides a useful starting-point in dealing with pastoral problems but makes it difficult to avoid generalization in the absence of precise data. Subjective attempts at more meaningful forms of classification have been attempted on a world basis. Davies and Skidmore (1966), noting that rainfall has a major influence, discuss tropical pastures on the basis of humid, sub-humid, semi-arid and arid zones. However, this simplified classification does not make allowance for tropical highland areas and the cooler subtropics, both areas having sown pasture species that do not occur elsewhere. Whyte *et al.* (1959) presented a table of grass species adapted for cultivation to the major climatic regions of the world, and their climatic classification of the tropics and subtropics consists of six units: tropical semi-arid, tropical restricted rainfall, tropical extended rainfall, subtropical semi-arid, subtropical seasonal rainfall and subtropical extended rainfall. However, many of the species listed are only suggestions as to what may be suitable and not what in fact is grown. Borget (1969) suggested six pastoral–climatic regions in the tropics: wet equatorial, dry equatorial, wet tropical, dry tropical, high-altitude equatorial and high-altitude tropical. He then outlined the major species combinations that characterize the regions. Unfortunately he does not discuss any areas > 18° latitude. Hutton (1970) distinguishes seven pastoral–climatic zones, comprising a wet equatorial zone with constant heat, rainfall and humidity, occurring between 5–10° N. and S. of the Equator. Four monsoonal areas are distinguished as follows: annual rainfall of 1000–2000 mm in two rainy seasons with short dry seasons; annual rainfall of 600–1500 mm in two short rainy seasons; annual rainfall of 750–1000 mm in one fairly long rainy season with a short dry season; and an annual rainfall of 500–1000 mm with one short rainy season and a long dry season. The subtropical zones are characterized as humid and dry respectively. Although this classification is more meaningful than those outlined above, it again lacks a provision for a tropical highland zone.

Reid (1973) attempted to use objective multivariate methods to define a classification that would be solely determined by the data rather than by the views of the individual investigators. To this end, 50 sites were selected covering a wide range of tropical and subtropical environments. Data for 46 of the sites were extracted from scientific papers and annual reports, those for the remainder were obtained by personal observation. These data referred to the actual pastoral situations and the many reports on species in plant-introduction gardens were not considered.

Site selection was quickly resolved for although there is an increasing interest in tropical pastures, very few of the many reports of site $\times$ species interactions provided any climatic or soil data. Only those sites which were characterized by the use of species for at least a period of five years were used in the analysis. This was taken as a measure of the adaptability of species to the environment and reduced the risk of fluctuation of species due to either a series of wet years or drought.

The major obstacle encountered in building a numerical classification of sown pasture species was the decision on which attributes could be profitably utilized. This was resolved when the data on each site had been collected and it was apparent that there was considerable unconformity between recorded characteristics. Whilst many sites had good climatic data, e.g. mean monthly temperature, mean monthly rainfall, evaporation readings, etc., others had little more than mean annual rainfall. Likewise with soil information, some sites were characterized by soil type, texture, structure, pH, drainage, etc., but the only common parameter over all was soil texture. Attributes finally selected were latitude, altitude, mean annual rainfall and soil texture. The length of humid season, length of dry season and a temperature factor were calculated using the method of Papadakis (1966).

Advice was then sought on which classificatory program to use and the methods selected were as follows. With the environmental data classification was effected by the intensely clustering strategy which is available in the Canberra program MULCLAS and the floristic data were processed by information analysis (Canberra program MULTBET). The mixed environmental and floristic set was classified by the 'flexible' sorting strategy and the group means at each hierarchical fusion were examined by the diagnostic program GROUPER.

The mixed classification of environmental and floristic data was arbitrarily truncated at the seven-group level (Fig. 3.12.1). It is of particular interest to ascertain the extent to which the attributes have contributed to the classification. It is impracticable to reproduce the results for all fusions, but those at the top of the hierarchy deserve attention.

The first dichotomy places major emphasis on the grass species *Cenchrus ciliaris* (A) and *Panicum maximum* (B). This division is strengthened when the major environmental attributes are considered, as the mean lengths of the dry season for A and B are $8 \cdot 1$ months and $2 \cdot 6$ months respectively, whilst the mean lengths of the humid season are 1·4 and 6·4 months. Whilst the two preceding grasses may be considered the primary species for separation, it should be noted that *Cenchrus setigerus* is found only in A and *Stylosanthes guyanensis* and *Centrosema pubescens* only in B.

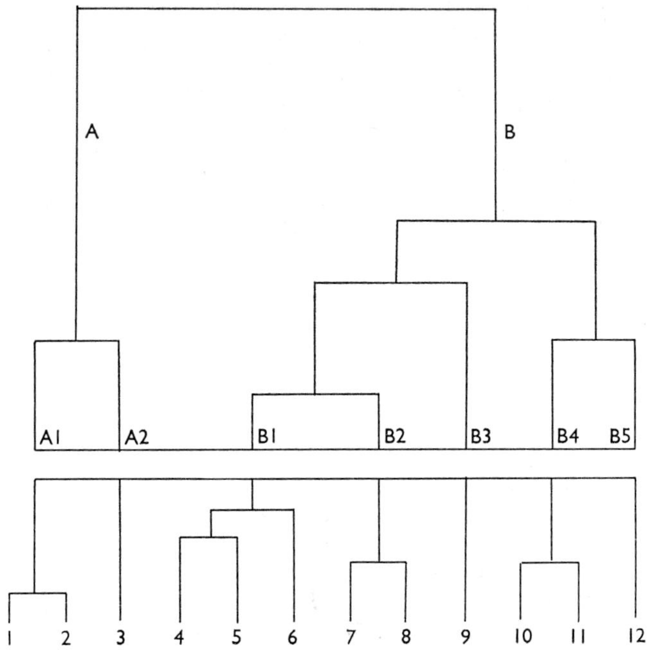

Fig. 3.12.1. Class hierarchy at 7- and 12-group levels.

We may usefully term group A1 *tropical semi-arid,* distinguished by having *Cenchrus setigerus* growing at all sites. Group A2 may be termed *subtropical seasonal rainfall* as the two dominant pasture species are *Medicago sativa* and *Panicum maximum* var. *trichoglume.* It is to be noted that *Cenchrus setigerus, Urochloa mosambicensis* and *Stylosanthes humilis* are absent.

Characterizing group B1, designated *tropical extended rainfall,* is *Digitaria decumbens* which occurs at 72% of the sites. Although *Stylosanthes guyanensis* and *Centrosema pubescens* occur they cannot be considered the major legumes. In group B2, designated *tropical seasonal rainfall, Stylosanthes guyanensis* occurs at 80% of the sites and *Centrosema pubescens* and *Melinis minutiflora* at 60% of the sites. The mean latitude is 16° and the dry season has a mean length of 3·8 months. *Pennisetum purpureum* is the primary species delineating group B3, termed *equatorial extended rainfall,* and is accompanied by a very small number of other grasses.

Groups B4/B5 are representative of the high-altitude tropics designated as *wet tropical highland* and *dry tropical highland* respectively. In B4 *Setaria sphacelata* is the only species common to all sites and the mean length of the dry season for the group is 1 month. *Desmodium uncinatum* is the only legume to occur at 80% of sites. Unlike B4, group B5 has a mean dry season length of 5·2 months and *Setaria sphacelata* is absent. *Chloris gayana* is the dominant grass species with *Glycine wightii* occurring at 50% of sites.

Further examination of the hierarchy at the 12-group level resulted in only minor clarification and revealed that whilst some groups could possibly be substantiated at a major group level, others were nonconformist. Thus group 7 may be considered as a very uniform unit, in that all sites are environmentally similar with a mean annual rainfall of 1250 mm and a dry season length of 4·3 months. This group falls naturally into the Sudanese savanna of previous geographical classifications and is characterized by the widespread use of *Andropogon gayanus* and *Stylosanthes guyanensis*. *Andropogon gayanus* has been successfully introduced into India at a site which falls within this group. A closely related group is group 8 which has a shorter dry season and in which *Andropogon gayanus* is replaced by *Hyparrhenia rufa*. *Brachiaria brizantha* is found at all sites within this group and the associated legume is *Calopogonium muconoides*. This may well be an indication of lower soil fertility levels to be found associated with this group as both these species, like *Hyparrhenia rufa*, have the ability to grow well in poor soils and are unlikely to suffer from competition from species demanding higher fertility (Buller 1960).

On the negative side, group 5 is open to question as it is constituted by the occurrence of only one species, *Brachiaria decumbens*, and the environmental data suggest that it should be reallocated to group 4. Likewise group 6 may be termed a nonconformist group as it includes sites that are markedly different from each other and from those in other groups. This nonconformity is reflected in the floristic component of these sites which have species occurring that are not evident elsewhere.

An inherent difficulty in regional classification is the lack of uniform attributes which enable valid comparisons between them to be made. Also, it is not possible to synthesize the general from the specific and a *de novo* approach is required. The previous classifications outlined above are all somewhat subjective and intuitive and none has used model numerical methods in the isolation of its respective units.

The purpose of this study was to introduce a technique for the delineation of sown tropical pasture regions, using existing published data as far as possible. Even with only 50 sites, and despite the very limited comparative information available for each, a clear-cut classification into seven major pastoral types has resulted. Moreover, this classification agrees well with the several subjective and intuitive classifications already in existence, while making provision for types which one or other of the previous classifications does not cover. The seven major groups may be designated as *tropical semiarid, subtropical seasonal rainfall, tropical seasonal rainfall, tropical extended rainfall, equatorial extended rainfall, wet tropical highland* and *dry tropical highland.* The groups are summarized in Table 3.12.1.

Table 3.12.1. Characteristics of species composition and environmental descriptions of the groups

Group	Site	Species	Environmental means	
A1	Jodhpur	*Cenchrus ciliaris*	Length of humid season	1·25 month
	Pali	*Cenchrus setigerus*	Length of dry season	8·6 months
	Lansdown	*Lasiurus sindicus*	Altitude	192 m
	Katherine	*Urochloa mosambicensis*	Mean annual rainfall	627 mm
	Charters Towers	*Panicum antidotale*		
	Bambey	*Stylosanthes humilis*		
	Barquisimeto	*Clitoria ternatea*		
	Ahmedabad			
A2	Biloela	*Medicago sativa*	Length of humid season	2 months
	Brian Pastures	*Panicum maximum* var.	Length of dry season	6 months
		trichoglume	Altitude	150 m
		Chloris gayana	Mean annual rainfall	711 mm
B1	Matao	*Digitaria decumbens*	Length of humid season	6·4 months
	South Johnstone	*Stylosanthes guyanensis*	Length of dry season	1·9 months
	Ingham	*Centrosema pubescens*	Altitude	655 m
	Sete Lagoas	*Melinis minutiflora*	Mean annual rainfall	1596 mm
	Dschang			
	Palmira			
	Popayan			
	El Capulin			
	Marrandellas			
	Corrientes			
	Cotaxtla			
B2	Shika	*Andropogon gayanus*	Length of humid season	5·8 months
	Ibadan	*Stylosanthes guyanensis*	Length of dry season	3·8 months
	Ranche	*Centrosema pubescens*	Altitude	616 m
	Bouake	*Leucaena leucocephala*	Mean annual rainfall	1350 mm
	Vientiane	*Brachiaria brizantha*		
	Dolisie			
	Serere			
	Bouar			
	Sardi			

Table 3.12.1 (*Continued*)

Group	Site	Species	Environmental means	
B3	Yangambi Gandajika M'vuazi Tanga Buenaventura Maha Kluang	*Pennisetum purpureum* *Panicum maximum*	Length of humid season Length of dry season Altitude Mean annual rainfall	8·4 months 1·1 month 271 m 2212 mm
B4	Entebbe Kairi Kitale Lyamungu Nioka Mulungu	*Setaria sphacelata* *Desmodium uncinatum* *Desmodium intortum* *Glycine wightii*	Length of humid season Length of dry season Altitude Mean annual rainfall	7·1 months 1 month 1441 m 1387 mm
B5	Kongwa Henderson Machakos Keyberg Ukiriguru Mt Makulu	*Panicum maximum* *Cenchrus ciliaris* *Digitaria milanjiana* *Chloris gayana* *Glycine wightii*	Length of humid season Length of dry season Altitude Mean annual rainfall	4·4 months 5·2 months 1240 m 854 mm

The method adopted is a new one in pastoral classification and it appears that further study is warranted in order to refine both the technique and the quality of the resultant classifications. The number of attributes could then be expanded to include a more detailed record of temperatures, details of rainfall on a monthly basis, soil data including pH and chemical analyses, and socio–economic factors related to the use of a particular pasture. Nevertheless, where environmental information is insufficient or too general, and where the selected environmental factors have not been correlated with the pastures, the pattern of pastoral species remains of potential value as an indicator of environmental differences. Provided that available information about the environment is not neglected, a study of species combinations offers a rapid means of classifying pastoral landscapes into units of similar habitat types.

With a further assemblage of detailed environmental data and more substantial floristic data it should be possible to construct a global classification that will aid in understanding sown pasture distribution and facilitate the profitable exchange of potential pasture species.

R. Reid

References

Borget, M. (1969). Results and present trends of fodder crop research at I.R.A.T. *Agron. Trop.* (2) **24**, 124–38.

Buller, R. E. (1960). Evolution of forage resources of the tropical grasslands of Mexico. Proc. 8th Int. Grassl. Congr., Reading, England, pp. 374–7.

Davies, W., and Skidmore, C. L. (1966). 'Tropical Pastures'. pp. 21–9. (Faber: London.)

Hutton, E. M. (1970). Tropical pastures. *Adv. Agron.* **22**, 1–73.

Papadakis, J. (1966). 'Climates of the World'. (Av. Cordoba 4564, Buenos Aires.)

Reid, R. (1973). A numerical classification of sown tropical pasture regions based on the performance of sown pasture species. *Trop. Grassl.* **7**, 331–40.

Whyte, R. O., Moir, T. R. G., and Cooper, J. P. (1959). 'Grasses in Agriculture'. (F.A.O.: Rome.)

Case 3.13. Pattern Analysis of Nitrogen Mineralization Potentials in Soil from a Ley–Pasture Crop Rotation Experiment

As part of investigations into the effects of ley pastures on soil fertility and wheat production, several methods for determining the nitrogen mineralization potential of soil were examined. Preliminary results indicated interactions between the methods for determining nitrogen mineralization potential and the field treatments; this complicated further investigation and interpretation of an already complex field experiment. A pattern analysis approach was adopted in an attempt to identify the major factors operating in this multivariate situation.

Materials and Methods

Description of the field experiment

The field experiment was a 17 treatments × 4 replications randomized block design on 15 hectares of black earth, Norillee association (Beckmann and Thompson 1960), at Norwin, Queensland. Treatments were 5 ley pastures (sward lucerne and 4 mixed pastures each comprising 1 perennial grass cultivar with lucerne) stocked at 3 rates (grazed with Merino wethers) and 2 cropping sequences (continuous wheat and a sorghum–wheat rotation). The grasses were sown in spring 1966, the lucerne was oversown in autumn 1967 and from June 1968 to April 1972 the established pastures were rotationally grazed by blocks with 2 weeks' grazing and 6 weeks' deferment, except during periods of extreme feed drought. The pasture plots were cultivated in June 1972 and subsequently all plots are being subjected uniformly to a fallow–wheat system over a 3- to 5-year period. Of the many variables measured during the grazing phase only pasture presentation data and animal grazing days are reported here. The dry-weight yields of the green grass or green lucerne component of pasture present at the commencement of each rotational grazing have been summed to provide indices of total production of grass and lucerne. Animal grazing days were determined from records for individual plots.

Laboratory experiment

Soil sampling. Soil was sampled in March 1973 when those plots previously under pasture had been in fallow for 9 months. Each plot was divided into five equal subplots, which for the pasture pretreatments lay across a possible fertility trend extending away from the original watering point. Five cores (0–15 cm layer × 4·5 cm diameter) were taken from each subplot and bulked. The soil was manually broken into *c.* 8-mm or

"

smaller aggregates and mixed thoroughly; subsamples from the five subplots were composited by weight. These field-moist composites were stored in sealed jars at $-15°C$, pending analysis. The remaining soil was air-dried in a forced-draught dehydrator at 35°C for 5 days and ground to pass a 2-mm sieve; subsamples from the five subplots were composited by weight. These air-dried composites were stored in sealed glass jars at room temperature, pending analysis.

Determination of nitrogen mineralization potentials. Each sample of field-moist and air-dried soil was analysed by 12 methods for determining nitrogen mineralization potential. These methods comprised a factorial array of different incubation conditions, namely aeration (aerobic or waterlogged), temperature (20, 30 or 40°C) and period of time (1 or 2 weeks). For aerobic conditions, 10 g soil was brought to pF 2 with distilled water and incubated in a wide-mouth bottle sealed with $2 \cdot 5$-μ-thick polyethylene film (Bremner 1965). For waterlogged conditions, the method of Waring and Bremner (1964) was used. Nitrogen mineralization potential was determined by the net change during incubation of the concentration of mineral nitrogen (NH_4-nitrogen + NO_3-nitrogen + NO_2-nitrogen) for samples incubated aerobically, or of ammonium-nitrogen for samples incubated under waterlogged conditions.

Pattern analysis. The program MULCLAS, with Burr standardized Euclidean metric as the similarity measure and flexible fusion with the cluster-intensity parameter β set at $-0 \cdot 25$, was employed for hierarchical classification. The field plots were grouped by treating nitrogen mineralization potential methods as attributes with both air-dried and field-moist samples included in the first instance (i.e. a 68 $\times$ 24 data matrix) and subsequently with field-moist samples only (i.e. a 68 $\times$ 12 data matrix). The relationships between groups of plots from the classification and field treatments and blocks were analysed by χ^2 tests of contingency tables. Analyses of variance based on the groups were carried out for lucerne production, grass production and animal grazing days. The contribution made by each attribute (nitrogen mineralization method) at each node of the hierarchy was determined using the program GROUPER (Lance *et al.* 1968).

The 68 $\times$ 12 data matrix was further analysed by principal component analysis (Pearce 1965). Correlation coefficients between each of the first five components and lucerne production, grass production and animal grazing days were calculated. The multiple correlation coefficient between principal component I on the one hand, and grass production and anima grazing days on the other, was also calculated.

Results

Analysis of data for both field-moist and air-dried soil samples

Hierarchical classification. A classification of the field plots based on the data for both field-moist and air-dried samples appeared quite incongruous at first glance. Closer examination revealed that in some sections of the classification the plots were grouped roughly into two groups corresponding to the two batches in which samples from the field had been subsampled, dried and composited in the laboratory. These batches differed markedly in pre-incubation ammonium concentration for the air-dried samples (means of 5 and 30 μg NH_4-N/g soil) but not for the field-moist samples (mean of $0 \cdot 7$ μg NH_4-N/g soil). This suggested that the classification was partly an artefact of sample preparation, which was probably associated with the drying of the soil and therefore absent from the data for field-moist samples.

Analyses of data for field-moist soil samples

Hierarchical classification. A grouping of the field plots based on the data for only field-moist soil samples did not reflect the two sample-preparation batches, thereby confirming that the artefact described above was associated with the soil-drying process. This grouping, truncated to show only the upper levels of the hierarchy, is given in Fig. 3.13.1. Statistically significant relationships between this grouping and external information are given in Tables 3.13.1–3.

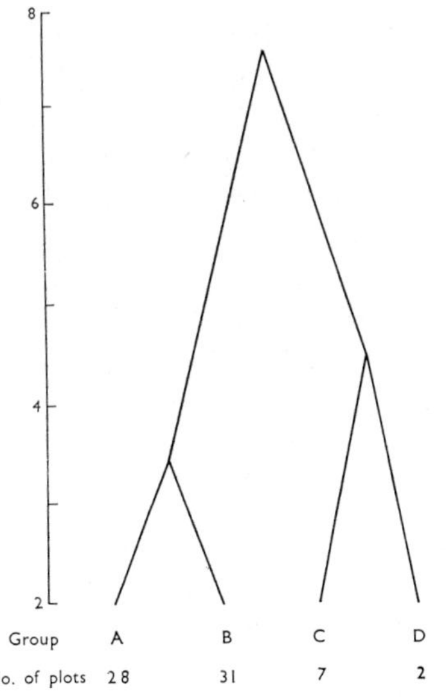

Fig. 3.13.1. Truncated hierarchical classification of the field plots.

Table 3.13.1. Relationship of the groups of plots to the previous cultural treatment of soil

| Group | Cultural treatment | | | Total |
	Crop sequences	Sward lucerne	Mixed pasture	
A	6	9	13	28
B	2	1	28	31
C	0	2	5	7
D	0	0	2	2
Total	8	12	48	68

$\chi^2 = 16 \cdot 1$* with 6 d.f.

Table 3.13.2. Relationship of the groups of plots to the previous production of grass and lucerne

| Group | Mean weight of grass (kg/ha) | | Mean weight of lucerne (kg/ha) | | |
	All treatments	Mixed-pasture treatments	All treatments	Sward lucerne and mixed-pasture treatments	Mixed-pasture treatments
A	1 850 a[A]	3 990 a	6 730 a	8 560 a	7 670 a
B	13 810 b	15 290 b	5 070 a	5 420 b	5 270 b
C	4 350 a	6 090 a	6 640 a	6 640 ab	5 080 b
D	18 760 b	18 760 b	4 080 a	4 080 b	4 080 b

[A] Means with the same letter do not differ significantly at $P = 0 \cdot 05$.

Table 3.13.3. Relationship of the groups of plots to the stocking rate of mixed-pasture treatments[A]

| Groups | Stocking rate (sheep/ac) | | | Total |
	3	6	9 reduced to 6 after 2 yr grazing	
(A + B)	15	16	10	41
(C + D)	1	0	6	7
Total	16	16	16	48

$\chi^2 = 10 \cdot 38$** with 2 d.f.

[A] Another analysis, including all cultural treatments, of an $r \times c$ contingency table composed of stocking rates $\times$ the four separate groups gave $\chi^2 = 24 \cdot 08$* with 12 d.f.

Groups A and B contain the majority of the 68 plots. Lower-level fusions indicated that group C containing only seven plots is a relatively diffuse group. Group D contains only two plots. The distribution of the plots in these four groups is not homogeneous with respect to previous cultural treatments (Table 3.13.1). Omission of group A and recalculation of χ^2 showed that group A, which contained three-quarters of both the crop sequence and sward lucerne plots, contributed most to the lack of homogeneity. The fact that group A also contained a reasonably large number (13) of the mixed-pasture plots suggested that variation not entirely associated with cultural pretreatment influenced the outcome of the analysis. Consequently relationships between the groupings based on mineralization and certain variables measured when the plots were under the previous cultural treatments were examined.

Analyses of variance, including either all treatments or only mixed pastures, showed significant differences among the groups in the total weight of grass produced (Table 3.13.2). The mixed-pasture plots that occurred in group A with most of the crop sequence and sward lucerne plots were low in grass. Similarly group C is identified as relatively low and groups B and D as relatively high in grass production. Lucerne production did not vary significantly among the groups in an analysis including all treatments (Table 3.13.2), indicating little effect of this variable on the nitrogen mineralization on which the grouping was based. This is exemplified by the fact that sward lucerne plots (high lucerne production) and crop sequence plots (no lucerne production) were grouped together in group A. Analyses of all pasture treatments or only mixed-pasture treatments did show significant differences among the groups for lucerne production (Table 3.13.2) but this is probably due to a generally inverse relationship between grass and lucerne productions reflecting the influence of grass production on the mineralization.

The distribution of plots in the four groups was not homogeneous with respect to stocking rate (crop sequences were included at zero stocking rate). Subdivision and analysis of further contingency tables showed the main contribution to the lack of homogeneity to be that six of the seven mixed-pasture plots in group (C + D) were from the highest stocking rate treatment (Table 3.13.3). To examine this result further with continuous data, total number of grazing days was analysed. However, F tests were significant only when crop sequence plots (zero grazing days) were included in the analyses of groups A, B, C and D or of (A + B) and (C + D).

The classification was statistically homogeneous with respect to field blocks although group A contained a somewhat higher proportion of plots from block 4. Likewise no marked relation between the groupings and position of plots in the field could be detected by inspection of the field plan.

Diagnosis of the classification. The results of GROUPER analysis of the truncated hierarchy are given in Tables 3.13.4–6.

Table 3.13.4. **Incubation conditions for determining nitrogen mineralization potential that contributed most to the grouping of field plots into groups A and B**

| | Incubation conditions | | Mean net nitrogen mineralization (μg N/g soil) | | Cumulative contributions to separation |
Aeration[A]	Temp. (°C)	Period (wk)	Group A (28 members)	Group B (31 members)	(%)
W	40	2	30·6	44·4	33·2
W	30	2	8·2	13·3	58·4
W	40	1	15·5	21·0	71·8
W	20	2	3·2	4·7	81·4
W	20	1	2·8	4·2	87·5
W	30	1	5·5	7·0	91·6

[A] W, waterlogged.

Table 3.13.5. **Incubation conditions for determining nitrogen mineralization potential that contributed most to the grouping of field plots into groups C and D**

| | Incubation conditions | | Mean nitrogen mineralization (μg N/g soil) | | Cumulative contributions to separation |
Aeration[A]	Temp. (°C)	Period (wk)	Group C (7 members)	Group D (2 members)	(%)
W	20	1	5·1	11·0	19·0
A	20	2	9·2	0·2	37·6
W	40	1	20·1	33·7	53·4
W	20	2	4·6	8·7	66·8
A	30	1	7·3	1·5	76·4
A	30	2	11·8	6·3	84·1
A	20	1	5·2	−0·2	91·5

[A]A, aerobic; W, waterlogged.

Table 3.13.6. **Incubation conditions for determining nitrogen mineralization potential that contributed most to the grouping of field plots into groups (A + B) and (C + D)**

| | Incubation conditions | | Mean net nitrogen mineralization (μg N/g soil) | | Cumulative contributions to separation |
Aeration[A]	Temp. (°C)	Period (wk)	Group (A + B) (57 members)	Group (C + D) (9 members)	(%)
A	40	1	10·1	18·9	15·9
W	20	1	3·5	6·4	29·0
A	20	2	2·5	7·2	41·5
A	20	1	−0·4	4·0	53·4
A	30	1	1·9	6·0	65·1
A	30	2	6·7	10·6	74·7
A	40	2	23·7	28·7	80·0
W	30	2	10·9	14·2	85·1
W	20	2	4·0	5·5	89·9

[A] A, aerobic; W, waterlogged.

Six of the 12 attributes (incubation conditions) accounted for 92% of contributions to the separation of groups A and B and these were the 6 waterlogged incubations. Mean net nitrogen mineralization in group A was less than in group B under waterlogged conditions, but greater under aerobic conditions.

Seven attributes accounted for 92% of contributions to the separation of groups C and D. Mean net nitrogen mineralization in group C was less than in group D under waterlogged conditions, but greater under aerobic conditions.

Nine attributes accounted for 90% of contributions to the separation of groups (A + B) and (C + D). Mean net nitrogen mineralization in group (A + B) was less than in group (C + D) under all incubation conditions.

Principal component analysis. The projections of the field plots onto principal components I and II are given in Fig. 3.13.2 with symbols to indicate how these plots were grouped in the hierarchical classification (Fig. 3.13.1). No major discrepancy between the hierarchical classification

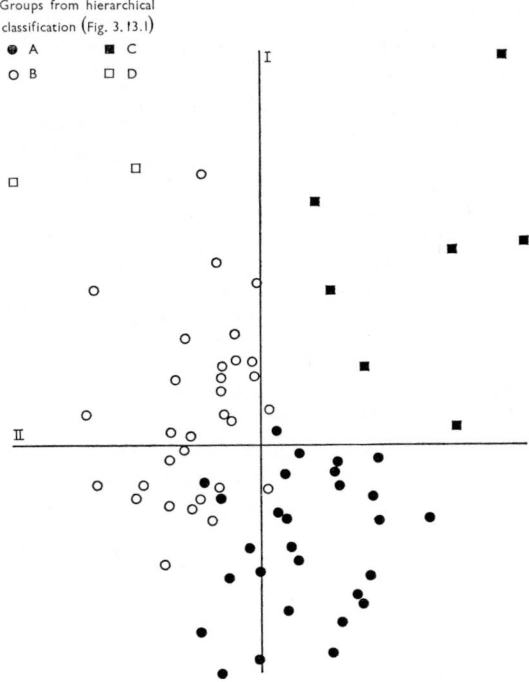

Fig. 3.13.2. Projections of the field plots onto principal components I and II.

and the principal component analysis is apparent. The latter does, however, suggest a closer relation between the high grass production groups B and D than was evident in the hierarchical classification. Correlation coefficients between the principal components of nitrogen mineralization potentials and external variables from the previous rotation phase of the field experiment are given in Table 3.13.7. Both principal components I and II were significantly correlated with grass production, the former being a positive correlation and the latter negative. Principal component I was also correlated with total number of grazing days; this variable contributed independently of total grass production at a low level of significance ($P<0\cdot1$) to a multiple correlation with principal component I. Similar correlations with principal components III, IV and V were statistically non-significant.

Table 3.13.7. **Correlation coefficients for principal components I and II with grass production, grazing days and lucerne production**

| Principal component | Simple correlation coefficient | | | Multiple correlation coefficient |
	Grass production	Grazing days	Lucerne production	Grass production and grazing days as two independent variables
I	0·518**	0·396**	−0·184	0·553[A]
II	−0·550**	−0·203	0·192	—

** $P < 0\cdot01$.
[A] $P < 0\cdot01$ for grass production; $P < 0\cdot1$ for grazing days.

The projections of the 12 incubation methods onto principal components I and II are given in Fig. 3.13.3. The methods are approximately disposed in three groups which can be identified on incubation conditions as (*a*) aerobic, 20 and 30°C, (*b*) aerobic, 40°C and (*c*) waterlogged. The disposition of these groups indicates a major discontinuity between the lower-temperature aerobic conditions and the waterlogged conditions.

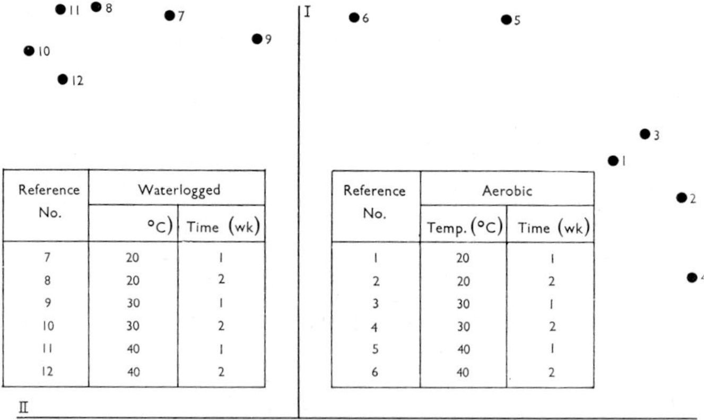

Fig. 3.13.3. Projections of incubation conditions, waterlogged or aerobic, onto principal components I and II.

Discussion

Pattern analysis proved to be a useful initial approach in this investigation. Firstly, it drew attention to an artefact in the data, associated with a batch difference in drying soil samples. The cause of the difference in initial ammonium concentration between batches of air-dried soil was not identified and whether it was initial ammonium concentration *per se* or some other change in the soil that influenced net nitrogen mineralization has not been explored. However, because of the possibility of a recurrence of this artefact in air-dried samples, air-drying has been rejected as a suitable soil preparation method for determining nitrogen mineralization potential in this field experiment.

Secondly, pattern analysis indicated that the major groupings in the field plots were related to the level of grass production under the previous cultural treatments. A further difference between the majority of the plots and certain mixed-pasture plots which had been grazed for the first 2 years at the highest stocking rate of 9 sheep per acre was also identified. Surprisingly, sward lucerne plots and crop sequence plots were grouped together and the level of lucerne production had no independent effect on the results obtained. Thus any effects that the growth of the legume may have had on the potential mineralization of soil nitrogen was not strongly expressed after nine months of bare fallow. It seems likely that any effect of sward lucerne on the nitrogen fertility of wheat crops in this field experiment will be associated with mineral nitrogen produced during the first fallow rather than with continued mineralization of legume-derived organic nitrogen in succeeding years.

Thirdly, pattern analysis indicated definite groupings among the methods employed to determine nitrogen mineralization potentials. Because these methods constituted a factorial array of incubation conditions, the main factors involved could be identified. Incubations under waterlogged conditions irrespective of temperature and time conditions were markedly contrasted with incubations at 20 or 30°C under aerobic conditions. High-temperature (40°C) aerobic incubations were situated between the waterlogged incubations and the lower-temperature aerobic incubations. Plots with previous high productions of grass mineralized more nitrogen than other plots under waterlogged conditions or high-temperature aerobic conditions but less under moderate-temperature aerobic conditions. These patterns have suggested a number of hypotheses on the microbiology of organic matter decomposition and net nitrogen mineralization which are currently being tested.

J. P. THOMPSON
D. L. LLOYD
A. W. MOORE

References

Beckmann, G. G., and Thompson, C. H. (1960). Soils and land use in the Kurrawa area, Darling Downs, Queensland. CSIRO Aust. Soils and Land Use Ser. No. 37.
Bremner, J. M. (1965). Nitrogen availability indexes. In 'Methods of Soil Analysis', Part 2. (Ed. C. A. Black.) pp. 1324–45. (Am. Soc. Agronomy: Madison.)
Lance, G. N., Milne, P. W., and Williams, W. T. (1968). Mixed-data classificatory programs. III. Diagnostic systems. *Aust. Comput. J.* **1**, 178–81.
Pearce, S. C. (1965). 'Biological Statistics: An Introduction'. (McGraw-Hill: New York.)
Waring, S. A., and Bremner, J. M. (1964). Ammonium production in soil under water-logged conditions as an index of nitrogen availability. *Nature (Lond.)* **201**, 951–2.

Case 3.14. Classification of Sequential Vegetation Data: A Three-dimensional Approach

There are many forms of agricultural and ecological research in which the analysis of vegetation is made in a sequence of time intervals. In grazing experiments, for example, pastures are commonly analysed at least annually for species composition. Despite the large amount of such data collection over the years by many experimentalists, methods of analysis of the data have never been standardized. A common method has been to analyse the data in two dimensions and to make intuitive comparisons over the third, e.g. to carry out a series of species $\times$ treatment analyses and compare the results between years. This method is unsatisfactory because it is subjective and omits vital information on the relative importance of the comparisons. For example in some cases it may be more meaningful to analyse treatment $\times$ years and compare the results between species.

Recently Williams and Stephenson (1973) described a numerical model which is useful for analysing such three-dimensional data. An agricultural example is given by Williams and Edye (1974). The mathematical background and proofs to the theorems are given in the original paper together with an example in marine ecology. The model is best suited to handling data based on absolute yields of individual species or groups of species rather than percentages. The use of raw percentage data eliminates the between-years dimension and thus reduces the effect of the model. In agricultural research we do not usually deal with a large number of species of animals, but in such cases number of animals would be suitable data for the model.

In this paper we shall discuss how the model was used to analyse botanical composition data from a grazing experiment. The original data consist of botanical composition obtained by the dry-weight rank method which gives percentage composition of the sward. Yields of the pasture were measured at the same time and thus the yield of dry weight of each of 9 species, or in some cases groups of species, could be calculated. The results were obtained for 8 paddocks over 6 years. The dimensions of the data matrix are thus 8 (paddocks) $\times$ 6 (years) $\times$ 9 (species groups).

Mean variance per comparison

For each dimension there is a matrix of the other two dimensions (e.g. each paddock has a matrix of species $\times$ years) and for each pair of, say, paddocks there are two such matrices. The difference between these matrices is subjected to analysis of variance. Each pair of paddocks has therefore a variance due to species, one to years and one for the interaction and these values are written away for later classification. If there are n paddocks there are $\frac{1}{2}n(n-1)$ such comparisons (28 in this case) and the mean of these will be the 'mean variance per comparison' due to species for paddock comparison.

The first part of the output from a program of this type is a table of 'mean variance per comparison' which is shown in Table 3.14.1. More than half the total information is contained in the paddocks × species comparison but years × species is also large. This means that we should pay most attention to these two comparisons and that although there is information in the years × paddocks comparison it is of relatively small importance. The interaction is derived for each comparison and involves all three dimensions; for each comparison it is different but the mean is the same for each pair of dimensions. If the interaction is small it is best regarded as 'noise'.

Table 3.14.1. Mean variance per comparison

	Paddocks	Years	Species	Interaction
Paddocks	—	2·325	7·951	1·024
Years	2·325	—	3·624	1·024
Species	7·951	3·624	—	1·024

Classification of the dimensions

As mentioned in the previous section, each comparison within a dimension has a variance due to each of the other dimensions and their interaction. These can be classified over the whole dimension using the variance as a dissimilarity measure. In this case the agglomerative 'flexible' fusion strategy of Lance and Williams (1967) was employed. Thus the output of the classification is in the form of a series of dendrograms; each pair of dimensions is classified in both directions, i.e. species are classified by paddocks and paddocks are classified by species. The use of these dendrograms will become clear as we proceed. Dendrograms for the paddock and species pair are given in Fig. 3.14.1.

Two-way tables

Knowing the relative importance of the comparisons and their classification, the experimenter needs to examine the data in order to interpret the results. The classification was not, however, made on original data but on its variances. These variances can be simulated by centring, i.e. paddocks are centred to a paddock mean and species to a species mean and so on.

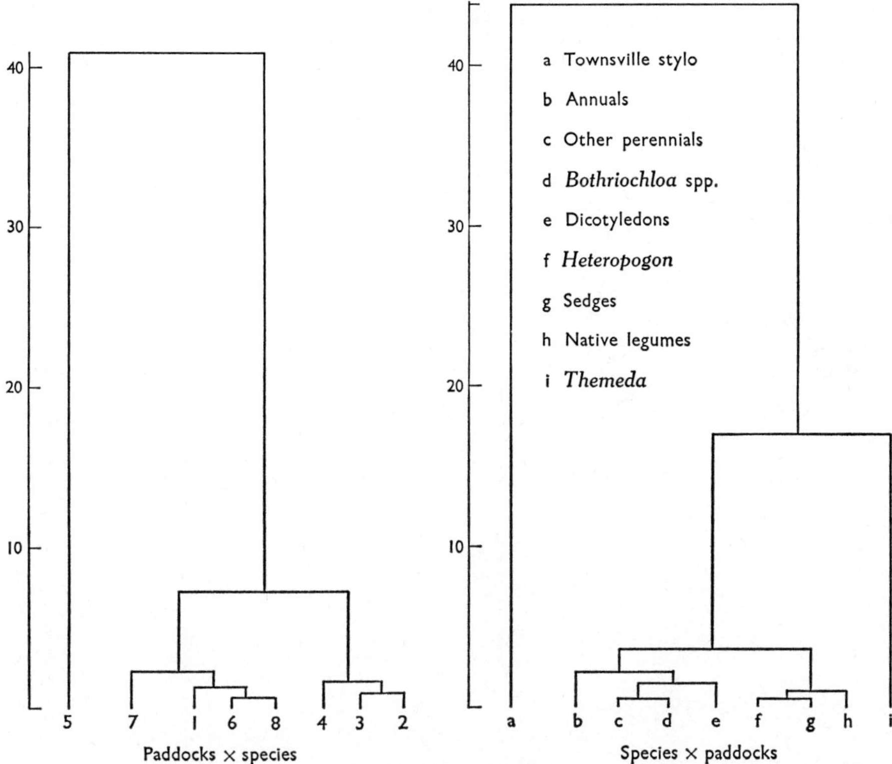

Fig. 3.14.1. Dendrograms for the paddock–species pair. Paddocks × species shows paddocks classified by species composition while species × paddocks shows species classified by their distribution between paddocks.

It is not difficult to prepare two-way tables which have been centred in this way. However, there are two entries in each cell; the top line adds to zero by columns and the bottom line adds to zero by rows. Most experimentalists would prefer to see only one entry per cell and in fact Williams and Stephenson (1973) reproduce the sum values of these entries only. This practice is not now recommended. Table 3.14.2 is the two-way table for the paddocks and species pair. Horizontal lines dividing the data represent the group classifications given in Fig. 3.14.1. Interpretation of this table is made by examining the distribution of values which sum to zero. Thus if we are interested in Townsville stylo we examine the bottom row of figures and note that it has a high negative value in paddock 5, positive values in paddocks 2, 3 and 4 and smaller positive and negative values in paddocks 7, 1, 6 and 8. If, however, we are interested in the paddocks, we examine the top figure by columns. For paddock 5 this tells us that blue grass,

Table 3.14.2. Two-way table, paddocks $\times$ species

Species	____			Paddocks					
	5	2	3	4	7	1	6	8	Total
Townsville	−21·6	8·0	9·5	8·7	1·5	3·6	1·5	2·8	14·0
stylo	−25·1	6·6	11·7	6·3	1·8	4·0	−3·7	−1·5	0
Themeda	−0·4	−25·1	−20·8	−21·6	−10·7	−16·6	−15·1	−15·2	−125·5
	13·5	−9·0	−1·2	−6·6	7·0	1·2	−2·8	−2·1	0
Annual grasses	−3·0	2·9	3·0	−2·0	−3·3	−3·0	−1·8	−2·6	−9·8
	−3·6	4·5	8·1	−1·4	−0·0	0·3	−4·0	−4·0	0
Blue grasses	13·0	11·7	11·2	13·7	8·9	15·1	10·8	9·0	93·4
	−0·5	0·4	3·4	1·5	−0·8	5·5	−4·3	−5·2	0
Other perennial	11·1	8·7	9·5	10·1	8·9	9·3	10·4	7·0	75·0
grasses	−0·1	−0·3	4·1	0·2	1·6	2·0	−2·4	−4·9	0
Dicotyledons	−1·5	2·2	−1·0	5·0	−2·5	0·8	2·8	1·8	14·8
	−4·3	1·7	2·0	3·4	−1·4	1·9	−1·5	−1·8	0
Native legumes	−4·2	−6·6	−9·3	−9·3	−4·4	−10·9	−8·2	−6·0	−58·9
	1·3	1·1	2·0	−2·6	5·0	−1·4	−4·2	−1·3	0
Heteropogon	12·4	8·0	9·4	5·7	8·5	9·4	9·4	9·8	72·6
	1·5	−0·7	4·3	−4·0	1·5	2·5	−3·1	−1·9	0
Sedges	−5·7	−10·0	−11·4	−10·3	−7·0	−7·7	−9·8	−6·7	−68·6
	1·1	−1·0	1·1	−2·3	3·6	3·0	−4·6	−0·8	0
	0	0	0	0	0	0	0	0	
Total	−16·2	3·3	35·5	−5·5	18·3	19·0	−30·6	−23·5	

Heteropogon contortus, and other perennials form the bulk of the sward, *Themeda* is present but not dominant and Townsville stylo forms the smallest part of the total composition. Examine the cell for *Themeda* in paddock 5, the upper value is −0·4 indicating a modest value for that column (which sums to zero) while the lower value is 13·5, the largest value for the row (which sums to zero). The value of −0·4 compared to the other values in the column indicates a modest amount of *Themeda* in the paddock but the higher value of 13·5 in the row indicates that it is paddock 5 that has almost all of the *Themeda* in all eight paddocks. A mean of the two values would reduce the information in this cell.

Data reduction

Interpretation of such large tables is at best clumsy and it would be desirable to reduce them for presentation.

Since we already have a classification for all combinations of pairs of data it is possible to group the data in Table 3.14.1 by averaging the appropriate values in each of the species–group/paddock–group cells. As we have already pointed out there is an information loss if averages of the upper and lower values are taken. Therefore two subtables are most useful, one containing the averages of the upper values in each macro-cell and one for the lower values. Such subtables are given in Tables 3.14.3(*a*) and (*b*). There is no doubt that the volume of data has been reduced but interpretation can be confusing because these subtables are unidirectional. Arrows mark the direction in which the values originally summed to zero. Thus to compare values for species within a paddock group one examines Table 3.14.3(*a*) in columns but to discover how the species are distributed between paddocks one examines Table 3.14.3(*b*) in rows.

Graphic representation

An experimentalist who wishes to publish his results analysed in this manner will most probably wish to reduce the data even further in order to

Table 3.14.3(*a*). Subtable of reduced data: paddocks × species

Species	Paddocks		
	5	234	7163
Townsville stylo	−21·6	8·7	2·4
Themeda	−0·4	−22·5	−14·4
Annuals	−3·0	1·3	−2·7
Blue grasses + other perennials + dicotyledons	7·5	7·9	6·7
Native legumes + *Heteropogon* + sedges	0·8	−3·3	−2·0

Sum to zero ↓

Table 3.14.3(*b*). Subtable of reduced data: species × paddocks

Species	Paddocks		
Townsville stylo	−25·1	8·2	0·2
Themeda	13·5	−5·6	0·8
Annuals	−3·6	3·7	−1·9
Blue grasses + other perennials + dicotyledons	−1·6	1·8	−0·9
Native legumes + *Heteropogon* + sedges	1·3	−0·2	−0·1

———→ Sum to zero

satisfy the editors and make interpretation easy for readers. One of the simplest ways of doing this is to reproduce the data in the form of figures. In pattern analysis the use of dendrograms has become well accepted and therefore the reproduction of these poses little difficulty. However, in the three-dimensional case we have two dendrograms for each pair of dimensions and it is likely that only one of these need be reproduced. For instance in the paddocks/species pair we could be most interested in classification of paddocks (by species) or of species (by paddocks). In my opinion it is the classification of paddocks which holds more interest and I find the inverse classification difficult to picture mentally. Each user of the model will be familiar with his own data and will have to decide which classifications are most useful to him.

Graphic representation can be improved by inclusion of the information in the subtables (3.14.3(a) and (b)) in histogram form below each group in the dendrogram. Here again there are two groups of values and the experimentalist could be easily confused in his choice of the appropriate subtable. Let us consider our example; if we have chosen the classification of the paddocks (by species) we shall want to represent values pertaining to the species below each group. This may be done by using the values in Table 3.14.3(a), in which case we are working in the direction of the arrow. Therefore the information represented tells us about the species composition in each paddock group. Alternatively the values in Table 3.14.3(b) could be used; this means we are using values *against* the direction of the arrow for each group, but with the arrow for each species between paddocks. Here the information is telling us how the total species are distributed between the paddocks. One important point to remember is that the classification of the dendrogram was made on information similar to that in Table 3.14.3(a) and therefore there is merit in using information contained in Table 3.14.3(b). Both sets of histograms are given in Fig. 3.14.2 and it is doubtful if radically different conclusions could be drawn from the two methods in this particular set of data.

Conclusions

The sort of data which so many agrostologists accumulate from botanical analysis is very difficult to analyse by conventional methods. Most experimentalists can recognize some of the patterns in their data but unless there is some numerical way of demonstrating this pattern they will be left unable to publish these observations unless they are satisfied with the 'there appeared to be a trend' type of statement. The use of a three-dimensional model gives substance to the easily observed patterns in the data; it will probably also show up other less obvious patterns and has the additional advantage of putting a value on the importance of the comparisons.

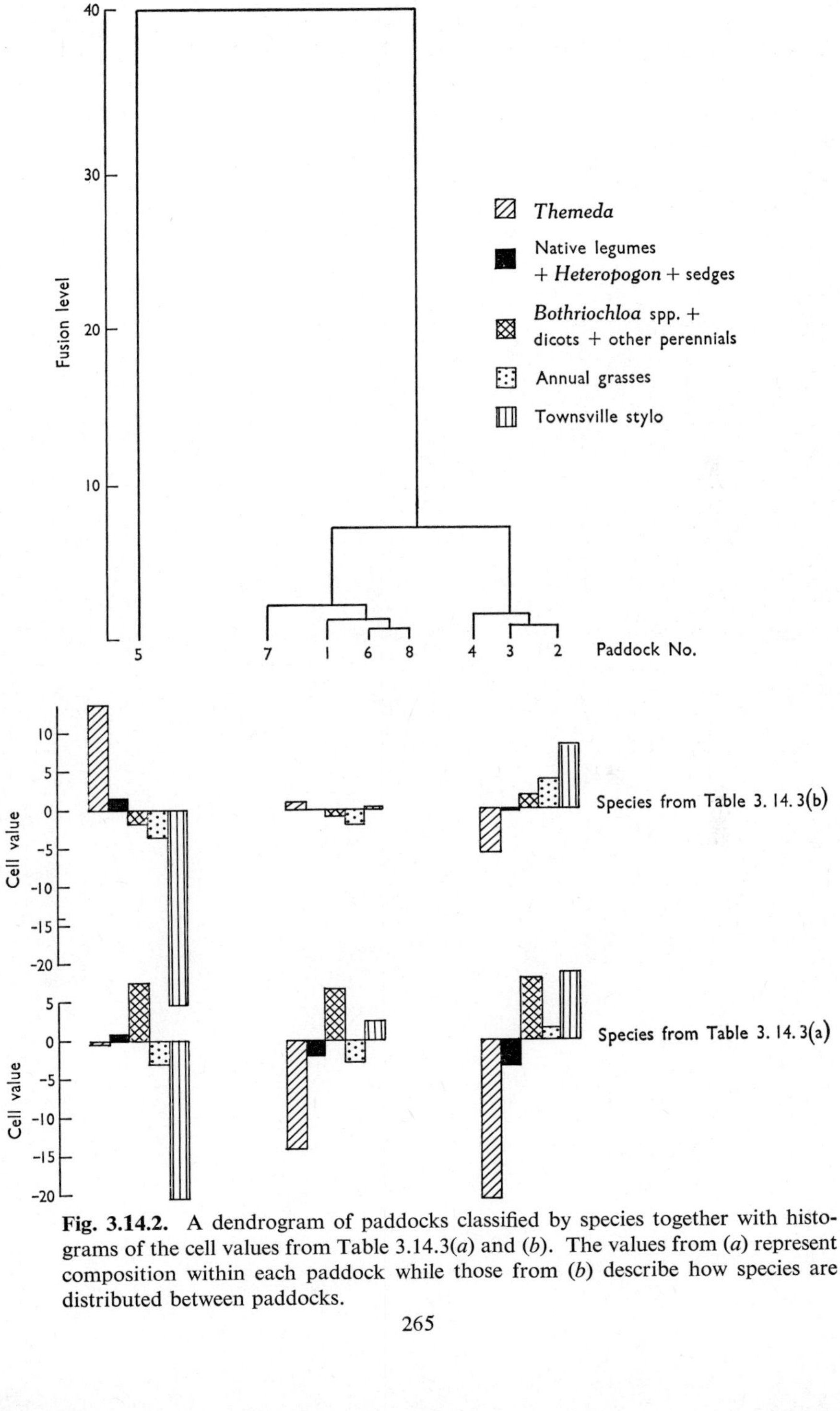

CASE 3.14

Fig. 3.14.2. A dendrogram of paddocks classified by species together with histograms of the cell values from Table 3.14.3(*a*) and (*b*). The values from (*a*) represent composition within each paddock while those from (*b*) describe how species are distributed between paddocks.

A disadvantage of the method is that we lose touch with the real data. Thus in the above example we cannot say what absolute yields of any species were. In certain instances this may be important and here it is the duty of the experimentalist to go back to his original data. This time he will have an advantage of hindsight in grouping species or treatments or years and he will be able to unravel more easily the maze of data that confronts him.

P. GILLARD

References

Lance, G. N., and Williams, W. T. (1967). A general theory of classificatory sorting strategies. I. Hierarchical systems. *Comput. J.* **9**, 373–80.
Williams, W. T., and Edye, L. A. (1974). A new method for the analysis of three-dimensional data matrices in agricultural experimentation. *Aust. J. Agric. Res.* **25**, 803–12.
Williams, W. T., and Stephenson, W. (1973). The analysis of three-dimensional data (sites × species × times) in marine ecology. *J. Exp. Mar. Biol. Ecol.* **11**, 207–27.

Case 3.15. Sequential Effects of Grazing, Burning and Fertilizing on a Native Pasture in South-east Queensland

This application exemplifies the use of pattern analysis in depicting sequential vegetation changes over a five-year period, using a large number of quadrat samples and involving a large number of species. The initial sampling was expected to identify intrinsic patterns associated with the site while subsequent samplings were expected to show increasingly the effects of the treatments that were superimposed on the area.

The area is situated in south-east Queensland, 48 km south-west of Maryborough (25°44'). The vegetation of the site is native grassland, typical of that which develops from grassy forest of *Eucalyptus maculata*, *E. crebra*, *E. intermedia* and *Acacia cunninghamii* following ring-barking (girdling), in this case more than 50 years previously. It has a gently undulating topography with little apparent soil variation.

The treatments were the presence or absence each of grazing, burning and fertilizers arranged in a factorial split plot design in two replications. Burning was carried out each year in spring after a fall of about 250 mm of rain was recorded. Fertilizer consisted of 500 kg/ha of molybdenized superphosphate (9·6% P, 10% S, 20% Ca, 0·03% Mo) applied in each of the first two years and half that amount in subsequent years as well as 300 kg ha^{-1} yr^{-1} of urea (46% N). The grazed treatments were open to uncontrolled continuous grazing by cattle in the area.

The botanical composition was estimated annually in autumn by means of presence/absence data obtained from a sampling grid of 1920 quadrats, each 20 cm square, or 80 quadrats per 0·2-ha plot. There were approximately 155 herbaceous plant species recorded from the site.

The results of the experiment have been reported elsewhere by Tothill (1974). They are considered to have considerable relevance to documenting the consequences of various agronomic applications to native pastures.

Sampling strategy

The first problem in studies of this kind is one of sampling. Williams (1971) makes a strong plea for maximum interaction between the ecologist and the analyst at all stages of designing and conducting such studies. While the ecologist must decide on the most suitable property of the vegetation to be measured, the analyst must decide the most suitable analysis to use and the most suitable values to be recorded for it. Finally, considerations of time, manpower and expense determine how much of this is practicable. In actual fact this experiment was designed with a conventional statistical, analytical approach in mind, which demonstrates there is flexibility in design while still satisfying basic analytical requirements.

Attributes sampled. Of the main properties of vegetation that may be measured (Tothill 1975) species composition by weight and by rooted frequency were chosen as these appeared to have the greatest relevance in describing the expected changes in botanical composition. The former could be readily estimated by means of the dry-weight rank procedure of 't Mannetje and Haydock (1963) and the latter simultaneously by enumerating all other species beyond the ranked set. Only the presence/absence data are considered here.

Sample distribution. A rectangular sampling grid was chosen for the economy of sampling which repeated reference to the same site allows. It is also a sample distribution recommended by Williams (1971) for pattern analysis.

Sample size. The 20-cm-square quadrat used here was found by 't Mannetje and Haydock (1963) to be adequate for dry-weight rank sampling of similar vegetation. Williams suggests a quadrat size of 1 m square because it increases the number of records obtained from each quadrat and the chance of detecting real positive associations or reducing the risk of generating false negative associations because one species, by virtue of its size, excludes others.

Sample number. Sample number and sample size are to a large extent alternatives, though it is generally better to have more smaller quadrats than fewer large ones. However, with 80 samples per plot, each sample providing a reasonable number of records, both number and size were considered adequate.

Environmental data. There may be no reasonable limit to the nature and amount of such data collected. It is not possible to measure backwards in time if one attribute is found, at a later date, to be valuable. Williams (1971) points to the difficulty of incorporating such data into one analysis. Soil chemical data were collected at the outset but appear to have no relevance to the pattern initially indicated, whereas topography showed a marked relationship.

Analytical tactics

At the time of designing these studies in 1963 it was proposed to use a conventional statistical approach for examining the composition by dry weight (dry-weight rank data) and by frequency percent. The main drawbacks to this form of analysis on such data are: (1) the repeated testing required to cover the set of species thought to be important; (2) the difficulty of identifying the important indicator species, particularly if some occur relatively infrequently; (3) the large number of species in the flora usually leading to species–site tabulations of quite unmanageable size. Where the vegetation is composed of but a few species, as in a planted pasture, this approach would have provided a satisfactory analysis of the situation.

Numerical techniques of pattern analysis have provided a very welcome means of handling large amounts of data derived from sampling many species over many sites. Not only is the task eased in a mechanical sense but also a better understanding of the full spectrum of species contributing to the pattern is obtained. When following sequential changes in botanical composition, due to a set of treatments imposed on the area, one expects some very significant changes to take place in some treatments. These are likely to be superimposed on a pre-existing pattern associated with soils, topography or some previous land use factor.

The choice of an analysis. The initial problem is whether to ordinate or whether to classify. Where the variation is likely to be discontinuous, i.e. where there are distinctive patterns, classification is obviously a pertinent procedure, and this is usually the case where agronomic treatments are involved. However, any background pattern in the vegetation that is associated with environmental factors prior to the institution of the treatments may be more continuous and therefore not so clear-cut. It may in this case be advantageous to carry out an ordination analysis to determine how distinctive the classification units are. Furthermore, the two procedures are complementary in the sense that by indicating similar patterns in the data, greater confidence in the analysis is gained.

Experimental analysis. Classification was carried out using an agglomerative, polythetic analysis entitled CENPERC 0 (Dale *et al.* 1971) which uses an information theory measure. The data input is binary (presence/absence) but is converted to numerical form by calculating the occurrences of species in each quadrat out of the total 80 per plot. The contribution of individual variables at each stage of the hierarchy of the classification was examined and arranged by the diagnostic procedure GROUPER (Lance *et al.* 1968).

Principal coordinate analysis was undertaken on the data by means of the ordination program GOWER (Gower 1966). The diagnostic program GOWECOR (Lance *et al.* 1968) was then used to produce correlation coefficients between vectors and the original attributes and to facilitate plotting.

Finally a two-parameter analysis (TWOP) (Dale and Anderson 1973) was undertaken. Though not necessary at this stage, it was found useful as a further diagnostic procedure in organizing the species relevant to the GROUPER groups.

Results and discussion

Both ordination and classification procedures indicated similar patterns in the data. The relevant data have been assembled in Fig. 3.15.1 which has been constructed from the classification and two-parameter analyses.

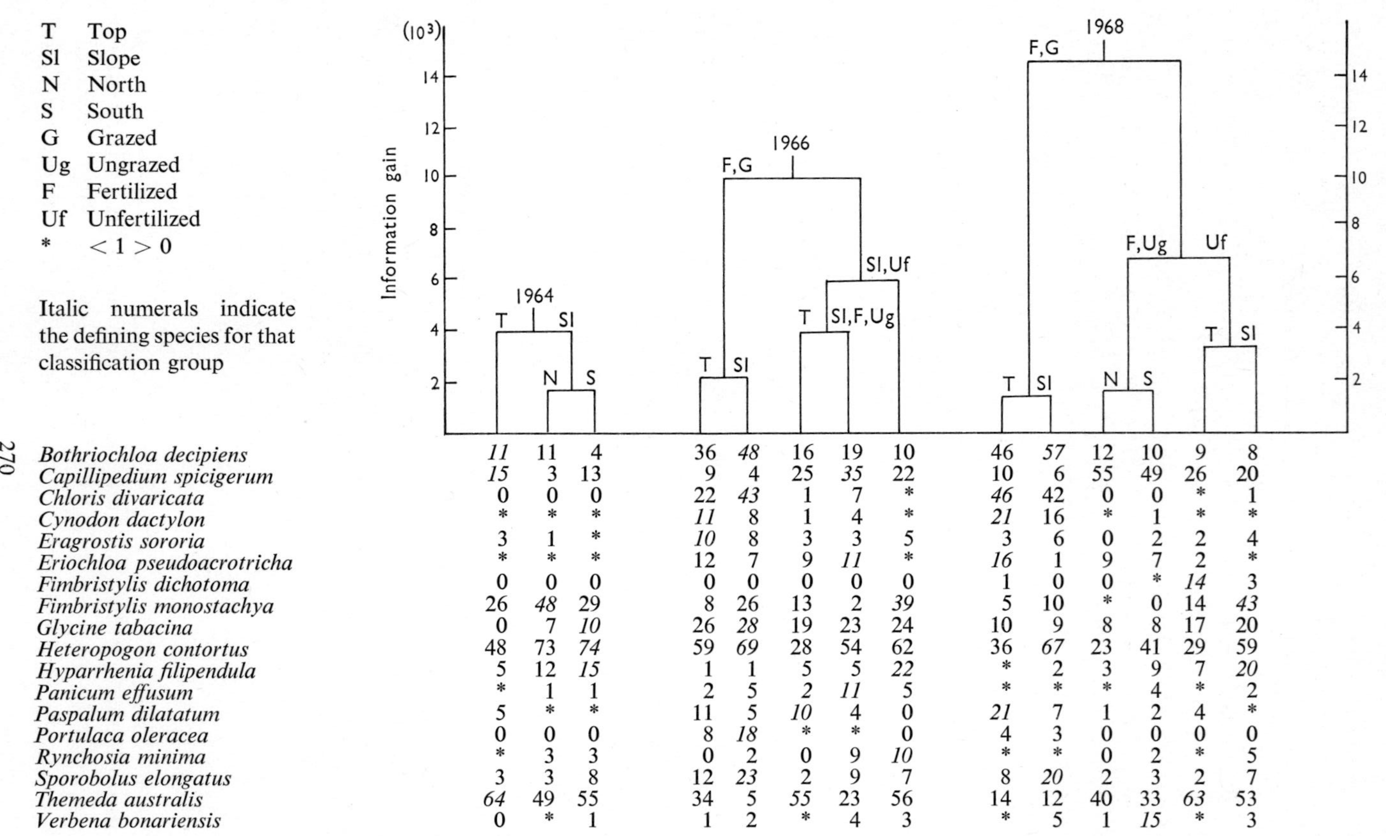

Species	1964			1966					1968					
Bothriochloa decipiens	*11*	11	4	36	*48*	16	19	10	46	*57*	12	10	9	8
Capillipedium spicigerum	*15*	3	13	9	4	25	*35*	22	10	6	55	49	26	20
Chloris divaricata	0	0	0	22	*43*	1	7	*	*46*	42	0	0	*	1
Cynodon dactylon	*	*	*	*11*	8	1	4	*	*21*	16	*	1	*	*
Eragrostis sororia	3	1	*	*10*	8	3	3	5	3	6	0	2	2	4
Eriochloa pseudoacrotricha	*	*	*	12	7	9	*11*	*	*16*	1	9	7	2	*
Fimbristylis dichotoma	0	0	0	0	0	0	0	0	1	0	0	*	*14*	3
Fimbristylis monostachya	26	*48*	29	8	26	13	2	*39*	5	10	*	0	14	*43*
Glycine tabacina	0	7	*10*	26	*28*	19	23	24	10	9	8	8	17	20
Heteropogon contortus	48	73	*74*	59	*69*	28	54	62	36	*67*	23	41	29	59
Hyparrhenia filipendula	5	12	*15*	1	1	5	5	22	*	2	3	9	7	20
Panicum effusum	*	1	1	2	5	2	*11*	5	*	*	*	4	*	2
Paspalum dilatatum	5	*	*	11	5	*10*	4	0	*21*	7	1	2	4	*
Portulaca oleracea	0	0	0	8	*18*	*	*	0	4	3	0	0	0	0
Rynchosia minima	*	3	3	0	2	0	9	*10*	*	*	0	2	*	5
Sporobolus elongatus	3	3	8	12	*23*	2	9	7	8	*20*	2	3	2	7
Themeda australis	*64*	49	55	34	5	*55*	23	56	14	12	40	33	*63*	53
Verbena bonariensis	0	*	1	1	2	*	4	3	*	5	1	*15*	*	3

Fig. 3.15.1. Classification dendrograms truncated at the useful group level for the three years 1964, 1966 and 1968 with frequency values for the collected defining species tabulated below.

The dendrograms were truncated at the lowest level of meaningful groupings. In 1964 the plot groupings, when superimposed on a contour plan, clearly indicated that topography alone determined the simple pattern at the outset, distinguishing top from slope and north-facing slope from south-facing slope. This topographic influence remained in all years at about the same level. However, the treatments imposed an increasing effect on the pattern of botanical composition with time.

The strength of the fertilized $\times$ grazed interaction is very evident, already being well established in 1966 and further strengthened by 1968. Also noticeable by its absence in all years was any effect of annual spring burning. The remaining treatments or their interactions had not established a distinct pattern by 1966, but the pattern of 1968 was beginning to emerge. By 1968 the groupings were clear and simple, viz. fertilized and grazed, fertilized and ungrazed, and unfertilized.

Tabulated below the dendrograms are the actual frequency values for all of the species that constituted defining species for any group in any year. The defining species for each group (frequencies in italics) have been taken from the two-way association table (species $\times$ groups) of beta values produced by the TWOP analysis with the lower defining threshold empirically set at $\beta = 0 \cdot 5$.

Ordination analysis helps us to understand the strengths of the classification groups. The first of the ordination plots (Fig. 3.15.2) shows the clear separation of the underlying topographic pattern in 1964. Vector 1, which accounts for 25% of the total variation, describes the top and slope pattern with Vector 2 accounting for 12% of the variation in the pattern associated with north and south. Plotting the ordination for the 1968 data, vector 1 accounts for 38% of the variation which is attributable to the presence or absence of fertilizer with vector 2 accounting for 20% of the variation due to the presence or absence of grazing. Grazing of fertilized grassland clearly results in a very marked and exclusive change in the pattern of vegetation but within this grouping topographic pattern is still discernible. The other treatment combinations show a fairly continuous pattern of variation without clear-cut boundaries. Vector 3, accounting for only 9% of the variation, describes the topographic pattern which again is more or less continuous.

Conclusion

These studies demonstrate that the recording of floristic composition by means of presence/absence data can be effectively employed to demonstrate site variability, even when this may be difficult to discern visually. They also show how effectively they can be used to follow changes in botanical composition over time as a result of the imposition of various treatments. There are a number of methods that may be used in handling such data.

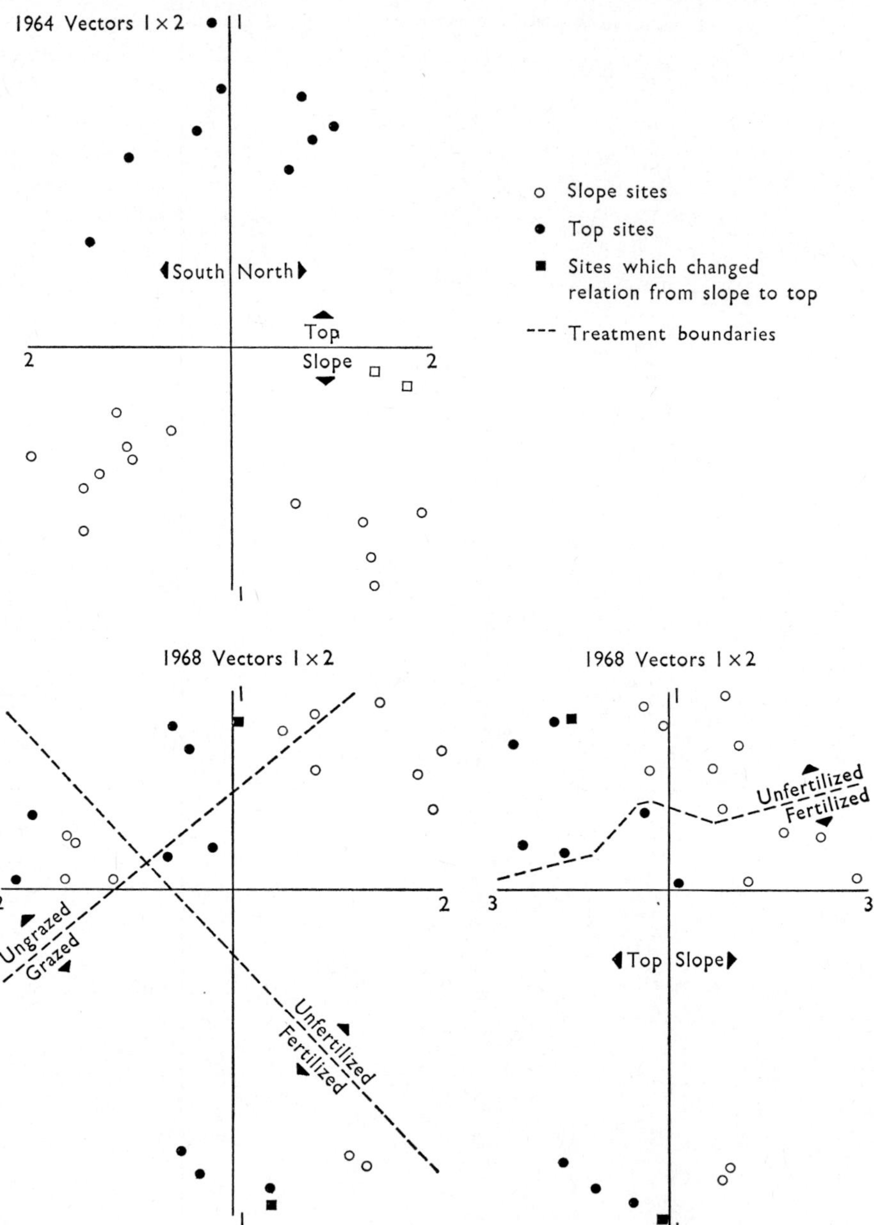

Fig. 3.15.2. The main Gower principal coordinate vectors for 1964 and 1968.

As well as lending confidence to the analysis they may complement each other since the patterns of change differ with the various treatments.

Presence/absence recording is rapid and objective. It has the advantage of being readily combined with the recording of other vegetation attributes without much further effort. It has been found that three operators can record about 2000 quadrats, with a potential total floristic richness of 200–300 species, in 5–8 days, providing they are familiar with the species, and, depending on the size of the area, very large-scale vegetation analyses can therefore be attempted.

J. C. TOTHILL

References

Dale, M. B., and Anderson, D. J. (1973). Inosculate analysis of vegetation data. *Aust. J. Bot*, **21**, 253–76.

Dale, M. B., Lance, G. N., and Albrecht, L. (1971). Extensions of information analysis. *Aust. Comput. J.* **3**, 29–34.

Gower, J. C. (1966). Some distance properties of latent root and vector methods used in multivariate analysis. *Biometrika* **53**, 325–38.

Lance, G. N., Milne, P. W., and Williams, W. T. (1968). Mixed-data classificatory programmes. III. Diagnostic systems. *Aust. Comput. J.* **1**, 178–81.

't Mannetje, L., and Haydock, K. P. (1963). The dry weight rank method for the botanical analysis of pasture. *J. Br. Grassl. Soc.* **18**, 268–75.

Tothill, J. C. (1974). The effects of grazing, burning and fertilizing on the botanical composition of a natural pasture in the sub-tropics of south-east Queensland. Proc. 12th Int. Grassl. Congr., Moscow. Section: Chemicalization of grassland farming, Pt 2, pp. 579–84.

Tothill, J. C. (1975). Botanical composition. In C.A.B. Bull. No. 150 (Ed. L. 't Mannetje).

Williams, W. T. (1971). Strategy and tactics in the acquisition of ecological data. *Proc. Ecol. Soc. Aust.* **6**, 57–62.

Case 3.16. Classification of Gilgai Data

In the southern brigalow region there are extensive areas of deep clay soils which show a strongly gilgaied pattern (Isbell 1962). Brigalow (*Acacia harpophylla*) forest is the main form of original vegetation on these soils and has been described by several workers, the latest of whom is Coaldrake (1970). In their natural state these forests are virtually unproductive in an agricultural sense since there is a very sparse ground layer of vegetation. However, the soils are inherently fertile and this has attracted development to both native and improved pastures by eliminating the forest. The strong microtopography of the gilgai has a very pronounced effect on patterning the herbaceous vegetation so induced, and this paper outlines the approach taken in analysing the vegetation pattern.

At the experimental site situated at Meandarra (27°23′ lat., 149°57′ long.) in southern Queensland, there exists an interesting series of vegetation types associated with gilgai on land that has received different cultural treatments. These range from uncleared forest through pasture of either native or introduced species on cleared land to pasture where gilgai have been levelled. The herbaceous vegetation in the virgin forest is sparse because of the overriding effect that trees have on the light and moisture supply. On the other hand levelling the gilgai, to overcome the physical and climatically related environmental problems which affect the use of these lands for crops and pastures, leads to a reduction in pattern. Between these situations the gilgai impart pronounced microtopographic effects to the herbaceous vegetation. Thus the following situations have been examined: (1) gilgai under forest; (2) gilgai cleared but within forest; (3) cleared gilgai in an extensive farm clearing with native pasture; (4) as for (3) but oversown with introduced grasses and legumes; (5) cleared and levelled gilgai oversown with introduced grasses and legumes.

The study was associated with one concerned with the agronomy of sheep and pasture production on cleared brigalow land. One of the more obvious features of the pastures, either native or introduced, is the very strong zonation of the vegetation into the topographic categories of top (or puff), slope and bottom. The bottom is characterized by being subject to occasional flooding during part of most wet seasons. This not only eliminates any resident vegetation but endows this zone with a large stored water reserve for the vegetation which establishes as the water recedes. The slope is characterized by a belt of dense permanent vegetation which is seldom inundated completely or for long but has access to a fair body of stored soil moisture. The top carries a rather sparse vegetation, largely of perennials but with some ephemeral species which establish under good conditions.

Since the region is characterized by very erratic climatic patterns and rain may occur in any season, the composition of the vegetation varies from season to season, though retaining the same general form.

This particular study was carried out in April 1972 following favourable summer and autumn rains. The floristic composition therefore does not include many of the true temperate components. One of each of the five gilgai types was selected as typical for its situation. A sixth gilgai, which duplicated the cleared native pasture type, has been examined elsewhere (see Dale, Case 3.21).

Sampling strategy

The sampling strategy was planned in conjunction with the analyst, Dr W. T. Williams. The gilgai were examined individually since they formed self-contained pattern units and their boundaries could easily be defined. Since the scale of pattern within the gilgai was small, a grid of contiguous quadrats, each 1 m square, was used. Floristic composition for each quadrat was obtained by means of percent cover estimates for each and all species present. This attribute was used because of the marked differences in growth form and distribution of the species between top and bottom. Slope, aspect, depth and topographic zone (top, slope, bottom) were also recorded as extrinsic attributes if they could be related to the pattern. Each gilgai was pegged out on a N.–S., E.–W. grid of cross strings at 2-m intervals. The highest string was considered the datum level for the depth measurements while the strings also ensured accurate placement of the quadrats on the uneven ground. A total of 125 species occurred over the area, the poorest gilgai having 39 species and the richest 66.

Analytical tactics

The visually striking zonation of the vegetation into top, slope and bottom indicated that a classification procedure would be an appropriate initial approach. Thus the agglomerative hierarchical classification CLASS (revision 15/6/72) was used. This is a combinatorial program which can use the space-conserving group average fusion strategy of Sokal and Michener (1958), referred to as Sokal sorting, and the Canberra metric as its measure. The diagnostic program GROUPER (Lance *et al.* 1968) was used in the examination and arrangement of the classification hierarchy.

Results

This analysis is a static one since it examines the vegetation pattern at one point in time, though over a range of situations. However, it can be interpreted in dynamic terms by imagining the series as a developmental one over time from a forest situation to a cleared and finally levelled one.

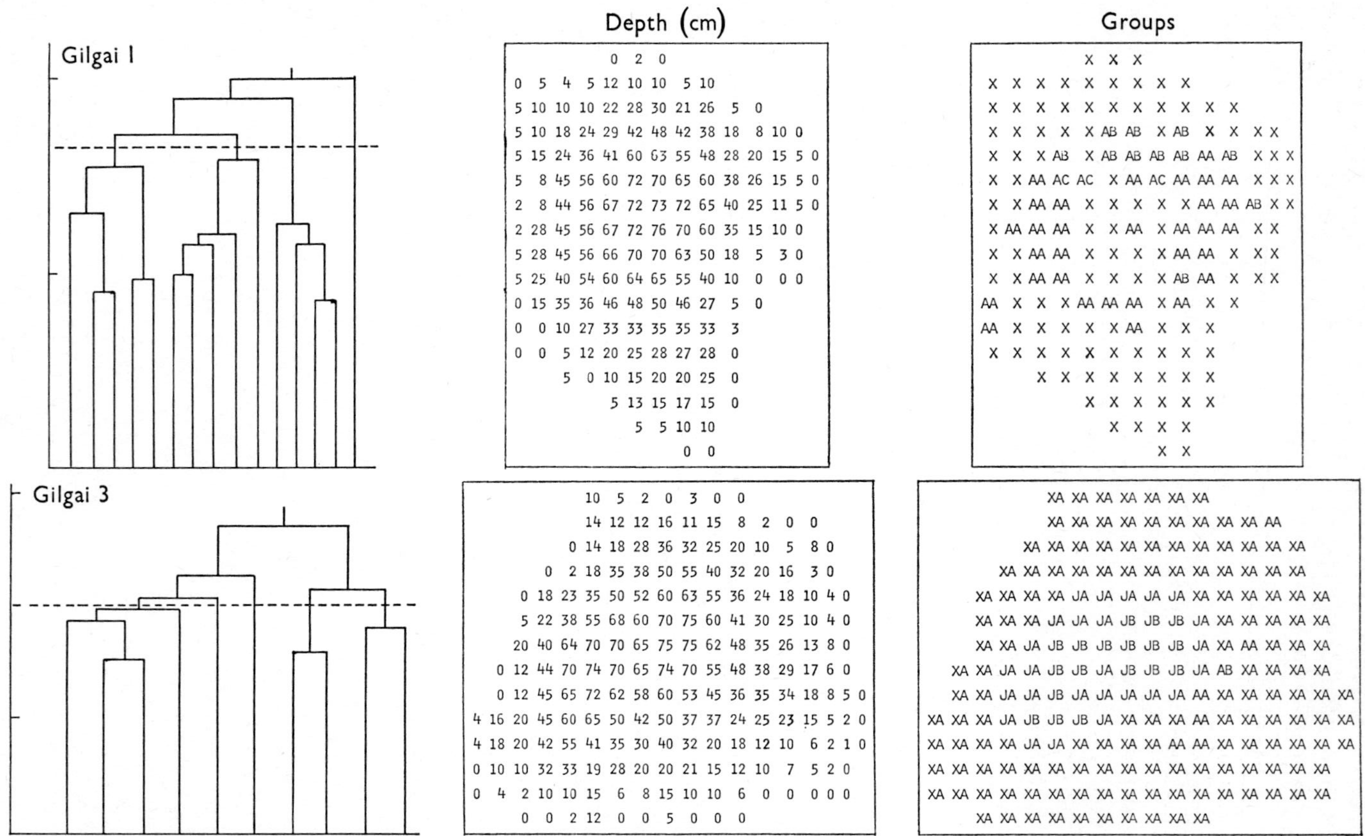

Gilgai 1
Gilgai 3
Depth (cm)
Groups

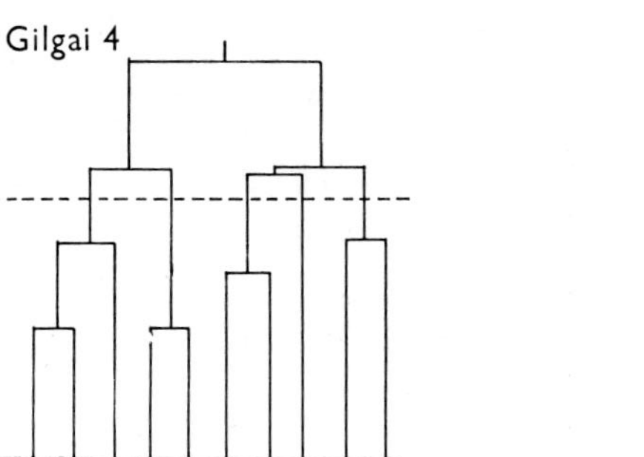

Fig. 3.16.1. Gilgai 1 (forest), 3 (cleared open native pasture) and 4 (cleared with introduced grasses and legumes) and for each (left) the classification dendrogram, (centre) topography as depth (cm) and (right) map of classification groupings. Dendrogram grouping codes, from left to right at level of dotted lines: gilgai 1, X, AA, AB, AC; gilgai 3, XA, AA, AB, JA, JB; gilgai 4, JA, XA, AA, AB, AC.

Fig. 3.16.1 gives for gilgai 1, 3 and 4 the classification dendrogram, the depths of the gilgai below datum, and a map of the groups which the dendrogram distinguishes plotted onto the quadrat layout.

The gilgai sequence

(1) *The forest.* The dendrogram does not indicate a clear separation into groups and most of the nodes are closely related. However, on mapping at the four-group level there is an indication of only two zones, viz. slope and the remainder. Apart from the main occurrence of the trees on the top and upper slope, the top and bottom both have sparse herbaceous vegetation and therefore do not differentiate. Most of the herbaceous vegetation occurs on the slope.

(2) *Cleared.* The dendrograms for the cleared gilgai 2, 3 and 4 initially divide into two groups, viz. gilgai bottom *v.* top + slope. This is followed by two or three more or less serial divisions which together suggest a rather indefinite slope zone and then a strong grouping representing the top. Some improvement in the fitting of these dendrogram groupings can be achieved by subjectively reallocating some of the lower-order groups on the basis of the occurrence of some species which seem to have a measure of site specificity.

(3) *Levelled.* In this gilgai all pattern has been destroyed and the occurrence of site groupings appears to be unrelated to any feature of the gilgai environment other than irregular establishment of the sown grasses or negative associations caused by large clumps excluding other species.

Discussion

All these classifications suggest that the classes are not as strongly expressed as might have been expected. This leads to a reassessment of the ecology of the gilgai vegetation and of the method of analysis.

The ecology. For gilgai 1 the sparseness of the vegetation in conjunction with the overriding effect of the trees probably determines the lack of pattern, while for gilgai 5 the almost complete destruction of the microtopography obviously eliminates the underlying cause of the patterns. The dendrograms for gilgais 2, 3 and 4 initially divide on the basis of gilgai bottom *v.* top + slope. This is ecologically meaningful since the bottom is clearly different from the other zones as the vegetation is either present or absent, depending on the season. The revegetation of this zone is partly genetically dependent on the slope and hence there is considerable relationship between bottom and slope in floristic composition.

The top and slope, on the other hand, are two zones of predominantly permanent vegetation. The top is subject to periods of considerable aridity and hence it maintains a sparse perennial vegetation with an ephemeral annual component in good seasons. The better water relations of the slope together with its relative security from flooding make it the most suitable site for both native or introduced perennial grasses. However, because of the strong element of 'top' species in this zone it bears more relationship to the top than to the bottom. Therefore, relating as it does to both the top and bottom zones of the gilgai, it can be understood how it is overshadowed by these zones in these analyses.

The particular attribute of the vegetation which is measured is likely to have a considerable effect on the outcome of the analysis. Had the attribute been composition on a weight basis, the slope zone, which carries the greatest bulk of herbage, would have been accentuated. However, since the interest in this instance is in abundance and distribution of species, composition by percent cover was considered to provide the best representation.

The analysis. Dale (in Case 3.21) has examined a large number of analytical methods on a further set of gilgai data. He found some improvement in fit to the gilgai topography and the visually distinct zones using a two-parameter model in an inosculate analysis (Dale and Anderson 1973). The analysis chose to group species in an inverse analysis rather than the normal analyses which have been provided in this paper and this leads to the formation of overlapping site groups. The fact that some improvement in the fit of the classification groups to the gilgai zones could be achieved by subjectively reallocating some of the lower-order dendrogram groups on the basis of species–site relationships lends some support to the suggestion that a single-parameter model may not have been entirely adequate for this situation.

Conclusion

The analysis suggests that the zonation of vegetation in the gilgai is not as distinctive as it at first appears. This has called for a re-examination of both the ecology and the analysis of the situation. A better understanding of the ecology of the slope zone in relation to the others has resulted. The fact that there appear to be strong species groupings as well as site groupings suggests there is more pattern information inherent in the data than is revealed by the analysis used.

Mapping out the groupings of quadrats suggested by the analysis provides a ready means of assessing the strength of the analysis.

J. C. Tothill

References

Coaldrake, J. E. (1970). The brigalow. In 'Australian Grasslands'. (Ed. R. M. Moore.) pp. 123–40. (A.N.U. Press: Canberra.)

Dale, M. B., and Anderson, D. J. (1973). Inosculate analysis of vegetation data. *Aust. J. Bot.* **21**, 253–76.

Isbell, R. F. (1962). Soils and vegetation of the brigalow lands, eastern Australia. CSIRO Aust. Soils and Land Use Ser. No. 43.

Lance, G. N., Milne, P. W., and Williams, W. T. (1968). Mixed data classificatory programs. III. Diagnostic systems. *Aust. Comput. J.* **1**, 178–81.

Sokal, R. R., and Michener, C. D. (1958). A statistical method for evaluating systematic relationships. Univ. Kansas Sci. Bull. No. 38, 1409–38.

Case 3.17. The Efficiency of Utilization of Forage Oats by Sheep: Use of Ordination and Canonical Coordinate Procedures

In agricultural experiments we are frequently interested in the interrelationship of two sets of variables. For example, we may record a number of measurements on both plants and animals in a grazing experiment and may wish to examine the effects of one set of variables on the other.

The conventional method for processing this type of data involves the use of regression analysis. However, difficulties like the following can arise with this method.

(1) In many situations it is not feasible to collect data in which the variables within sets are not correlated. If the predictor variables are highly correlated, the regression coefficients may be unstable and the data difficult to interpret.

(2) Regression analysis is a univariate method and we can consider only one dependent variable at a time. If we have more than a few dependent variables the method becomes tedious and in the situation where the dependent variables are themselves correlated or we want an overall or combined measure of the response, multivariate analysis is more appropriate.

(3) In some instances the data matrix may be overdefined (i.e. we have more variables than we have observations) and the variance–covariance matrix singular.

In this paper an interpretation based on the results of ordination and canonical coordinate procedures is presented. It is demonstrated that these methods can be applied where the data matrix is highly overdefined; that it is possible to consider combinations of both predictor and dependent variables; and that the problem of high correlation between predictor variables may be largely overcome. Regression analysis is also used and some discussion of the different approaches is given.

Experimental procedure

The experiment to be analysed has been reported in detail elsewhere (Hedges *et al.* 1973). Briefly the experiment was planned to examine the hypothesis that the quantity of forage available at the commencement of grazing would influence the efficiency of utilization; it being supposed that the greater the quantity available, the lower the efficiency. A range of forage availabilities was produced by sowing pairs of $0 \cdot 2$-ha plots on 6 occasions at intervals of 3–19 days. It was intended to have 6 levels of availability, each replicated twice. However, rainfall during the growth period was well below average and consequently the range in forage availability was not as extensive as planned. When the dry matter initially available for grazing (DMIA) was estimated 2 days before grazing began, 5 of the 12 plots had approximately equal quantities (range 3100–3300 kg/ha). It was decided to divide the 12 plots into 2 series. In the series to be considered in this paper there were 8 plots in which the DMIA ranged from 2730 to 4850 kg/ha and the plots were stocked at approximately 1 sheep/40 kg DMIA. Thus the experiment examined the effects of different quantities of DMIA per unit area when initial stocking pressure was nearly constant.

The oats were grazed with Merino wethers aged 20 months and grazing began about 3·5–5 months after sowing. The sheep were weighed at intervals of 3–8 days and grazing continued until a fall in liveweight was detected. The maximum average liveweight of the sheep ('peak weight') and the time from commencement of grazing at which this occurred ('days to peak weight') were estimated from quadratic regressions fitted to the three highest mean liveweights for each plot; the attainment of peak weight was an objective indication that the crop had been effectively utilized. The quantity of residual forage was estimated on 16 August (at about the time of peak weight) and at the end of grazing (3–6 days after peak weight).

Forage intakes were estimated by the total faecal collection method and forage digestibility for each of the six sowings was determined *in vivo* on material harvested from adjacent ungrazed areas. To provide an integrated estimate of the efficiency of utilization the total metabolizable energy harvested (TMEH) by the sheep was estimated as the sum of the energy required for liveweight gain plus that for maintenance.

Analysis and results

A total of 32 attributes were recorded for each plot and these are numbered in Table 3.17.1. This numbering is used to refer to the variables in subsequent tables and in the text. The raw data matrix is available from the author or on application to the Editor-in-Chief, Editorial and Publications Service, CSIRO, 372 Albert Street, East Melbourne, Vic. 3002.

Table 3.17.1. Attributes recorded for each plot, classified on the basis of their plant or animal origin

Plant attribute	No.	Animal attribute	No.
Total DM[A] initially available (kg/ha) (DMIA)	P1	Average peak bodyweight (kg)	A1
Dead DM initially available (kg/ha)	P2	Average bodyweight gain to peak weight (kg/head)	A2
Green DM initially available (kg/ha)	P3	Number of days taken to reach peak weight	A3

[A] DM, dry matter.

Table 3.17.1 (*Continued*)

Plant attribute	No.	Animal attribute	No.
Percentage nitrogen in total DMIA	P4	Average daily gain to peak weight (g head^{-1} day^{-1})	A4
Percentage nitrogen in dead DMIA	P5	Total weight gain to peak weight (kg/ha)	A5
Percentage nitrogen in green DMIA	P6	Grazing days per hectare to peak	A6
Dead DM as percentage of total DMIA	P7	Total ME harvest to peak (Mcal/ha)	A7
Percentage digestibility of total DMIA	P8	Total faecal nitrogen (kg/ha)	A8
Percentage digestibility of forage nitrogen	P9	Total faecal dry matter (kg/ha)	A9
Growth of DM during grazing (kg/ha)	P10	Average daily faecal nitrogen (g head^{-1} day^{-1})	A10
DM available on 16 Aug. (kg/ha)	P11	Average daily faecal DM (g head^{-1} day^{-1})	A11
DM available at end of grazing (kg/ha)	P12	Average daily nitrogen intake (g head^{-1} day^{-1})	A12
Day of sowing	P13	Total nitrogen intake (kg/ha)	A13
Days from sowing to grazing	P14	Average daily DM intake (g head^{-1} day^{-1})	A14
DM available on 16 Aug. (% of total DMIA)	P15	Total DM intake (kg/ha)	A15
DM available at end of grazing (% of total DMIA)	P16	Weight of initial forage per sheep (kg)	A16

Regression analysis. Initially, an attempt was made to analyse the data using regression procedures. However, many of the forage attributes (predictors) were highly correlated, an increase in the age of the crop being associated with larger quantities of total DMIA (P1) ($r = 0.79$; $p < 0.05$), with a decrease in the nitrogen content (P4) ($r = -0.82$; $p < 0.05$) and an increase in the proportion of dead material (P7) ($r = 0.77$; $p < 0.05$). Thus DMIA and the proportion of DMIA present as dead material were highly correlated (Fig. 3.17.1(*a*)). Unexpectedly, however, DM digestibility was not linearly related to forage age or quantity of DMIA (Fig. 3.17.1(*b*)) but instead was highest at the second and third sowings. Despite these peculiarities the relationship between total metabolizable energy harvested and DMIA (Fig. 3.17.1(*c*)) showed no departure from linearity and gave no indication of undue wastage of energy at the higher levels of DMIA.

As the sheep were confronted with an initially equal amount of feed per head and as the proportion harvested was fairly constant, it might be expected that all plots would have reached peak weight at the same time. Fig. 3.17.1(*d*) shows that this was not so and that the rate of forage utilization was not constant; together with Fig. 3.17.1(*b*) it implies that the rate of utilization was to some extent related to forage digestibility. This suggested that the simple description of the efficiency of utilization in terms of the linear regression of TMEH on DMIA might be inadequate.

Various multiple regression equations were fitted. However, partly because of the high degree of correlation and partly because of the small number of observations (particularly for digestibility where there were only 5 separate estimates for the 8 plots), it was not possible to clarify the situation further by regression analysis. Consequently it was felt that a multivariate approach might lead to a clearer understanding of the effects of the various

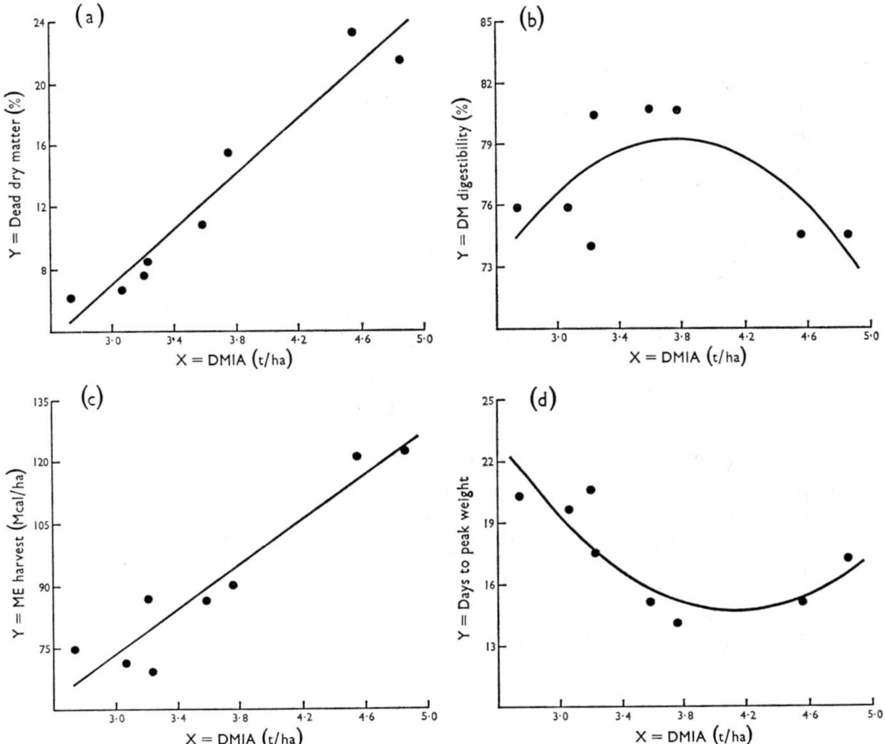

Fig. 3.17.1. Relationships between dry matter initially available (DMIA) and (*a*) percentage of dead material in DMIA (P7), (*b*) forage dry-matter digestibility (P8), (*c*) total metabolizable energy harvested (A7), and (*d*) the number of days taken to reach peak weight (A3) for experiment 1. Regression equations (X = DMIA, t/ha):

$$[P7] = 8 \cdot 94X - 19 \cdot 9, \qquad (r = 0 \cdot 97; \ P < 0 \cdot 001) \qquad (1a)$$

$$[P8] = 14 \cdot 8 + 34 \cdot 5X - 0 \cdot 0046X^2, \qquad (R = 0 \cdot 69; \ P \simeq 0 \cdot 10) \qquad (1b)$$

$$[A7] = 2711X - 778 \cdot 0, \qquad (r = 0 \cdot 95; \ P < 0 \cdot 001) \qquad (1c)$$

$$[A3] = 75 \cdot 9 - 29 \cdot 6X + 0 \cdot 00358X^2. \qquad (R = 0 \cdot 85; \ P < 0 \cdot 05) \qquad (1d)$$

forage attributes on grazing efficiency and thereby facilitate the generation of more precise hypotheses for future examination.

Principal component analysis. It is of interest to disregard temporarily the experimental treatments (i.e. sowing dates) and simply consider the system as consisting of 8 individuals (plots) each defined by 32 continuous attributes. The inter-attribute correlation matrix can then be submitted to a principal component analysis to produce a set of orthogonal vectors reflecting the various linear relationships between the attributes.

In Table 3.17.2 are given the latent roots and the 12 most important direction cosines for the first three vectors. The inter-attribute correlation matrix was singular and so there is no test of significance for this analysis. However, it can be seen from Table 3.17.2 that the first 3 vectors accounted for $84 \cdot 1\%$ of the total variance, i.e. 84% of the total variance was accounted for by 3 linear combinations of the original 32 variables. The roots corresponding to subsequent vectors were small and have been disregarded.

Examination of the direction cosines with the original axes (Table 3.17.2) shows that the first vector reflects the close association between time of sowing (P13), green (P3) and total (P1) DMIA and total animal production per hectare as estimated by metabolizable energy harvested (A7) and total liveweight gain (A5). This vector also shows quite clearly that the use of

Table 3.17.2. **Principal component analysis of the inter-attribute correlation matrix**

Vector I		Vector II		Vector III	
Root: $16 \cdot 726$		$6 \cdot 513$		$3 \cdot 688$	
Variation: $52 \cdot 3\%$		$20 \cdot 3\%$		$11 \cdot 5\%$	
No.	PC	No.	PC	No.	PC
P4	$0 \cdot 220$	P11	$0 \cdot 361$	A3	$0 \cdot 255$
A12	$0 \cdot 213$	P15	$0 \cdot 285$	A6	$0 \cdot 195$
P6	$0 \cdot 213$	A8	$0 \cdot 264$	A9	$0 \cdot 153$
A1	$0 \cdot 210$	P10	$0 \cdot 228$		
P13	$0 \cdot 207$	A3	$0 \cdot 220$		
		A9	$0 \cdot 211$	P11	$-0 \cdot 156$
		P13	$0 \cdot 200$	P8	$-0 \cdot 159$
A15	$-0 \cdot 220$	A2	$0 \cdot 184$	A10	$-0 \cdot 192$
P3	$-0 \cdot 232$	P16	$0 \cdot 178$	A13	$-0 \cdot 221$
A5	$-0 \cdot 233$			P12	$-0 \cdot 274$
A7	$-0 \cdot 233$			P5	$-0 \cdot 312$
P7	$-0 \cdot 234$	P14	$-0 \cdot 200$	P16	$-0 \cdot 357$
P2	$-0 \cdot 235$	A13	$-0 \cdot 294$	A2	$-0 \cdot 369$
P1	$-0 \cdot 238$	P8	$-0 \cdot 363$	A4	$-0 \cdot 433$

different sowing dates to produce a range of levels of DMIA led to substantial differences in forage quality as indicated by the amount (P2) and proportion (P7) of dead material and the percent nitrogen (P4 and P6) in the forage. This vector therefore is consistent with equations (1*a*) and (1*c*) and does not call for any modification to the conclusions drawn from them.

The direction cosines of vector II indicate that high dry-matter digestibility (P8) was associated with increased total nitrogen intake (A13) and a reduction in the time taken to reach peak weight (A3). There is also an indication that low digestibility was associated with an increase in the amount of wastage both as undigested material voided as faeces (A8 and A9) and as residual forage of low quality and/or acceptability (P11, P15 and to a lesser extent P16). This vector thus supports the tentative conclusions drawn from equations (1*b*) and (1*d*), and indicates that this relationship was independent of the level of DMIA and accounted for *c*. 20% of the total variation.

The presence of the third vector shows that after accounting for the above two effects, there remained a relatively weak effect associated with individual sheep liveweight gain (A4 and A2) and accounting for 11·5% of the total variation.

Principal coordinate analysis. The variables were first standardized to common variance by the Euclidean method of Burr (1968) to remove the effects of scale. Two matrices of inter-plot Euclidean distances were then calculated and separately subjected to principal coordinate analysis (Gower 1966, 1967). This produced two sets of orthogonal vectors. The first set (Table 3.17.3(*a*)) reflected the relationships between the plots in terms of the

Table 3.17.3. Configuration of the plots on the Gower vectors

		(*a*) Plant vectors (Trace = 2·955)			(*b*) Animal vectors (Trace = 3·742)		
		1	2	3	1	2	3
	Latent roots:	1·721	0·732	0·237	1·822	0·768	0·675
	Variation:	58·2%	24·8%	8·0%	48·7%	20·5%	18·0%
Plot no.	DMIA (kg/ha)			Principal coordinates			
1	4850	−0·716	−0·104	0·209	−0·750	0·113	0·338
2	4552	−0·580	−0·369	−0·046	−0·753	−0·030	−0·116
3	3583	−0·125	0·390	−0·102	0·047	−0·282	−0·435
4	3756	−0·133	0·045	−0·290	0·012	−0·404	−0·202
5	3228	0·085	0·535	0·035	0·401	−0·314	0·454
6	3203	0·225	−0·043	0·173	0·016	0·527	−0·045
7	2735	0·574	−0·369	−0·170	0·521	0·022	0·238
8	3065	0·672	−0·086	0·191	0·507	0·368	−0·230

forage attributes, and the second (Table 3.17.3(*b*)) in terms of animal attributes. The first two plant vectors accounted for 83% of the variation between plots in terms of plant attributes and the first three animal vectors accounted for 87·2% of the variation in terms of sheep attributes. The correlation coefficient between the points on these vectors and the original data (Williams and Lance 1968) are given in Table 3.17.4. They suggest that:

(1) Plant vector 1 strongly reflects the quantity of DMIA (total P1 and green P3) and the substantial increase in the quantity (P2) and proportion (P7) of dead material associated with the increase in DMIA as the time from sowing to grazing (P14) increased.

(2) Plant vector 2 illustrates the association between DM digestibility (P8) and the quantity of residual forage at about the time of peak weight (P11) and at the end of grazing (P12) 3–6 days after peak weight; seen in this context this vector indicates that reduced forage digestibility was associated with higher levels of residual or wasted forage.

(3) Animal vector 1 reflects total animal production per hectare as estimated by total metabolizable energy harvested (A7), total liveweight gain (A5), etc.

(4) The second and third animal vectors appear to be associated with days to peak weight (A3), average daily gain (A4) and faecal output.

Canonical coordinate analysis. The Gower ordination procedure has produced two sets of orthogonal vectors in which most of the variation in the original data is contained in the first 2 or 3 vectors. As these vectors are reasonably meaningful it is of interest to see how closely the two vector sets are related. The appropriate procedure is canonical coordinate analysis (Williams and Lance 1968).

Table 3.17.4. Correlation coefficients between the points on the Gower vectors and the original attribute values in experiment I

Plant 1		Animal 1		Plant 2		Animal 2		Animal 3	
No.	*r*	No.	*r*	No.	*r*	No.	*r*	No.	*r*
P13	0·92	A16	0·87	P8	0·77	A3	0·78	A3	0·35
P4	0·90	A12	0·87	P9	0·56	A14	0·63		
						A8	0·62	A4	−0·48
P2	−0·92	A9	−0·87	P10	−0·39	A11	0·51	A12	−0·48
P14	−0·92	A6	−0·89	P15	−0·64	A9	0·49	A14	−0·58
P7	−0·92	A15	−0·91	P12	−0·71			A13	−0·63
P3	−0·94	A5	−0·97	P16	−0·75	A13	−0·64	A10	−0·77
P1	−0·95	A7	−0·99	P11	−0·95	A4	−0·68	A11	−0·83

The canonical correlation coefficients and the direction cosines relating the canonical vectors to the Gower vectors are given in Table 3.17.5(a). The configuration of the plots on the vectors is shown in Table 3.17.5(b), and the correlation coefficients between the points on the canonical vectors and the original data are given in Table 3.17.6.

The canonical correlation coefficients show that it was possible to find two vectors in each set that were orthogonal to each other and also very closely related to the corresponding vector in the other set. It can be seen from the cosines in Table 3.17.5(a) that in both the plant and the sheep sets the first canonical vector is almost identical with the first Gower vector. The order of the plots on the vectors (Table 3.17.5(b)) and the correlation coefficients in Table 3.17.6 again show the very close association between the quantity of dry matter available and the energy harvested (A7) and total liveweight gain (A5).

The cosines (Table 3.17.5(a)) indicate that the second canonical sheep vector is almost identical with the second Gower sheep vector whereas the

Table 3.17.5. Canonical analysis between Gower plant and animal vectors in experiment I

(a) *Direction cosines relating canonical vectors to Gower vectors*

Canonical vector:	I	II	III
Root:	0·999	0·918	0·027
Correlation:	0·999	0·958	0·164

Gower vectors	Cosines		
Plant 1	0·984	0·095	0·153
Plant 2	0·180	−0·544	−0·819
Plant 3	0·005	0·833	−0·553
Animal 1	0·968	−0·221	0·120
Animal 2	0·240	0·952	−0·188
Animal 3	−0·073	0·211	0·975

(b) *Canonical coordinates*

Plot no.	Canonical 'plant' vector		Canonical 'sheep' vector	
	I	II	I	II
1	−0·722	0·163	−0·723	0·344
2	−0·637	0·107	−0·728	0·113
3	−0·053	−0·309	0·009	−0·371
4	−0·124	−0·279	−0·071	−0·430
5	0·180	−0·254	0·280	−0·292
6	0·214	0·189	0·145	0·489
7	0·497	0·114	0·492	−0·044
8	0·647	0·270	0·596	0·190

Table 3.17.6. **Correlation coefficients between canonical vectors and the original data in experiment I**

| First canonical vector | | | | Second canonical vector | | | |
| Plant | | Animal | | Plant | | Animal | |
No.	r	No.	r	No.	r	No.	r
P4	0·90	A12	0·88	P11	0·79	A8	0·78
P13	0·87	A16	0·85	P15	0·62	A9	0·74
P6	0·87	A1	0·84	P13	0·54	A6	0·69
				P10	0·49	A3	0·62
P14	−0·87	A11	−0·78				
P3	−0·94	A6	−0·83				
P2	−0·96	A15	−0·89				
P7	−0·96	A5	−0·97	P14	−0·54	A4	−0·59
P1	−0·97	A7	−0·97	P8	−0·94	A13	−0·81

second plant canonical vector is a combination of the second and third Gower plant vectors. These new vectors can again be interpreted by reference to the order of the plots on the vector and the correlation coefficients between them and the original data. It would seem that the second canonical vectors are reflecting the non-random external influence associated with the contrast of the three plots with high DM digestibility (plots 3, 4 and 5; mean digestibility 81%) against the rest (mean 75%). The correlation coefficients (Table 3.17.6) suggest higher levels of residual forage (P11 and P15) and faecal output (A8 and A9) and lower intakes of nitrogen (A13) on the plots with forage of lower DM digestibility; they also indicate that this effect is not associated with the quantity of dry matter available.

Discussion

The original hypothesis was tested by regression analysis. The regression of total energy harvested on DMIA showed no departure from linearity and implied that the sheep harvested a constant proportion of energy regardless of the quantity of dry matter initially available. Further examination by regression analysis, however, revealed that the rate of forage utilization was not constant and appeared to be related to digestibility which itself was not linearly related to DMIA. This raised some doubts as to the general validity of the regression between energy harvest and DMIA and suggested that further examination of the data was warranted.

Both the principal component and the Gower principal coordinate procedures indicated that there was, in fact, a second non-random effect in the data which accounted for about one-quarter of the total variation. The ordination and canonical coordinate procedures suggested that this effect was associated with a higher level of wastage and reduced acceptability of the forage on the plots of lower digestibility. The effect was not associated with DMIA.

The analyses also indicated that approximately half of the variation in the data was associated with the original planned comparison and that this was independent of digestibility. Therefore, they support the conclusions drawn from regression but suggest that the efficiency of utilization and DMIA will be unrelated only if DMIA and digestibility (or forage acceptability) are unrelated. It is of interest to note that the relative contributions of the two effects could not be established by regression analysis.

As was pointed out by Hedges *et al.* (1973) there are no tests of significance which can be applied to the above results and obviously the suppositions that have been made need to be tested by further experimentation.

It is considered that the lack of orthogonality, the high degree of internal correlation, the possible errors in estimation of variables and the relatively small number of points precluded the satisfactory interpretation of these data by regression analysis. Obviously, there will be many cases where standard regression techniques should be used and it will not be necessary, or profitable, to proceed to the alternative methods used here. But this will not always be so. In agricultural situations even the most carefully controlled experiments sometimes do not proceed as planned, an unsuspected or unplanned source of variation may enter into the results or the data may not conform to the assumptions of the standard techniques.

While a considerable number of words have been used to present the multivariate analyses which are unfamiliar to most biologists, the time actually taken to compute the results was not great. It is suggested that these are basically simple methods which can often be very useful for the examination of data by reducing them to a few vectors which represent the major sources of variation. Care in interpreting these vectors is clearly necessary, but it is often possible to establish useful hypotheses from them.

D. A. HEDGES

References

Burr, E. J. (1968). Cluster sorting with mixed character types. I. Standardization of character values. *Aust. Comput. J.* **1**, 97–9.

Gower, J. C. (1966). Some distance properties of latent roots and vector methods used in multivariate analysis. *Biometrika* **53**, 325–38.

Gower, J. C. (1967). Multivariate analysis and multidimensional geometry. *Statistician* **17**, 13–28.

Hedges, D. A., Wheeler, J. L., and Williams, W. T. (1973). The efficiency of utilization of forage oats in relation to the quantity initially available. *Aust. J. Agric. Res.* **24**, 257–70.

Williams, W. T., and Lance, G. N. (1968). Choice of strategy in the analysis of complex data. *Statistician* **18**, 31–44.

Case 3.18. Some Relationships between Voluntary Feed Consumption and Feed Characteristics

In a recent study of 70 silages Wilkins *et al.* (1971) reported significant regression relationships between silage composition and voluntary intake by sheep. The data used in that study were obtained over several years and under a considerable range of conditions, the aim of the study being to generate hypotheses for future tests seeking to determine causal relationships between silage characteristics and intake. However, several of the silage characteristics were highly correlated and also some significant differences were found between regressions within types of silage material (viz. ryegrasses, other grasses and legumes). These features made the data difficult to interpret and the authors draw attention to the possibility that some of the relationships might be spurious.

It has been suggested elsewhere in this text that some multivariate procedures, by providing clearer insight, can sometimes offer considerable assistance in the generation of hypotheses, particularly from difficult data sets. In this paper we wish to examine the usefulness of several approaches in providing insight into (i) the way in which the various silage attributes were related to voluntary intake and (ii) the extent to which these relationships were influenced by any discontinuities in the data.

The Data Matrix

The data for the voluntary dry-matter intake of silage by sheep and for 7 silage characteristics for 91 silages were collected by the Grassland Research Institute, Hurley, Britain, during the period 1961–66. Voluntary dry-matter intake (DMI) is expressed as grams of dry matter consumed per day per kilogram of liveweight to the power of 0·73. The 70 silages reported by Wilkins *et al.* (1971) were a subset of the current data and the methods of ensiling, chemical analysis and feeding were detailed by those workers, though not all of the silage characteristics reported by them were available for this analysis. The seven attributes used in this study were: (i) percentage acetic acid (Ac); (ii) percentage nitrogen (N); (iii) apparent dry-matter digestibility *in vivo* (DMD); (iv) percentage lactic acid (La); (v) percentage total acid (Ta); (vi) pH of the silage; and (vii) lactic acid as a percentage of total acid (LaT). Variables (i) to (v) are expressed as percentage of dry matter.

Methods of Analysis and Results

Relationships between variables

The correlation matrix. The simplest method of indicating which of the seven silage characteristics were related to DMI is to calculate the correlation coefficients between all variables; these are presented in Table 3.18.1.

Table 3.18.1. Correlation matrix

	N	DMD	La	Ta	pH	LaT	DMI
Ac	0·132	0·223*	0·114	0·691***	0·083	−0·541***	−0·460***
N		0·323***	0·240*	0·188	0·207*	0·183	0·516***
DMD			0·370***	0·455***	−0·294**	0·077	−0·072
La				0·735***	−0·705***	0·672***	0·227*
Ta					−0·424***	0·029	−0·219*
pH						−0·623***	−0·018
LaT							0·585***

In this and in all subsequent tables in this paper * $P < 0·05$; ** $P < 0·01$; ***$P < 0·001$.

Their examination shows that highly significant ($P < 0·001$) correlations existed between silage consumption and acetic acid, nitrogen content and LaT; the correlations between DMI and La and Ta content were also significant ($P < 0·05$). However, Ac and LaT are also significantly correlated ($P < 0·001$) with each other and there is, in fact, a high degree of inter-correlation between most of the predictors; this poses problems for interpretation as it is difficult to assess which relationships arise merely because of correlations with other variables.

Multiple regression analysis. The data used by Wilkins *et al.* (1971) did not include all the silages available for this study and in order that the results could be more appropriately compared with those of subsequent analyses, the raw data matrix was reanalysed by multiple regression analysis. A stepwise elimination procedure was adopted whereby the regression including all predictor variables was calculated and then the predictor with the lowest non-significant *t*-value was successively deleted until only significant terms remained. The results are presented in Table 3.18.2 and the equation involving four variables accounted for 65% of the variation in DMI and is highly significant ($P < 0·001$). The equation suggests that Ac and LaT had independent and opposite influences on DMI and that high nitrogen content was associated with higher intake. The negative association between DMI and digestibility was surprising and it is later shown to be spurious.

However, it is possible to derive several other equations which fit the data almost as well as the one selected by the stepwise procedure. For example, LaT or Ac can be replaced by either Ta or La and the multiple correlation coefficient only varies from 0·77 to 0·80 (Table 3.18.2). All

Table 3.18.2. Multiple regression equations for predicting dry matter intake (g day^{-1} kg$^{-0.73}$)

Stepwise regression	
$47 \cdot 0 + 0 \cdot 24$ LaT $- 2 \cdot 21$ Ac $+ 14 \cdot 1$ N $- 0 \cdot 46$ DMD	R $= 0 \cdot 80$
Alternative regressions	
$36 \cdot 8 - 0 \cdot 84$ Ta $+ 0 \cdot 38$ LaT $+ 13 \cdot 1$ N $- 0 \cdot 35$ DMD	R $= 0 \cdot 80$
$58 \cdot 8 + 0 \cdot 66$ Ta $- 4 \cdot 55$ Ac $+ 15 \cdot 9$ N $- 0 \cdot 49$ DMD	R $= 0 \cdot 77$
$30 \cdot 0 - 1 \cdot 27$ La $+ 0 \cdot 52$ LaT $+ 13 \cdot 0$ N $- 0 \cdot 39$ DMD	R $= 0 \cdot 79$
$60 \cdot 9 + 0 \cdot 94$ La $- 3 \cdot 69$ Ac $+ 15 \cdot 1$ N $- 0 \cdot 51$ DMD	R $= 0 \cdot 78$

these equations have similar predictive power and all suggest two independent and opposing effects of the acid components of the silages on DMI. It is, however, rather difficult to attach a biological identity to these effects because the signs of the coefficients for particular variables change depending on which other variables are included.

The authors of the original paper proceeded to an interpretation from regression analysis. However, we are here concerned to assess the usefulness of alternative procedures which may add to the interpretation.

Reduction of the dependent variable (DMI) to states. The 91 silages were ordered on the basis of their DMI values and then divided by the method of Dale (1964) into groups. This method successively divides (or chops) the data at the point that will generate maximal between-group sums of squares for the variable in question. That is, it is a useful way of reducing a continuous variable, in our case DMI, to a series of ordered states; it provides a simple procedure for examining the extent to which differences in predictor variables occur over the whole range of the dependent variable. While the method optimizes the successive splitting of groups, the choice of the final number of states to be used is somewhat subjective and is made from an examination of the differences in the sums of squares of the groups produced by each division; in this particular instance four were chosen.

We can now consider each silage as belonging to one of the four states for DMI, and can calculate the means and Student t-values for the differences between means for each of the seven silage characteristics. The values for the comparisons state 1 *v.* state 2, state 3 *v.* state 4 and states 1 + 2 *v.* states 3 + 4 are presented in Table 3.18.3(*a*).

Examination of Table 3.18.3(*a*) reveals that LaT and the nitrogen content of the silages were significantly different in all three comparisons. Differences in lactic acid content were significant for the low range of dry-matter intake (state 1 *v.* 2) but not for the upper range (state 3 *v.* 4). In the extended comparison (1 + 2 *v.* 3 + 4) differences in acetic acid content were significant.

This implies that the relationships between both LaT and nitrogen content and DMI held over the whole range of intake, that lactic acid concentration was important only at lower levels of intake, and that acetic acid content was important only at the extremes.

It is also of interest to examine the distribution of the silage parent materials between the states. The results (Table 3.18.3(b)) show that the distribution was not uniform, there being no legume silages in the lowest state and no cereal silages in the highest two states. This implies that some of the observed differences in DMI could be due to the particular type of material ensiled.

Principal component analysis. In situations where interpretation of regression analysis is made difficult by high correlations between predictor variables it is often useful to carry out a principal component analysis to derive an efficiently reduced number of new orthogonal variables. The new variables are linear combinations of the original ones and can be considered as representing the major linear trends or effects in the data. Quite frequently

Table 3.18.3. Mean values within each of the four DMI states (a) and number of silages from each type of parent material (b)

(a)	Dry-matter intake state				Combined states	
	1	2	3	4	(1+2)	(3+4)
DMI	28·4	48·3	63·1	79·8	42·0	67·4
		—		—		—
Ac	4·8	4·4	3·4	2·2	4·5	3·0
	(0·65)[A]		(1·72)		(3·33**)	
N	1·8	2·5	2·5	3·2	2·2	2·7
	(4·05***)		(4·54***)		(3·47***)	
DMD	71·0	69·0	69·2	69·2	69·4	69·5
	(0·77)		(0·03)		(0·30)	
La	3·2	6·0	6·5	6·3	5·1	6·5
	(2·57*)		(0·19)		(1·76)	
Ta	12·1	11·9	11·4	9·1	11·9	10·8
	(0·13)		(1·22)		(1·13)	
pH	4·6	4·5	4·5	4·6	4·5	4·5
	(0·66)		(0·58)		(0·01)	
LaT	25·7	49·5	56·5	67·8	41·9	59·4
	(3·76***)		(2·01*)		(4·12***)	
(b)						
Grass	9	18	25	6	27	31
Legume	0	9	8	5	9	13
Cereal	6	5	0	0	11	0

[A] *t*-Values, shown in parentheses, apply to the difference between the two means above them.

it will be found that most of the variation in the original data matrix can be accounted for by the first two or three vectors. When standardized, the vectors contain the cosines of the angles between the vector and the original axes. The size and sign of the direction cosines show which variables have been important in determining the path of the vector and whether they have had similar or opposing influences.

The results of the principal component analysis carried out on the correlation matrix for the seven predictor variables are presented in Table 3.18.4. The first vector accounted for 42% of the total variation in the predictor variables and appears to reflect differences in total acid content and pH; it may be considered as the general acidity index. The second vector reflects what is normally considered as silage quality as indicated by a high proportion of lactic acid and a low acetic acid content; it accounted for 28% of the variation. Vector 3 (16·5% of the variance) largely reflects differences in nitrogen content and vector 4 (9·6%) differences in DMD. In all, the first four vectors accounted for 96% of the total variation which suggests that subsequent vectors can be ignored.

It should be noted that often vectors such as the third and, particularly, the fourth which account for a relatively small proportion of the variance would not be easily interpreted. Their interpretation in this case was greatly facilitated by the fact that both are substantially reflecting a large proportion of the variation in only one variable.

Regression of DMI on principal component variables. As the vectors appear to have meaningful interpretations we can proceed to calculate the

Table 3.18.4. Principal component vectors derived from the correlation matrix of the seven silage attributes

Vector 1 Variation 41·7		Vector 2 Variation 28·3		Vector 3 Variation 16·3		Vector 4 Variation 9·6	
Variable	Cosine[A]	Variable	Cosine	Variable	Cosine	Variable	Cosine
La	0·555	Ac	−0·643	Ac	−0·196	La	−0·260
Ta	0·464	Ta	−0·372	Ta	−0·184	N	−0·252
DMD	0·343	pH	−0·278	La	−0·018	Ac	−0·215
LaT	0·337	DMD	−0·210	LaT	0·183	Ta	−0·197
N	0·166	N	−0·161	DMD	0·255	LaT	−0·171
Ac	0·137	La	0·097	pH	0·372	pH	−0·162
pH	−0·445	LaT	0·540	N	0·831	DMD	0·853
(1·8%)[B]		(21·5%)		(35·4%)		(6·3%)	

[A] The direction cosines have been ordered.
[B] The variation in DMI accounted for by regression on each vector is presented here for convenience.

value of the four new orthogonal variables for each individual and use these values as predictors in regressions with DMI. The values or scores are calculated by performing the following matrix multiplication:

$$A \ (91, \ 7) \ C \ (7, \ 4) \ \rightarrow \ Z \ (91, \ 4),$$

where A is the standardized data matrix of the 91 silages each specified by 7 predictor variables; C is the principal components matrix of 4 column vectors of direction cosines with each of the 7 predictors; and Z is the matrix of 4 new orthogonal variates for each of the 91 silages. The regression coefficients relating DMI to the Z variables are presented in Table 3.18.5. These can be regarded as four separate independent linear equations explaining the deviation of Y (DMI) about its mean value of 54·2 g, or as multiple regression equations with 2, 3 or 4 terms in Z. In any case the constant term will always be the mean observed value of Y (i.e. intake).

The multiple regression on all four orthogonal variates accounted for 65% of the total variation in intake; variable 3 reflecting nitrogen content accounted for 35·4%, variable 2, or silage quality, for 21·5%, variable 4, the DMD, for 6·3% and variable 1, silage acidity, less than 2%.

The relationship between vector 2 and DMI suggests that low acetic acid content and a high proportion of the total acid as lactic appear to have enhanced high voluntary intakes. It is, of course, not possible to say if the combination of high acetic and low lactic acid is necessary to depress intake or, in fact, whether both variables are merely general indicators of the type of fermentation process which has occurred. The possible interaction of total acid content also cannot be ignored since this variable had a reasonably high direction cosine on the vector. Subsequent work by Hutchinson and Wilkins (1971) tends to confirm that this vector is really reflecting the type and extent of the fermentation pathway of the particular silages. From experiments where acetic acid or acetate was added to the silage diet or infused into the rumen, they concluded that high acetate level *per se* is

Table 3.18.5. Regression coefficients between DMI and orthogonal component variables

	β	t^{A}	r^{B}	Variation in DMI (%)
Vector 1	1·24	2·08*	0·13	1·8
Vector 2	5·24	7·22***	0·46***	21·5
Vector 3	8·78	9·24***	0·59***	35·4
Vector 4	−4·89	3·93***	−0·25*	6·3
			R = 0·80	R² = 65·0

[A] t-Values apply to partial correlation coefficients in multiple regression involving all 4 vectors.
[B] r-Values are simple correlation coefficients between DMI and each vector.

unlikely to result in low intake of silage. However, it is worth noting that the lactic acid content (6·1 to 5·4% of dry matter) of the ryegrass silage fed in those experiments was in the range (Table 3.18.3) where it was not correlated with intake in the data used in our study, i.e. they were working with a 'lactic acid' silage.

The results of the principal component analysis suggest that in silages containing low levels of lactic acid (i.e. in 'acetic acid' silages) the acetic acid content *per se* might appear to influence voluntary consumption; it might well be, however, that the Ac was merely reflecting the extent to which fermentation had proceeded in the acetic acid direction. While in these cases the acetic acid content would be a quite good predictor of voluntary intake, the true biological meaning of the relationship would need to be borne in mind. Further experimentation is necessary to clarify this point.

The importance of the vector reflecting nitrogen content was anticipated from the results of Tables 3.18.1 and 3.18.3. The lower level of correlation with other attributes (Table 3.18.1) and its relationship over the whole range of intake (Table 3.18.3) together with the results in Table 3.18.5 imply that it was the best single predictor of intake in this data set. As we have already pointed out for vector 2, this does not necessarily imply a causal relationship. Subsequent work by Hutchinson *et al.* (1971) in fact suggests that the relationship may have been induced by the correlation of nitrogen content with other unmeasured chemical or physical properties of the silage.

It is interesting that the major dimension in the silage composition data (i.e. acidity) had little effect on silage intake. The simple correlation between points on this vector and DMI was not statistically significant, but when the residual error term was reduced by incorporating the other vectors into the equation the partial correlation coefficient became significant ($P < 0·05$). Hence although this does imply an independent effect of silage acidity on intake, the effect was very small and of somewhat doubtful significance.

Again, as in the multiple regression analysis, the relationship between DMD and intake was negative; here it was only detected after the variation due to other vectors was removed.

Discontinuities in the data

Ordered principal component variables. A simple method of examining the data for discontinuity is merely to order and inspect the principal component scores; major discontinuities will show as large gaps in the values along the vector and if the vector is reflecting differences between types of silage material then this should be apparent from the distribution of the various silage types on the vector. Because the vectors are orthogonal each can be considered separately.

With 91 silages it is impractical to show the complete ordered vectors. None of the variables showed major discontinuities in its values, and it is necessary to show only the highest scores at each end of each variable to demonstrate the points to be made. These are given in Table 3.18.6.

Examination of Table 3.18.6 reveals that the positive end of vector 1 contains a high preponderance of ryegrasses, and that the negative ends of vectors 3 and 4 contain a high preponderance of cereals and legumes respectively. From this we can conclude that ryegrasses tended to make more acid silages, that the cereals produced silages of low nitrogen content and that the legume silages were of lower DMD.

The polar concentrations of silage types on some of the vectors suggest that the regression relationships between DMI and these vectors need to be re-examined. In the case of vector 4 it is fairly obvious from an examination of the points on the vector that the negative relationship between DMI and DMD was in fact spurious. Despite the lower DMD of the legumes, they tended to be eaten more readily than other silages and this produced an apparent negative relationship between intake and digestibility. However, an examination of the negative end of this vector (Table 3.18.6) indicates that within the legume silages there was in fact a tendency for increased DMD to be associated with increased DMI. The overall negative relationship should therefore be disregarded. Examination of the other vectors did not reveal any further discontinuity or heterogeneity.

Table 3.18.6. Six highest and six lowest values for each principal component variable

Vector 1 (Acidity)		Vector 2 (Quality)		Vector 3 (Nitrogen)		Vector 4 (Digestibility)			
Value	Type[A]	Value	Type	Value	Type	Value	Type	DMD	DMI
3·50	RG	2·22	RG	3·23	RG	1·92	OG	79·5	23·4
3·28	RG	1·98	RG	2·72	OG	1·70	OG	80·0	19·9
3·04	RG	1·94	Leg	2·27	RG	1·60	OG	79·3	25·0
2·62	RG	1·93	RG	2·00	OG	1·49	RG	73·5	53·5
2·57	RG	1·91	OG	1·82	RG	1·30	OG	73·9	36·3
2·37	RG	1·74	Leg	1·75	Leg	1·17	OG	73·7	19·0
−2·53	OG	−2·42	OG	−1·63	C	−1·34	Leg	62·1	49·8
−2·56	OG	−2·57	C	−1·78	C	−1·36	Leg	69·6	70·4
−2·63	OG	−2·70	OG	−2·05	C	−1·47	Leg	56·6	62·4
−3·06	OG	−2·88	RG	−2·06	C	−1·48	Leg	55·7	46·0
−3·50	Leg	−3·73	Leg	−2·12	C	−1·59	Leg	57·7	30·1
−3·50	Leg	−3·94	Leg	−2·17	C	−2·03	Leg	57·5	45·6

[A] Type of parent material: RG, ryegrass; OG, other grass; Leg, legume; C, wheat or barley.

Within-group analyses. A more precise means of examining the data for discontinuity is to repeat the above analyses within each of the different types of silage material; alternatively one could estimate the effects of the various types by including suitable code variables in the regression analyses. In general this will only be feasible in the situation where the likely sources of discontinuity are known or suspected *a priori.*

Repeating the principal component and regression analyses within the various types of silages indicated that the relationship between DMI and DMD was the only one that was essentially reflecting differences between types which were not attributable to or closely related to continuous variation in the predictor variables. Within both grass and legume silages, the relationship between DMI and the vector representing DMD was positive though only within the legume silages was it statistically significant. The only other major difference found between groups was that the relationship between the nitrogen vector and intake was much weaker within the legume silages than it was within other silages.

Classification. In situations where there is no *a priori* basis for separate within-group analyses, or where it is desired to look for unsuspected discontinuities, the data may be subjected to some form of classification and diagnostic procedure to see how the population is divided and which variables are important in each division. In the type of situation with which we are here concerned it is usual to base the classification only on the predictor variables. This is because we are usually interested in assessing whether the various divisions based on known predictors have produced groups that differ substantially in their values for the dependent variable.

If the classification indicated major discontinuity then the results of an analysis of the combined data may be meaningless or uninterpretable. In this case it may be profitable to proceed to separate analyses within the groups produced by the classification.

The 91 silages were considered as being specified by the 7 silage characteristics; these variables were standardized to common variance by the Euclidean method of Burr (1968) and the silages then classified by the incremental sum of squares strategy (Burr 1970) using the polythetic agglomerative program MULCLAS (Lance and Williams 1967*a, b*). The resulting hierarchy is shown in Fig. 3.18.1 with the composition of the groups. The contribution of each attribute to each of the fusions in Fig. 3.18.1 was elucidated using the diagnostic program GROUPER (Lance *et al.* 1968) and the results are shown in Table 3.18.7(*a*). The DMI values were not used to produce the classification and so we can calculate Student *t*-values for the between-group differences in intake; the *t*-values and appropriate degrees of freedom are given in Table 3.18.7(*b*).

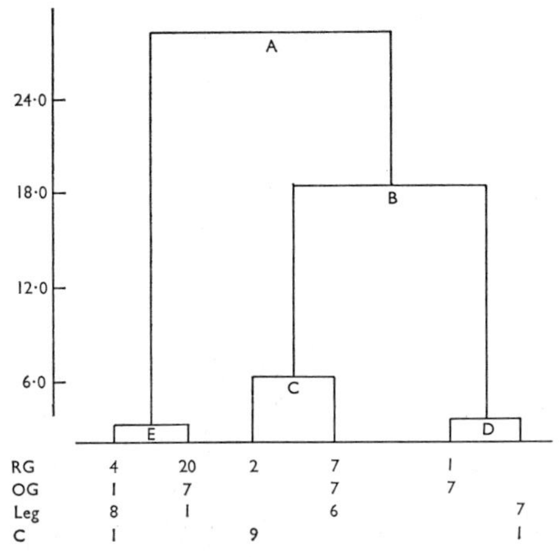

Fig. 3.18.1. Dendrogram for the classification of the 91 silages with respect to the 7 silage characteristics, and the number of silages of each type in each group.

All fusions in Fig. 3.18.1, except the highest at A, involved the merging of groups that differed significantly in DMI. Examination of the magnitude of the contributions of the various attributes to the fusions shows that the attributes, or combinations thereof, which were important here were the same as those isolated by the principal component analysis.

The fusions at D and E were both associated with DMD and, as foreshadowed by the ordination procedure, lower DMD was associated with higher intakes in groups containing a high proportion of legume silages.

Table 3.18.7. Contributions of the silage characteristics to the fusions in Figure 3.18.1 and the group mean DMI values

(*a*)

| Variable | Contribution to fusion at | | | | |
	A	B	C	D	E
Ac	0·9	*5·3*	0·3	0·3	0·9
N	1·3	0·0	*3·4*	0·6	0·3
DMD	2·8	0·2	0·8	*2·3*	*2·1*
La	*7·8*	0·5	0·1	0·0	0·0
Ta	*6·6*	1·4	0·2	0·0	0·2
pH	3·2	2·2	0·9	0·5	0·0
LaT	2·0	*6·5*	0·2	0·0	0·0

(*b*)   Mean DMI values of the groups fused

	A	B	C	D	E
Mean 1[A]	56·5	58·9	43·6	28·4	62·5
Mean 2	52·1	38·9	67·4	49·5	53·7
t	1·31	4·17	6·10	3·51	2·14
d.f.	89	45	28	14	42
	n.s.	***	***	**	*

[A] Mean 1 refers to the left-hand group fused and mean 2 to the right-hand group.

Conclusions

It has been possible to find four orthogonal variables which accounted for 65% of the variation in voluntary intake of the silages; most of this variation was accounted for by only two of these variables. Thus an index of silage quality (vector 2 reflecting Ac, LaT and to a lesser extent Ta) accounted for 21·5% of the variation in DMI and vector 3 (the nitrogen content) accounted for 35·4%.

A similar proportion of the variation in DMI could be accounted for by a multiple regression involving the original variables Ac, LaT, N and DMD. Thus the results of both methods indicate that the same four variables were related to DMI and both methods derived equations with the same predictive power. However, some important differences occurred in the biological interpretations which the two methods suggested. It is considered that the indication by the principal component analysis that Ac and LaT were reflecting a common effect provides the more plausible and meaningful biological interpretation.

The orthogonality of the principal component variables allowed a simple, yet effective, examination of the data for discontinuities. Each vector can be examined separately without reference to the others and if it is required, some vectors can be dropped from the regression equation without necessitating the recalculation of the remaining coefficients.

Although the results have not been presented here, there was also some advantage in using the principal component approach within the various groups of silages, viz. grasses, cereals or legumes. With this method, the biological interpretations of the relationships for the different groups were quite consistent and in line with the interpretation based on the analysis of the complete data matrix. The stepwise regression procedure, however, tended to select a different number and combination of variables within the three groups.

It is suggested that the biological interpretation of data in which the predictor variables are highly correlated can sometimes be substantially facilitated by the principal component and associated regression procedures that have been described. It should be stressed that the above methods, including conventional regression analysis, only estimate mathematical associations. The biological interpretation of these associations is a subjective process which requires considerable care.

In practice the examination of the data for discontinuity or heterogeneity is an integral part of the interpretation of the relationships existing within the data. The two have been separated and discussed in the above order for convenience of presentation only.

I am extremely grateful to Dr R. J. Wilkins of the Grassland Research Institute, Hurley, Britain, and to Dr K. J. Hutchinson of the Pastoral Research Laboratory, CSIRO, Armidale, for making these data available and also to Dr Hutchinson for helpful discussion on the interpretation of the results.

D. A. HEDGES

References

Burr, E. J. (1968). Cluster sorting with mixed character types. I. Standardization of character values. *Aust. Comput. J.* **1**, 97–9.

Burr, E. J. (1970). Cluster sorting with mixed character types. II. Fusion strategies. *Aust. Comput. J.* **2**, 98–103.

Dale, M. B. (1964). Multivariate analysis of heterogeneous data. Ph.D. Thesis, University of Southampton.

Hutchinson, K. J., and Wilkins, R. J. (1971). The voluntary intake of silage by sheep. II. The effects of acetate on silage intake. *J. Agric. Sci.* **77**, 539–43.

Hutchinson, K. J., Wilkins, R. J., and Osbourn, D. F. (1971). The voluntary intake of silage by sheep. III. The effects of post-ruminal infusions of casein on the intake and nitrogen retention of sheep given silage *ad libitum*. *J. Agric. Sci.* **77**, 545–7.

Lance, G. N., Milne, P. W., and Williams, W. T. (1968). Mixed data classificatory programs. III. Diagnostic systems. *Aust. Comput. J.* **1**, 178–81.

Lance, G. N., and Williams, W. T. (1967a). Mixed data classificatory programs. I. Agglomerative systems. *Aust. Comput. J.* **1**, 15–20.

Lance, G. N., and Williams, W. T. (1967b). A general theory of classificatory sorting strategies. I. Hierarchical systems. *Comput. J.* **9**, 373–80.

Wilkins, R. J., Hutchinson, K. J., Wilson, R. F., and Harris, C. E. (1971). The voluntary intake of silage by sheep. I. Interrelationships between silage composition and intake. *J. Agric. Sci.* **77**, 531–7.

Case 3.19. Classification of Sheep Body Composition Data

An experiment was carried out in an animal house to study the effect of compensating growth on the body composition of sheep. Compensatory growth is usually observed after an animal has experienced a period of restricted growth and may be defined as the exhibition of a greater growth rate than normally expected at that particular bodyweight. The explanation of this phenomenon is of little concern in the present context.

Other experiments dealing with this subject have suffered from deficiencies in design, management and detailed measurement. In some experiments the diet has been altered during the period of weight loss and again for the weight gain period. In others animals have been slaughtered after a specified time and thus groups have been compared without regard to animal size. Body composition has been determined in some experiments by indirect methods such as tritiated water or specific gravity—the validity of treatment comparisons rests on the assumption that conversion equations apply equally to starved animals, compensating animals and continuously growing animals. Some experimenters have chosen to look at only half the animal—the carcass—and thus have ignored changes in the other body components which may in turn have affected the carcass composition.

Data from two experiments are reported in this paper. The first experiment was designed to overcome many of the deficiencies mentioned above. Pairs of animals were slaughtered at intervals over an extensive bodyweight range in each treatment (Fig. 3.19.1). At slaughter the components of the body were weighed and later they were chemically analysed, giving 46 pieces of

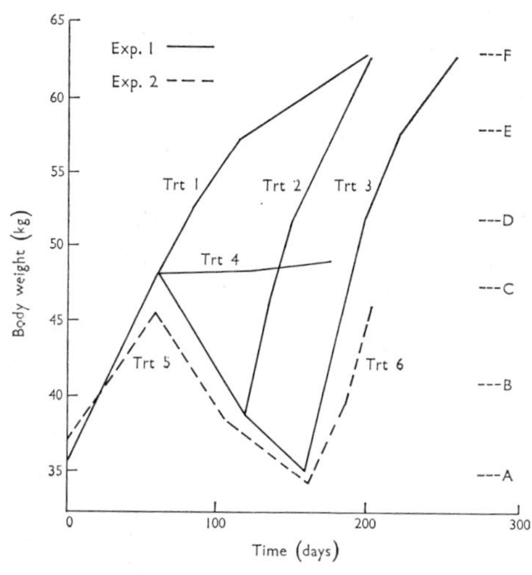

Fig. 3.19.1. Growth paths and slaughter weights A–F of animals in the first and second experiments.

information per sheep (Table 3.19.1). A second experiment was carried out to study the digestive efficiencies of animals following a similar growth pattern to treatment 3 of the first experiment. The same measurements were taken at slaughter and so the total number of animals became 57. Further details of the design, animal management etc. can be obtained from Winter (1971).

Table 3.19.1. The weights of the following attributes[A] were recorded for each of the 57 animals slaughtered

No.	Name	No.	Name	No.	Name
1	Body	17	Large intestine	33	Visceral protein
2	Empty body	18	Rumen contents	34	Visceral water
3	Carcass	19	Abomasum contents	35	Visceral fat
4	Head	20	Omasum contents	36	Visceral ash
5	Hide	21	Small intestine contents	37	Hide
6	Limbs	22	Large intestine contents	38	Hide protein
7	Blood	23	Total contents	39	Hide water
8	Liver	24	Carcass protein	40	Hide fat
9	Heart	25	Carcass water	41	Hide ash
10	Lungs	26	Carcass fat	42	Head and limbs
11	Spleen	27	Carcass ash	43	Head and limbs protein
12	Caul fat	28	Empty body protein	44	Head and limbs water
13	Rumen	29	Empty body water	45	Head and limbs fat
14	Abomasum	30	Empty body fat	46	Head and limbs ash
15	Omasum	31	Empty body ash		
16	Small intestine	32	Viscera		

[A] Attributes were expressed as proportions by division as follows, viz. $2 \div 1$; $3–17 \div 2$; $18–22 \div 23$; $24–27 \div 3$; $28–31 \div 2$; $33–36 \div 32$; $38–41 \div 37$; $43–46 \div 42$.

As with most sets of data there are several methods of analysis but the most useful in this field is comparison of linear regression 'growth' equations using bodyweight, carcass weight etc. as the independent variate and other measured attributes such as carcass fat weight, small intestine weight etc. as the dependent variate (Table 3.19.2). This technique is based on the logarithmic expansion of Huxley's (1932) allometric equation $y = ax^b$.

This technique is excellent for comparing the relative growth of one component with the whole, such as the percentage of fat, protein and water in the the carcass relative to the carcass weight. When only a small number of variables are studied, treatment differences are easy to comprehend. Interpretation is often made easier here as the attributes are invariably correlated, e.g. negative correlation of percentage fat and water in a muscle. However, when dealing with a large number of attributes it is almost impossible to gain an overall view of the treatment effects. Many of the attributes are not correlated with the chosen independent variate and others exhibit negative correlations for one treatment and positive correlations for another. It is not possible to gauge the overall significance of these effects since each attribute is examined independently of all others. Thus this

Table 3.19.2. Constants in the regression equations in the analysis of covariance of the chemical components with log EBW as the independent variate in the equation $y = a + bx$

Dependent variate	Constants, experiment 1						Constants, experiment 2			
	a			$b \pm$ S.E.			a		$b \pm$ S.E.	
(log wt)	G 1	G 2	G 3	G 1	G 2	G 3	G 5	G 6	G 5	G 6
Protein	0·007	−0·227	−0·305	0·457^A ±0·101 r	0·607^A ±0·119 rs	0·665^A ±0·093 s	−0·018	−0·146	0·461^A ±0·165	0·549 ±0·231
Water	0·817	0·554	0·214	0·305^A ±0·162	0·475^A ±0·111	0·680^A ±0·121	0·302	0·235	0·620^A ±0·103	0·667 ±0·275
Fat	−2·201	−2·041	−1·732	2·059^B ±0·197	1·950^B ±0·210	1·758^B ±0·229	−2·001	0·001	1·958^B ±0·230 t	0·674 ±0·292 u
Ash	−1·302	−0·280	−0·182	0·896 ±0·113 r	0·277^A ±0·141 s	0·236^A ±0·105 s	−0·712	−0·238	0·496^A ±0·193	0·263^A ±0·239

^{A} Value of b significantly < 1.
^{B} Value of b significantly > 1.
r, s, t, u. Within experiment, values on the same line with different subscripts differ significantly ($P < 0.05$).

regression technique is good at describing differences in relative growth rates of each attribute but is of little use in any judgment of the importance of these differences or in the assessment of overall treatment effects.

It was this problem that encouraged the use of pattern analysis. Instead of each attribute being considered independently, all the attributes could be considered together without bias to give an overall view of each animal. Treatment differences could be determined by examination of the fusion groups and these differences could be defined by the attributes contributing to the fusion.

The data were classified on three occasions, twice by the agglomerative strategy available from the Canberra program MULCLAS and once during the development of the divisive program REMUL. For each MULCLAS run the group means at each hierarchical fusion were examined by the diagnostic program GROUPER.

The first MULCLAS classification describes the population using the raw data, i.e. the weight of each physical and chemical component (Fig. 3.19.2).

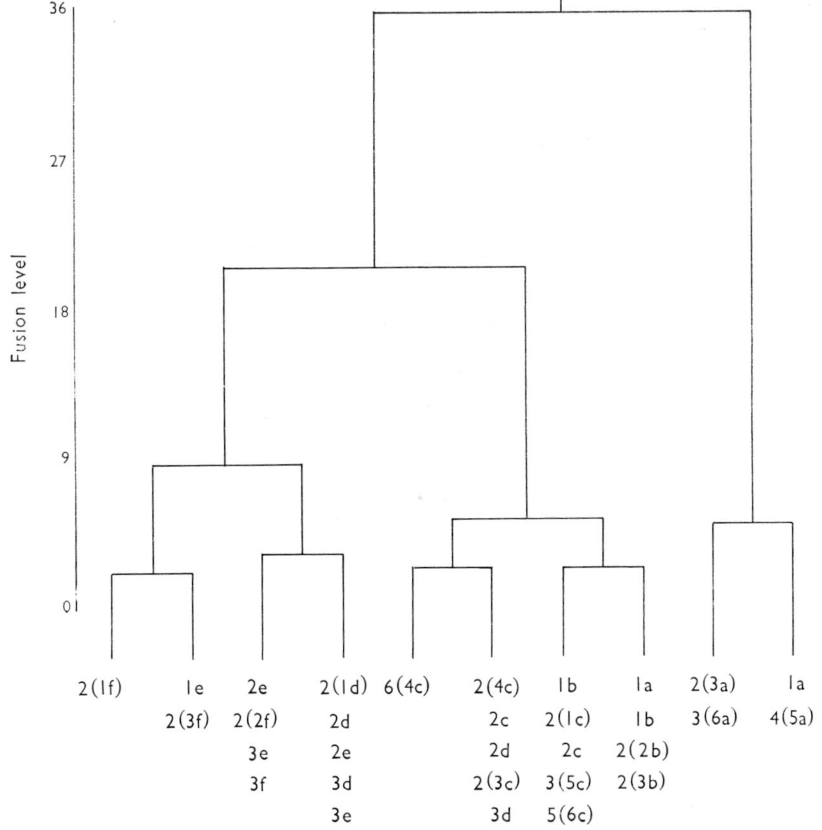

Fig. 3.19.2. Classification of sheep using weights of 46 measured variates.

Four major groups are defined which can be best described using the slaughter weight notations A, B and C, D and E, F. It is of interest to note that liveweight, empty bodyweight and carcass weight do not contribute at any fusion point. The major separation of A from the rest is based on differences in empty body and carcass *water* and *protein* weights. The next division of B and C from D, E and F is due to differences in empty body and viscera *fat* weights. The separation of F from D and E is due to differences in liver and small intestine size.

In general, animals at the same slaughter weight are in the same or a closely related fusion group. This classification indicates that treatment had only minor effects on body composition. Apart from the last two fusions most of the group differences are confined to the visceral components.

A few interesting points from the classification are:
the fattest animals are those which are grown continuously at the D, E and F slaughter weights;

the leanest are those slaughtered at the start of the experiment;

animals starved to slaughter weight A have more ash ($\equiv$ bone);

animals maintained at a constant bodyweight have less abomasal contents than growing animals of similar size.

It could be argued that this classification is inevitable since many of the attributes are highly correlated with bodyweight, e.g. carcass fat. Also not all animals given the same notation (A–F) were killed at exactly the same bodyweight, so slaughter weight comparisons of different treatments may be slightly in error. To overcome these criticisms the data were re-expressed as proportions (Table 3.19.2), then reclassified by MULCLAS.

This classification (Fig. 3.19.3) is similar to the first with a few notable changes in attributes contributing to fusion and in the positioning of certain fusion groups. The major contributing attributes to the last fusion are viscera water, fat, ash and protein compared with empty body and carcass water and protein in the first classification. However, empty body and carcass water and fat are the major contributors at the next fusion level. The main difference between the classifications is the separation of the animals at slaughter weight A. It is now apparent that animals which have been starved (3a and 6a) are fatter than the animals at the start of the experiment (1A and 5A). Some other interesting points from the classification are:
when animals are starved either to lose weight or to maintain it (groups 5 and 6, Fig. 3.19.3) the spleen continues to grow whilst the liver and rumen tend to atrophy (cf. groups 3 and 4);

animals which have grown continuously (most of groups 7 and 8) have less gut contents and a lower proportion of the contents in the rumen than animals regrown after a period of weight loss (groups 9 and 10);

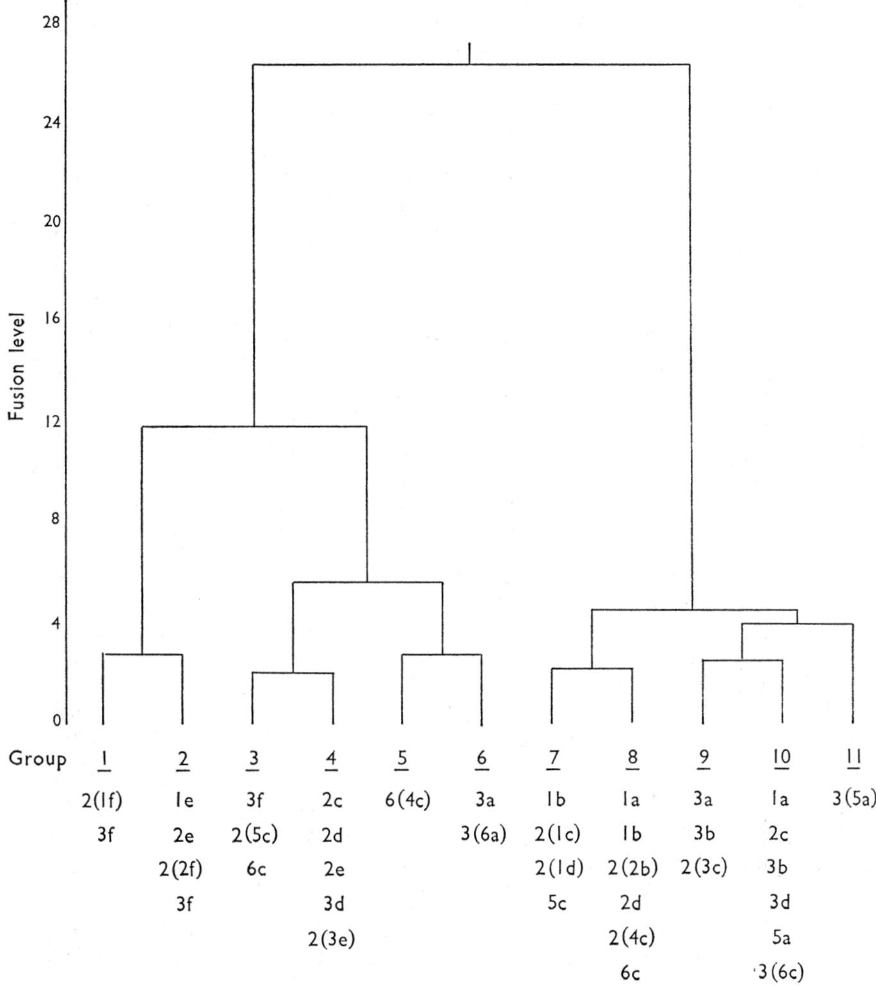

Fig. 3.19.3. Classification of sheep using proportions of 46 measured variates.

group 11 is distinguished from groups 9 and 10 by having lower proportions of ash and more hide protein (this is a source of labile protein and had not been fully restored in the regrowing animals).

This second classification has been of some benefit. Although the integrity of some fusion groups has been lost, some additional important facts have been revealed. The initial conclusions that treatment had only minor effects on body composition in terms of fat, protein and water have been confirmed but there are some differences attributable to gut contents, visceral components and amount of ash.

The data were used as a test batch for REMUL. This program is divisive and division is not dependent on group size. The classification is shown in Fig. 3.19.4. Of the 57 animals considered, 37 were deemed similar whilst the remainder were further subdivided. The major contributor at the first and second division is empty body water. Fusion groups 2, 3 and 4 (animals starved back to slaughter weight A) are separated from group 5 because of a higher proportion of ash. The remaining groups have animals from slaughter weights E and F. Groups 9 and 10 are distinguished from the others because of reduced small intestine weights.

From the experimenter's viewpoint too much information is lost in the first division in this classification. The reason for this may lie in the presentation of the data. It might have been better in this case to express the data

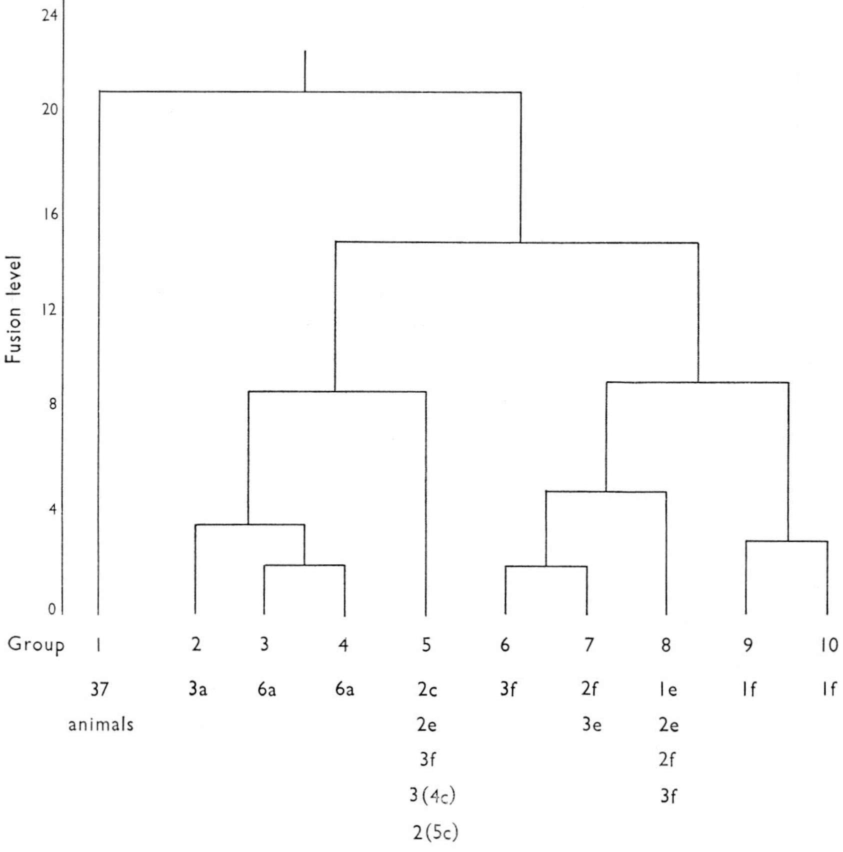

Fig. 3.19.4. Divisive classification of sheep by REMUL using proportions of 46 measured attributes.

in logarithmic form since many relationships such as empty body weight and empty body water are curvilinear and this program has divided at about the point of inflexion, leaving the majority in one large group.

In summary it can be stated that pattern analysis of this type of data is most useful for gaining an overall judgment of treatment effects and if used in conjunction with traditional techniques will make a significant contribution to the interpretation of such data.

W. H. WINTER

References

Huxley, J. S. (1932). 'Problems of Relative Growth'. (Methuen: London.)

Winter, W. H. (1971). A study of weight loss and compensatory gain in sheep. Ph.D. Thesis, University of Melbourne.

Case 3.20. The Influence of Attributes on the Classification of the Grasses (Poaceae)

Any numerical taxonomic analysis begins with a set of entities to be classified and a set of attributes derived from these entities. Though not commonly recognized it is clear that since the number of attributes available is usually much less than the number possible, any classification of the entities will be highly dependent upon the nature of the set of attributes employed. In this regard the classification of the grasses is of particular interest. Modern classifications of the family incorporate a very diverse set of attributes (Prat 1960) amongst which are a range of anatomical, biochemical and cytological properties not available to earlier taxonomists. They relied instead upon gross morphology and in particular spikelet structure for the construction of their classifications.

It has been argued that provided a reasonable sample of attributes is employed their nature is of little importance. The basis of this viewpoint is the linkage of the genes into chromosomes. Since most attributes are dependent for their expression on several genes and these are linked together on relatively few chromosomes (ignoring polyploidy where, nonetheless, individual chromosomes are replicated) it would appear that any reasonable set of attributes would be a reasonable sample of the genotype. Hence it would be expected that classifications based upon different sets of attributes would be congruent. However, the hypothesis of congruency has not been well supported by previous studies (Ehrlich and Ehrlich 1967; Clifford and Lavarack 1974) and does not receive unqualified support from the studies reported upon below.

Whereas previous numerical taxonomic studies of the family (Clifford 1965; Clifford and Goodall 1967; Clifford *et al.* 1969) employed a wide range of attributes, the present study investigates the classifications arising from restricted attribute sets.

One set appertained to vegetative, the other to reproductive attributes. The sample of grass genera was that previously reported upon (Clifford *et al.* 1969) and the attributes were likewise unchanged save for being divided into two subsets. There were, however, fewer unscored data than heretofore. The genera are listed alphabetically in Table 3.20.1 and are numbered serially for ease of reference. Classifications were generated using an information-gain strategy, fusion of groups being determined on the basis of the minimum information-gain (ΔI) at each step in the fusion cycle

Table 3.20.1. List of genera studied

1	*Agropyron*	24	*Centotheca*	47	*Jouvea*	70	*Phalaris*
2	*Agrostis*	25	*Chloris*	48	*Lecomtella*	71	*Pharus*
3	*Ammophila*	26	*Chusquea*	49	*Lepturus*	72	*Phragmites*
4	*Ampelodesmos*	27	*Coleanthus*	50	*Lolium*	73	*Phyllorachis*
5	*Anomochloa*	28	*Cortaderia*	51	*Lygeum*	74	*Poa*
6	*Anthephora*	29	*Cynodon*	52	*Melica*	75	*Polypogon*
7	*Anthoxanthum*	30	*Danthonia*	53	*Melinis*	76	*Pommereulla*
8	*Aristida*	31	*Diarrhena*	54	*Melocanna*	77	*Saccharum*
9	*Arthropogon*	32	*Dichanthium*	55	*Micraira*	78	*Sesleria*
10	*Arthrostylidium*	33	*Ehrharta*	56	*Milium*	79	*Sorghum*
11	*Arundinaria*	34	*Elytrophorus*	57	*Miscanthus*	80	*Spartina*
12	*Arundinella*	35	*Enneapogon*	58	*Molinia*	81	*Sporobolus*
13	*Avena*	36	*Eragrostis*	59	*Monerma*	82	*Stipa*
14	*Bambusa*	37	*Eriachne*	60	*Nardus*	83	*Streptochaeta*
15	*Boivinella*	38	*Erianthus*	61	*Olyra*	84	*Streptogyna*
16	*Bothriochloa*	39	*Euchlaena*	62	*Orcuttia*	85	*Thysanolaena*
17	*Brachyelytrum*	40	*Festuca*	63	*Oryza*	86	*Trachys*
18	*Brachypodium*	41	*Garnotia*	64	*Oryzopsis*	87	*Triticum*
19	*Bromus*	42	*Glyceria*	65	*Panicum*	88	*Uniola*
20	*Brylkinia*	43	*Heteropogon*	66	*Pappophorum*	89	*Vulpia*
21	*Buergersiochloa*	44	*Hubbardia*	67	*Pariana*	90	*Zea*
22	*Calamagrostis*	45	*Imperata*	68	*Perotis*	91	*Zizanea*
23	*Capillipedium*	46	*Isachne*	69	*Phaenosperma*	92	*Zoysia*

(MULTBET, Lance and Williams 1967). The resultant hierarchies are presented in Figs 3.20.1 and 3.20.2.

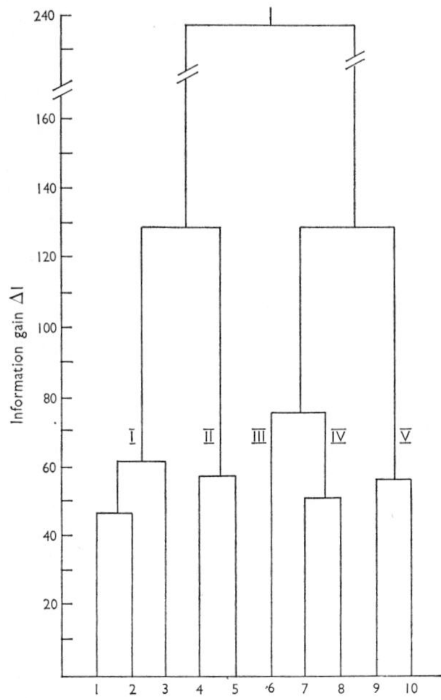

Fig. 3.20.1. Classificatory dendrogram resulting from the clustering of 92 grass genera employing reproductive attributes only.

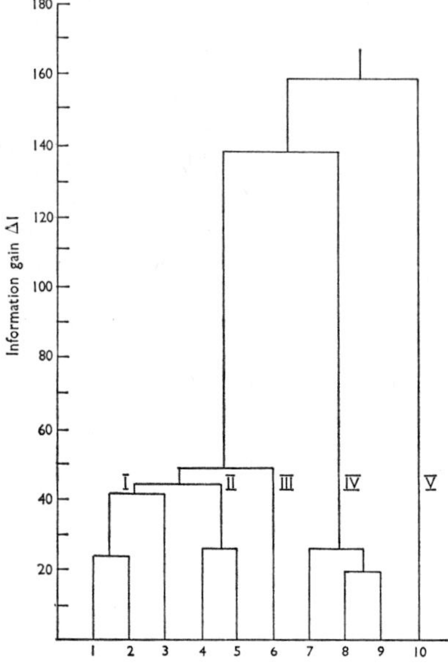

Fig. 3.20.2. Classificatory dendrogram as in Fig. 3.20.1 but employing vegetative attributes only.

A comparison of Figs 3.20.1 and 3.20.2 reveals that different hierarchies result from the employment of different attribute sets. There is, for instance, more tendency for chaining in the dendrogram arising from data that pertain to vegetative attributes than in the dendrogram based upon reproductive attributes. Furthermore, the generic compositions of the groups in each dendrogram differ in that at the 10-group level there is no group with identical generic compositions in both hierarchies. This may be confirmed by reference to Fig. 3.20.3 where the two classifications are expressed in terms of a two-way table. To conserve space in the table each genus is quoted in terms of its code number.

From Fig. 3.20.3 it is clear that whilst there is an overall correspondence between the two classifications, the agreement is not close. Thus whilst the members of groups 4 and 5 of the 'reproductive classification' are mostly to be found in group 10 of the 'vegetative classification', the remaining genus *Brachyelytrum* (17) is placed in group 3 of that classification. Such a close correspondence, however, is not reflected in the majority of groups. Indeed, there is no correspondence between the two classifications with respect to group 1 of the 'reproductive classification', for there each of the six genera in the group appears in a separate group of the 'vegetative classification'.

One of the reasons for the discrepancies observed between the two classifications could be that the dendrograms are truncated at what are

Reproductive

Vegetative		I			II		III	IV		V	
		1	2	3	4	5	6	7	8	9	10
I	1						90			6,39, 77	16,23, 32
	2						12,24, 49		55	38,47	
	3	31				17		34	27,51, 60	15,45, 68,86	41,57, 79
II	4	88						8,35, 58	30	9,48, 53,65	43
	5	37		73			72,85	28		46	
III	6						62,81, 92	25,29, 36,66, 76			
	7			63,67, 91							
IV	8	21	10,11, 14,26, 54,84	71							
	9	44		5,83							
V	10	69		59	20,22, 2,3, 40,50, 56,64, 70,74, 75,82, 89	1,4, 13,18 19,78 87			7,42, 52		

Fig. 3.20.3. The 92 grasses of Table 3.20.1 arranged into 10 groups according to both vegetative and reproductive attributes.

biologically different levels. Such an influence is likely to be less significant at higher levels in the hierarchy and so the classifications were also compared at the five-group levels. At this level there appeared the five major groups recognized in earlier analyses. The distribution of these five groups in a two-way table is shown in Fig. 3.20.4.

From Fig. 3.20.4 it can be seen that whereas the bambusoid and pooid grasses are relatively homogeneous groups with respect to both vegetative and reproductive attributes the other three groups are heterogeneous. Thus the panicoid grasses, defined broadly, are uniform with respect to reproductive attributes but are subdivided into two groups, panicoid proper and

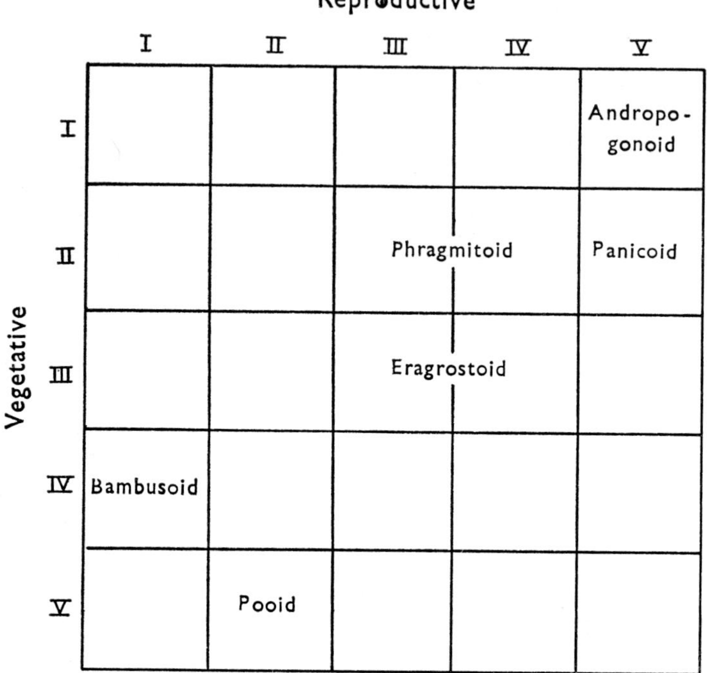

Fig. 3.20.4. The major groups of grasses with respect to the 5-group levels of Fig. 3.20.3.

andropogonoid, on the basis of their vegetative attributes. In contrast the phragmitoid and eragrostoid grasses are relatively uniform with respect to their vegetative but diverse with respect to their reproductive structures.

Most classifications will be based on a mixture of attributes, some vegetative and some reproductive in nature, and so the marked differences resulting from the separate consideration of these two types of attribute will not be evident. Nonetheless it is clear from the above analyses that the proportions of the two types of attribute will influence the classification. Hence it cannot be stressed too strongly that in all taxonomic published work it is incumbent upon the writer to expose fully the data employed.

H. T. CLIFFORD

References

Clifford, H. T. (1965). The classification of the Poaceae: a statistical study. *Univ. Qd Pap. Dep. Bot.* **4**(15), 243–53.

Clifford, H. T., and Goodall, D. W. (1967). A numerical contribution to the classification of the Poaceae. *Aust. J. Bot.* **15**, 499–519.

Clifford, H. T., and Lavarack, P. S. (1974). The role of vegetative and reproductive attributes in the classification of the Orchidaceae. *Biol. J. Linn. Soc.* **6**, 97–110.

Clifford, H. T., Williams, W. T., and Lance, G. N. (1969). A further numerical contribution to the classification of the Poaceae. *Aust. J. Bot.* **17**, 119–31.

Ehrlich, P. R., and Ehrlich, A. M. (1967). The phenetic relationships of butterflies. I. Adult taxonomy and the non-specificity hypothesis. *Syst. Zool.* **16**, 301–17.

Lance, G. N., and Williams, W. T. (1967). Mixed-data classificatory programs. I. Agglomerative systems. *Aust. Comput. J.* **1**, 15–20.

Prat, H. (1960). Vers une classification naturelle des Graminées. *Bull. Soc. Bot. Fr.* **107**, 32–79.

Case 3.21. A Comparison of some Recently Developed Methods

Since the methods to be discussed here will, almost by definition, be unfamiliar to most, an attempt has been made to link them by considering only a single set of data. These consist of records for the percent cover of 52 species, taken from 72 stands arranged in a 9×8 grid. The stands were arranged to cover a single gilgai feature as will be apparent from Fig. 3.21.1(a) where the topography of the area is indicated using depths below an arbitrary datum. The grid display also permits a slightly easier examination of the results of the methods, whereas with stands that cannot be topographically related it is often more difficult to provide an adequate interpretation. Three classes of results will be examined:

(1) One-parameter classification methods

(2) Two-parameter classification methods

(3) A seriation method.

Unfortunately space does not permit an adequate examination of factor analysis, multidimensional scaling or extrinsic classification methods.

One-parameter methods

The first two methods to be discussed have no figures to illustrate the results. Goodall (1969) proposed a method for isolating stands with an undue preponderance of rare species, while Ling (1972) has suggested a clustering algorithm based on k-linkage (a presumed improvement on nearest-neighbour methods). The reason for the absence of results is simple. Both methods indicate that the data are homogeneous, Goodall's even providing a significance test for the result. Ling's method, it is true, does isolate a couple of very small clusters but then provides no evidence of major groupings. However, neither method indicates if any structure is present or how to identify it if it does exist. I have not yet tried Gower's (1974) method with these data but it too might well indicate homogeneity. The next six methods are all in the traditional mode of numerical classification, being derived from various search procedures but all primarily concentrating on the measure of similarity. The results are shown in Figs 3.21.1(b–d) and 3.21.2(a–c) and the analyses will be discussed in order.

(1) POLYDIV (Fig. 3.21.1(b))

This is a polythetic divisive method. The stand scores on the first principal component are calculated and then they are divided into two subgroups so as to maximize between-group variance. Each such subgroup is then regarded as a new population and subjected to a similar division procedure.

Noy-Meir (1973) has suggested a cross-classification based on divisions of successive components from all the data rather than the recalculated components used in POLYDIV. There is a difficulty in knowing when to stop, although a new measure seems to be proving more sensitive than the one at present in the computer program.

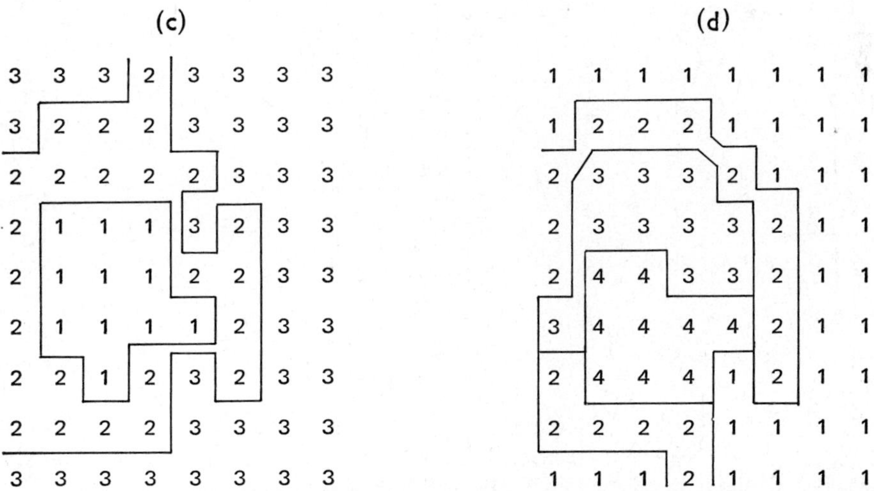

Fig. 3.21.1. (a) Topography of the test area showing depth below arbitrary datum. (b) POLYDIV analysis, 4 groups. (c) CENTPERC 0 extension, 3 groups. (d) CENTPERC 2 extension, 4 groups.

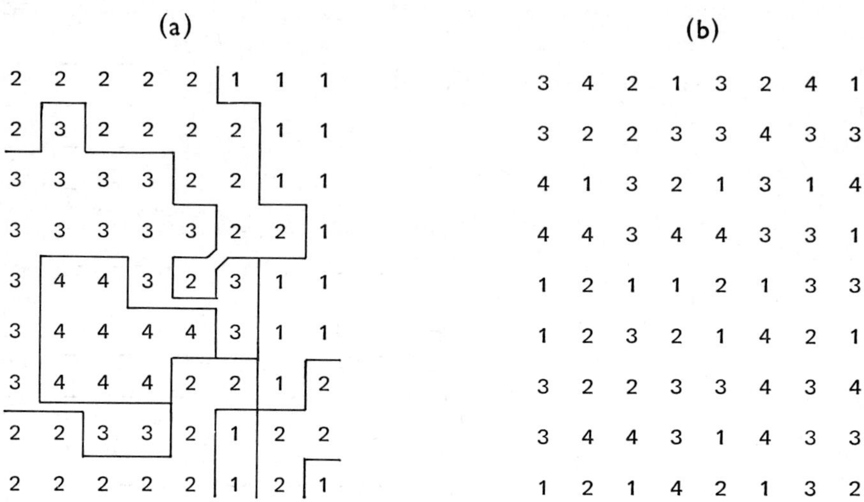

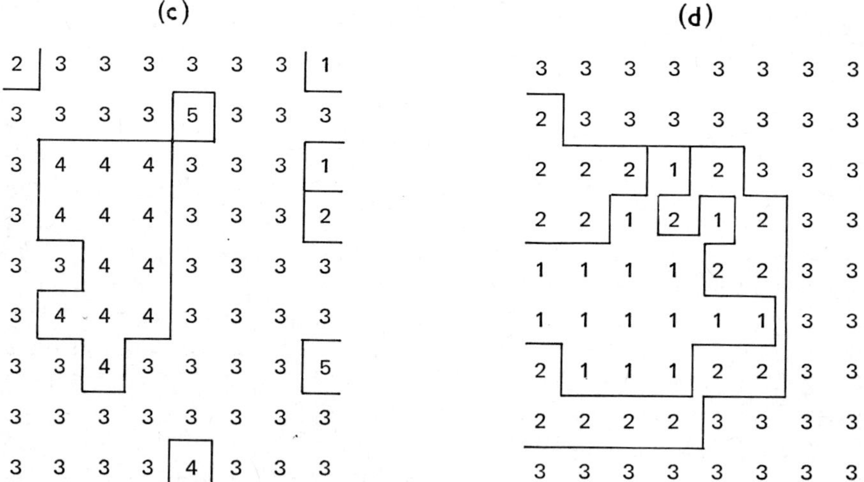

Fig. 3.21.2. (*a*) CENTPERC 2 qualitative information only, 4 groups. (*b*) CENTPERC 2 extension quantitative information only, 4 groups. (*c*) CENTPERC 4, 5 groups. (*d*) Two-parameter normal analysis 3 groups determined by stopping rule.

The results are tolerably indicative of the gilgai structure, showing concentric patterns, even if the separation of groups 2 and 3 is a little odd and might suggest some asymmetry, perhaps due to aspect effects.

(2) CENTPERC 0 extension (Fig. 3.21.1(*c*))

(3) CENTPERC 2 extension total information (Fig. 3.21.1(*d*))

(4) CENTPERC 2 extension qualitative information only (Fig. 3.21.2(*a*))

(5) CENTPERC 2 extension quantitative information only (Fig. 3.21.2(*b*))

Dale *et al.* (1971) proposed the various CENTPERC methods to include quantitative information in a similarity measure based on information statistics. Williams (1972, 1973) has shown how a more rigorous examination suggests an extension to the original statistics used. Here we examine the two complete extended statistics, and also the partition of one of them into its constituent qualitative and quantitative components. Note that the quantitative information is that left after removal of qualitative effects.

The results for the complete extensions suggest, though not strongly with these data, that the CENTPERC 0 statistic may be preferable to the CENTPERC 2 statistic, although both lead to the suggestion of some sort of concentric structure. The qualitative information statistic clearly closely resembles its parent, although it is perhaps less clear, while the quantitative data provide no useful pattern at all. For small areas quantitative data often simply increase the noise level, because the factors causing variation are often historical and genetic and not easily identified.

(6) CENTPERC 4 (Fig. 3.21.2(*c*))

There has been much concern within numerical classification with the problem of double-zero matches. It is indeed a difficult question to decide if two items should be regarded as similar because they lack a large number of properties. Finally the decision must be subject, or even user, specific, although there is some evidence to suggest that suppressing double-zero matches does avoid some problems with non-linearity of relationships. In any case there is a need for an information measure of similarity which allows suppression, and CENTPERC 4 is an attempt to provide such a measure. I refrain from presenting the method because the results suggest that the attempt failed, being unduly sensitive to minor aberrations.

Most of these examples required only a few minutes of computer time using the Control Data CD3600 computer of the CSIRO Division of Computing Research. However, none of the agglomerative methods would be tolerable with more than 400 stands, while POLYDIV would not be very usable with more than 200 species. None of these hold out much hope for very large-scale problems.

(a)

—	—	—	—	—	—	—	—
—	—	—	—	—	—	—	—
—	—	—	172	—	—	—	—
—	—	222	—	115	—	—	—
196	112	224	115	—	—	—	—
131	149	215	182	99	106	—	—
—	94	215	161	—	—	—	—
—	—	—	—	—	—	—	—
—	—	—	—	—	—	—	—

(b)

—	—	—	—	—	—	—	—
115	—	—	—	—	—	—	—
149	79	169	—	192	—	—	—
93	57	—	176	—	117	—	—
—	—	—	—	90	142	—	—
—	—	—	—	—	—	—	—
131	—	—	—	57	140	—	—
106	115	102	144	—	—	—	—
—	—	—	—	—	—	—	—

Fig. 3.21.3. (*a*) *a* Values for group 1 stands, two-parameter analysis. (*b*) *a* Values for group 2 stands, two-parameter analysis. (*c*) *a* Values for group 3 stands, two-parameter analysis.

(c)

88	64	75	80	69	40	88	110
—	159	261	164	137	197	64	99
—	—	—	—	—	129	40	32
—	—	—	—	—	—	80	52
—	—	—	—	—	—	195	53
—	—	—	—	—	—	239	123
—	—	—	—	—	—	200	59
—	—	—	—	132	195	96	24
102	129	169	69	88	115	51	24

Two-parameter classifications

In this section only one method will be considered, the two-parameter method of Macnaughton-Smith (1965). Because of the many interesting features which two-parameter methods introduce into classification an extended discussion will be given for this method. The first analysis is a simple normal analysis. The search method used is monothetic-divisive, although the nature of the groups sought suggests that this will be a close approximation to the 'best' polythetic grouping method. A reallocation procedure for 'cleaning up' groups is known but remains to be tested. An empirically derived stopping rule can be used, and in the present case suggests three groups, shown in Fig. 3.21.2(*d*). The concentric pattern is again clearly visible, but in addition for each group the stands have *a* parameter values, while the species *b* parameters are known for all three groups. In Fig. 3.21.3(*a–c*) the *a* values for the three stand-groups have been mapped to indicate variation within the classes. The *a* values would also permit the identification of depauperate stands and sometimes enable a minor misplacement to be corrected, because a stand representing a rare type, and hence with few similar stands in the data set, will show low *a* values even in the group in which it is placed.

The *b* values for the species can be used to provide information usually obtained from an inverse analysis. The results of such a sorting are shown in Table 3.21.1. The procedure is simply to identify for each species the stand group with which it has most affinity, and to group species that have a common affinity for the same stand group. Sometimes a stand group appears with no associated species, and in this case the group may be regarded as transitional.

Table 3.21.1. Derived species classes from normal two-parameter analysis

Species	Stand group 1	Stand group 2	Stand group 3
Echinochloa colonum	476	47	1
Cyperus difformis	90	50	—
Diplachne parviflora	40	10	0+
Senecio lautus	20	10	0+
Leptochloa debilis	1	—	—
Cyperus gracilis	1	—	—
Senecio spp.	1	—	1
Dichanthium sericeum	9	90	58
Chloris divaricata	20	66	33

Table 3.21.1 (*Continued*)

Species	Stand group 1	Stand group 2	Stand group 3
Euphorbia drummondii	26	54	7
Cynodon dactylon	45	52	—
Centipeda minima	11	17	1
Chloris truncata	3	12	5
Panicum subzerophyllum	1	9	0+
Oxalis corniculata	4	8	4
Eclipta platyglossa	5	7	1
Sonchus oleraceus	5	7	3
Salvia reflexa	2	6	—
Paspalidium caespitosum	—	5	2
Panicum queenslandicum	2	4	1
Vittadinia triloba	—	3	2
Plantago varia	0+	2	1
Alternanthera denticulata	—	2	—
Sporobolus contiguus	1	2	1
Eragrostis parviflora	—	1	—
Sorghum almum	—	1	—
Sporobolus scabridus	—	1	—
Rhagodia nutans	—	1	—
Verbena officinalis	—	1	—
Cucumis myriocarpus	0+	1	—
Cyperus betchei	0+	0+	—
Cyperus dactylotes	—	0+	—
Amaranthus mitchellii	—	0+	—
Damasonium australe	—	0+	—
Bassia tetracuspis	6	36	125
Atriplex semibaccata	6	37	73
Enchylaena tomentosa	2	7	31
Atriplex muelleri	—	3	14
Salsola kali	0+	—	10
Paspalidium constricta	—	—	7
Portulaca oleracea	4	4	5
Zygophyllum apiculatum	—	—	5
Sporobolus caroli	—	1	4
Chloris aciculata	—	—	1
Eriochloa pseudoacrotrichum	—	—	1
Panicum buncei	—	—	1
Paspalidium gracilis	—	—	1
Rhagodia linifolia	0+	—	1
Boerhavia diffusa	—	—	1
Brunoniella australe	—	0+	1
Lepidium bossopifolium	—	—	1
Atriplex spp.	—	—	1

An inverse analysis would provide groups of species, and through the use of the *a* values could also group the stands. However, of more interest is the inosculate search method where at each division the analysis can choose between an inverse and a normal division. In the present case an interesting result ensues. Only inverse divisions are found, indicating that there are no major stand groups. This of course ties in with the homogeneity found in Goodall's and Ling's methods but with the two-parameter method the *a* values can be used to order the stands, providing an ordination. In Fig. 3.21.4(*a–c*) the results for three inverse groups are shown. The gilgai can be regarded as homogeneous, but each stand can be regarded as possessing a mixture of 2 or 3 groups of species. Dale and Webb (1975) have argued that the point of change from normal to inverse divisions which can be found in analyses of phytosociological data marks a 'termination' level independent of any arbitrary stopping rules and which should have phytosociological significance. However, inosculate analysis may be less well suited to taxonomic problems.

Seriation

In examining the topological dimensions of the data the first requirement is for some estimate of the number of dimensions. Using Trunk's (1968) method the result obtained was 4, which is, it is hoped, not too unlike the three groups of the two-parameter method. Trunk's method assumes that the variables are uncorrelated and correlation seems likely to increase the dimensionality somewhat. Reasonable results can be obtained using Gower's (1966) principal coordinate analysis throughout in four dimensions especially with double-zero-match suppression, but it is perhaps more interesting to examine Shepard and Carroll's (1965, see also Noy-Meir 1974) parametric mapping method and Hill's (1973) reciprocal averaging method. Parametric mapping proved disappointing. After considerable computation the program showed no signs of converging and the intermediate answers were quite unsatisfactory. This has been found to be typical of the performance of the present algorithm but it is hoped that others can be found which will prove better.

Hill's method poses something of a problem. It is not clear that a series of stands should lie in anything other than one dimension yet we know that the present data probably lie in three or four. In Fig. 3.21.5 the results for the first two axes are shown. While the first axis is almost a perfect reproduction of the topography, the second has little obvious meaning, although it might reflect aspect differences. The difficulty with seriation is that there is no means of knowing when a sequence due to one cause finishes and that due to another starts. However, the success of the first axis is undeniable. Obviously further studies are needed before a real appreciation of the role of seriation can be made.

(a) *Echinochloa colonum* group

94	94	0	94	94	0	94	94
379	94	284	94	94	94	94	0
959	1756	6533	4311	764	94	0	0
764	1959	10743	7635	3215	379	189	0
2370	4763	11316	4763	3322	959	94	189
2789	6901	10042	8955	3867	1655	189	189
1554	3322	10743	6901	379	379.	0	0
475	379	764	959	94	0	94	0
0	0	189	189	0	0	0	0

(b) *Bassia tetracuspis / Atriplex semibaccata* group

1735	1242	758	376	1145	1340	2541	3056
1735	758	662	662	1537	1934	2135	3686
1934	662	94	0	188	2439	1145	951
567	0	0	0	282	1934	2849	1636
1340	0	0	0	94	1438	6464	1537
758	0	0	0	0	1242	6117	3369
951	188	0	94	1145	1735	6117	1537
1438	471	282	2135	3369	5774	2337	854
3056	2439	2849	1242	2952	3899	1242	758

Fig. 3.21.4. (*a*) *a* Values from first group of species in inverse two-parameter analysis. (*b*) *a* Values from second group of species, in inverse two-parameter analysis. (*c*) *a* Values for third group of species, in inverse analysis two-parameter method.

(c) *Dichanthium sericeum / Chloris divaricata* group

513	366	735	919	477	37	220	330
1105	1850	3211	1963	1253	698	37	0.
1365	403	513	1253	2830	808	110	73
1031	220	146	330	587	1068	73	73
1775	146	37	183	256	1439	330	73
808	73	220	0	220	624	993	366
1216	256	37	220	366	1514	550	220
1031	1588	1270	1216	513	587	403	0
256	845	1068	403	110	146	220	37

First axis

−248	−425	−891	−84	−399	−313	−481	−498
−18	47	312	−36	−623	−584	−177	−613
334	501	732	788	124	−203	−325	−531
508	811	927	936	111	−99	−119	−503
370	718	800	687	421	246	−666	−384
667	484	758	442	598	−289	−406	−834
337	250	513	308	−51	−134	−716	−400
98	199	488	−34	−237	−269	−694	−315
−216	−402	−472	−313	−323	−502	−406	−322

Second axis

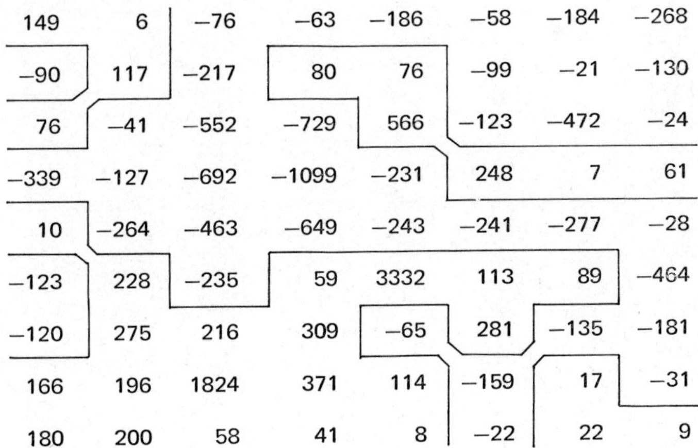

149	6	−76	−63	−186	−58	−184	−268
−90	117	−217	80	76	−99	−21	−130
76	−41	−552	−729	566	−123	−472	−24
−339	−127	−692	−1099	−231	248	7	61
10	−264	−463	−649	−243	−241	−277	−28
−123	228	−235	59	3332	113	89	−464
−120	275	216	309	−65	281	−135	−181
166	196	1824	371	114	−159	17	−31
180	200	58	41	8	−22	22	9

Fig. 3.21.5. Hill seriation method, stand loadings on first and second axes.

Conclusions

The applications presented here have shown both successes and failures. The two-parameter method and seriation both have much promise for the future while the extended CENTPERC models seem to provide useful methods. Some of the other methods are much less successful and either require considerable modification before they can be widely used, as with the Shepard–Carroll parametric mapping, or can be rejected completely, as with CENTPERC 4. I hope, though, that they all show that the prophecies made here and in other places are based on some contact with real data, real problems and real needs for answers.

M. B. DALE

References

Dale, M. B., Lance, G. N., and Albrecht, L. A. (1971). Extensions of information analysis. *Aust. Comput. J.* **3**, 29–34.

Dale, M. B., and Webb, L. J. (1975). Numerical methods for the establishment of associations. *Vegetatio* **30**, 77–87.

Goodall, D. W. (1969). A procedure for the recognition of uncommon species combinations in sets of vegetation samples. *Vegetatio* **18**, 19–35.

Gower, J. C. (1966). Some distance properties of latent root and vector methods used in multivariate analysis. *Biometrika* **53**, 325–38.

Gower, J. C. (1974). A maximal predictive classification. *Biometrics* **30**, 643–54.

Hill, M. O. (1973). Reciprocal averaging: an eigenvector method of ordination. *J. Ecol.* **61**, 237–50.

Ling, R. F. (1972). On the theory and construction of *k*-clusters. *Comput. J.* **15**, 326–32.

Macnaughton-Smith, P. (1965). Some statistical and other numerical methods of classifying individuals. Home Office Res. Unit Rep. No. 6. (H.M.S.O.: London.)

Noy-Meir, I. (1973). Divisive polythetic classification of vegetation by optimized division on ordination components. *J. Ecol.* **61**, 753–60.

Noy-Meir, I. (1974). Multivariate analysis of the semiarid vegetation in south-eastern Australia. II. Vegetation catenae and environmental gradients. *Aust. J. Bot.* **22**, 115–40.

Shepard, R. H., and Carroll, J. D. (1965). Parametric representation of nonlinear data structures. In 'Multivariate Analysis'. (Ed. P. R. Krishnaiah.) pp. 561–92. (Academic Press: New York.)

Trunk, G. V. (1968). Statistical estimation of the intrinsic dimensionality of data collections. *Inf. Control* **12**, 508–25.

Williams, W. T. (1972). Partition of information. *Aust. J. Bot.* **20**, 235–40.

Williams, W. T. (1973). Partition of information: the CENTPERC problem. *Aust. J. Bot.* **21**, 277–82.

Index to Parts 1 and 2